MW00966301

Chemical Oxidation and Reactive Barriers

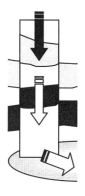

Remediation of
Chlorinated and
Recalcitrant
Compounds

EDITORS
Godage B. Wickramanayake, Arun R. Gavaskar,
and Abraham S.C. Chen
Battelle

The Second International Conference
on Remediation of Chlorinated and
Recalcitrant Compounds

Monterey, California, May 22–25, 2000

 BATTELLE PRESS
Columbus • Richland

Library of Congress Cataloging-in-Publication Data

International Conference on Remediation of Chlorinated and Recalcitrant Compounds (2nd : 2000 : Monterey, Calif.)
    Chemical oxidation and reactive barriers : remediation of chlorinated and recalcitrant compounds (C2-6) / edited by Godage B. Wickramanayake, Arun R. Gavaskar, Abraham S.C. Chen.
        p.   cm.
    "The Second International Conference on Remediation of Chlorinated and Recalcitrant Compounds, Monterey, California, May 22–25, 2000."
    Includes bibliographical references and index.
    ISBN 1-57477-100-0 (alk. paper)
        1. Organochlorine compounds--Environmental aspects--Congresses. 2. Oxidation--Congresses. 3. In situ remediation--Congresses. 4. Hazardous waste site remediation--Congresses. I. Wickramanayake, Godage B., 1953– II. Gavaskar, Arun R., 1962– III. Chen, Abraham S.C., 1951– IV. Title.

TD1066.O73 I58 2000c
628.1'683--dc21

                                                                                00-034232

Printed in the United States of America

Copyright © 2000 Battelle Memorial Institute. All rights reserved. This document, or parts thereof, may not be reproduced in any form without the written permission of Battelle Memorial Institute.

Battelle Press
505 King Avenue
Columbus, Ohio 43201-2693, USA
614-424-6393 or 1-800-451-3543
Fax: 1-614-424-3819
E-mail: press@battelle.org
Website: www.battelle.org/bookstore

For information on future environmental remediation meetings, contact:
Remediation Conferences
Battelle
505 King Avenue
Columbus, Ohio 43201-2693, USA
Fax: 614-424-3667
Website: www.battelle.org/conferences

CONTENTS

Advanced Oxidation Processes for Destruction of Recalcitrant Compounds

Innovative Applications of Permeable Barrier Systems

Long-Term Performance of Permeable Reactive Barriers

FOREWORD

Recent developments in chemical oxidation technologies have stimulated renewed interest in applying these technologies at real-world sites. At sites where contaminated groundwater plumes pose a risk to human health and the environment, reactive barriers are finding increasing use by remediation professionals. *Chemical Oxidation and Reactive Barriers: Remediation of Chlorinated and Recalcitrant Compounds* brings together the latest ideas and applications in areas such as treatability studies and modeling for and field applications of in situ permanganate oxidation; the use of Fenton's Reagent; advanced oxidation processes for destruction of recalcitrant compounds; innovative applications of permeable barrier systems; the use of innovative media in permeable barriers; and long-term performance of permeable reactive barriers.

This is one of seven volumes resulting from the Second International Conference on Remediation of Chlorinated and Recalcitrant Compounds (May 22–25, 2000, Monterey, California). Like the first meeting in the series, which was held in May 1998, the 2000 conference focused on the more problematic contaminants—chlorinated solvents, pesticides/herbicides, PCBs/dioxins, MTBE, DNAPLs, and explosives residues—in all environmental media and on physical, chemical, biological, thermal, and combined technologies for dealing with these compounds. The conference was attended by approximately 1,450 environmental professionals involved in the application of environmental assessment and remediation technologies at private- and public-sector sites around the world.

A short paper was invited for each presentation accepted for the program. Each paper submitted was reviewed by a volume editor for general technical content. Because of the need to complete publication shortly after the Conference, no in-depth peer review, copy-editing, or detailed typesetting was performed for the majority of the papers in these volumes. Papers for 60% of the presentations given at the conference appear in the proceedings. Each section in this and the other six volumes corresponds to a technical session at the Conference. Most papers are printed as submitted by the authors, with resulting minor variations in word usage, spelling, abbreviations, the manner in which numbers and measurements are presented, and formatting.

We would like to thank the people responsible for the planning and conduct of the Conference and the production of the proceedings. Valuable input to our task of defining the scope of the technical program and delineating sessions was provided by a steering committee made up of several Battelle scientists and engineers – Bruce Alleman, Abraham Chen, James Gibbs, Neeraj Gupta, Mark Kelley, and Victor Magar. The committee members, along with technical reviewers from Battelle and many other organizations, reviewed more than 600 abstracts submitted for the Conference and determined the content of the individual sessions. Karl Nehring provided valuable advice on the development of the program schedule and the organization of the proceedings volumes. Carol

Young, with assistance from Gina Melaragno, maintained program data, corresponded with speakers and authors, and compiled the final program and abstract books. Carol and Lori Helsel were responsible for the proceedings production effort, receiving assistance on specific aspects from Loretta Bahn, Tom Wilk, and Mark Hendershot. Lori, in particular, spent many hours examining papers for format and contacting authors as necessary to obtain revisions. Joe Sheldrick, the manager of Battelle Press, provided valuable production-planning advice; he and Gar Dingess designed the volume covers.

Battelle organizes and sponsors the Conference on Remediation of Chlorinated and Recalcitrant Compounds. Several organizations made financial contributions toward the 2000 Conference. The co-sponsors were EnviroMetal Technologies Inc. (ETI); Geomatrix Consultants, Inc.; the Naval Facilities Engineering Command (NAVFAC); Parsons Engineering Science, Inc.; and Regenesis.

As stated above, each article submitted for the proceedings was reviewed by a volume editor for basic technical content. As necessary, authors were asked to provide clarification and additional information. However, it would have been impossible to subject more than 300 papers to a rigorous peer review to verify the accuracy of all data and conclusions. Therefore, neither Battelle nor the Conference co-sponsors or supporting organizations can endorse the content of the materials published in these volumes, and their support for the Conference should not be construed as such an endorsement.

Godage B. Wickramanayake and Arun R. Gavaskar
Conference Chairs

SIMULATION OF OXIDATIVE TREATMENT OF CHLORINATED COMPOUNDS BY PERMANGANATE

Hubao Zhang (Duke Engineering & Services, Albuquerque, New Mexico)
Franklin W. Schwartz (The Ohio State University, Columbus, Ohio)

ABSTRACT: Computer models are very important in optimizing the oxidzation remediation process because detailed process knowledge reduces the cost of the operation and potential problems. A computer model (ISCO3D) was developed to incorporate NAPL dissolution, chemical reactions, and solute mass transport, which would accompany an in-situ chemical oxidization scheme. The Strang operator-splitting method, coupling the different physical and chemical processes, and an exponentially-expressed solution of the kinetic reaction equations has been implemented to speed up the solution process. The code has been demonstrated with column experiments, small-scale field experiments, and a field demonstration. These examples are representative of published experimental work for the oxidization of chlorinated compounds by permanganate in a variety of settings. In all cases, the simulation results matched well with the measurements. The computer model simulated the concentration behavior with time for permanganate, various contaminants (TCE, PCE, and others), a reactive aquifer material (mainly organic carbon), Cl^-, CO_2, H^+, and MnO_2 during these experiments. The computer model also tracked changes in porosity and permeability due to mineral precipitation. The computer model is proving to be a useful tool for assisting the design and the prediction of the oxidization processes under in-situ field conditions.

INTRODUCTION

This paper describes a new computer code (ISCO3D), which was developed to simulate the coupled processes of NAPL dissolution, chemical reactions, and solute mass transport related to an *in-situ* oxidization scheme. The code is also capable of simulating key reactions between aquifer material and MnO_4^- and the kinetic sorption of chemicals, and has been optimized for computational efficiency. The code fills a gap between the laboratory studies and field operations by providing a numerical model capable of assisting with the design of systems and the interpretation of field observations.

MATHEMATICAL FORMULATION

In developing the computer code, it is assumed that there exists a zone of residual NAPL or a dissolved plume of chlorinated compounds in the saturated zone. MnO_4^- is injected into an aquifer as an oxidant and assumed to follow some flow path. The MnO_4^- is assumed to react with natural organic carbon and other oxidizable aquifer materials, as well as the dissolved chlorinated compounds in the aqueous phase. With the injection of MnO_4^- into a NAPL zone, the utilization of contaminants in the aqueous phase accelerates the dissolution of the NAPL.

Assuming that the NAPL is immobile and the saturation of the NAPL in the medium is small, the ground-water flow equation can be written as

$$\nabla(\mathbf{K} \nabla h) + Q = S_s \frac{\partial h}{\partial t} \tag{1}$$

where $\mathbf{K}$ is the hydraulic conductivity tensor (L/T), h is the water head (L), Q is the source (L^3/T), t is time (T), and S_s is the specific storage (L^{-1}). The transport equation is expressed in operator notation as:

$$\frac{\partial c_i}{\partial t} + L(c_i) = r_i \tag{2}$$

where $L = \nabla \bullet (n v - n D \nabla)$ \hfill (3)

where n is porosity, $\mathbf{D}$ is the dispersion coefficient tensor ($L^2 T^{-1}$), v is the linear groundwater velocity vector (LT^{-1}), c_i is the concentration for species i (ML^{-3}), and r_i is the reaction rate for species i ($ML^{-3} T^{-1}$).

Equations representing chemical reactions between MnO_4^- and tetrachloroethene (PCE), trichloroethene (TCE), and dichloethene (DCE), and aquifer material (for example, natural organic carbon) are summarized as follows:

(a) PCE

$$C_2Cl_4 + 2MnO_4^- \Rightarrow 4Cl^- + 2CO_2 + 2MnO_2(s) \tag{4}$$

(b) TCE

$$C_2HCl_3 + 2MnO_4^- \Rightarrow 3Cl^- + 2CO_2 + H^+ + 2MnO_2(s) \tag{5}$$

(c) DCE

$$C_2H_2Cl_2 + 2MnO_4^- \Rightarrow 2Cl^- + 2CO_2 + 2H^+ + 2MnO_2(s) \tag{6}$$

(d) Natural organic carbon

$$C + MnO_4^- \Rightarrow CO_2 + MnO_2(s) \tag{7}$$

(5) Sorption reaction

$$C_2Cl_4 \Leftrightarrow C_2Cl_4(s) \tag{8}$$

Equations (4) through (6) describe the oxidation of PCE, TCE, and DCE by MnO_4^-. Equation (7) simplifies the oxidization of aquifer materials (natural organic carbon and others) by MnO_4^-. Batch experiments showed that the oxidization processes (4) through (6) are irreversible reactions (Yan and Schwartz, 1999). Yan and Schwartz (1999) have determined the order and rate constants for reactions in batch experiments (Table 1). In the experiments, initial TCE concentration varied from 0.031 to 0.083 mM while initial MnO_4^- concentration changed from 0.37 to 1.2 mM.

In this study, a modified Strang-splitting approach was developed to couple the mass transport and chemical reaction solutions. In the calculation of chemical reactions, a first-order loss rate was assumed so that the reaction calculation can be expressed exponentially. The approach significantly reduces the solution time.

Table 1. Kinetic rate constants for PCE, TCE, and DCE oxidation by MnO_4^-.

Chlorinated Ethylenes	k $(M^{-1}s^{-1})$
PCE	0.045
TCE	0.65
Cis-DCE	0.92
Trans-DCE	30
1,1-DCE	2.38

SIMULATION RESULTS
Simulation of Permanganate Treatment of PCE in Column Experiments

Schnarr et al. (1998) undertook column experiments with an initial PCE saturation of 1%. They flushed the columns with a 10 g/L MnO_4^- solution for 214 h, and then with deionized water to 680 hours (Figure 1). First, the MnO_4^- oxidized organic matter and other reactive materials in the column, before it reached the NAPL zone. With time, destruction of PCE in the aqueous phase enhanced the solubilization of the DNAPL. Simulations indicated that about 2.62 g/kg of soil materials, including solid organic carbon in the column, was oxidized to match the experimental result. This consumption of MnO_4^- in reactions is supported by the late arrival time in the measured breakthrough curve of MnO_4^-.

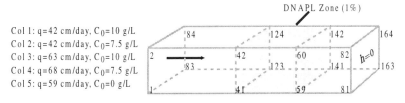

Col 1: q=42 cm/day, C_0=10 g/L
Col 2: q=42 cm/day, C_0=7.5 g/L
Col 3: q=63 cm/day, C_0=10 g/L
Col 4: q=68 cm/day, C_0=7.5 g/L
Col 5: q=59 cm/day, C_0=0 g/L

Δx = 0.01 m, Δy = Δz = 0.045 m, α = 0.02 m, n = 0.41

**Figure 1. Model setup for simulating the column experiments by
Schnarr et al., (1998)**

In each column experiment, there were three kinetic reactions of interest. Reaction rate constants from batch experiments (Table 1) (Yan and Schwartz, 1999) were used to describe oxidation of PCE by MnO_4^-. A large reaction rate constant of 450 $M^{-1}s^{-1}$ is assumed for reactions between aquifer materials and MnO_4^-. The approximate magnitude was estimated by the relative rate of migration of MnO_4^- as compared to the linear ground-water velocity. Modeling the kinetic dissolution of pure-phase PCE was similar to that formulated by Power et al. (1994).

The simulated breakthrough curve of the MnO_4^- is similar to the measured one in terms of the arrival time. The simulated breakthrough curves for PCE are comparable to the measurements (Figures 2a through 2e). We were able to reproduce the dramatic concentration decrease in the aqueous phase once MnO_4^- was applied to the NAPL zone. When the MnO_4^- was replaced by water in the injection fluid the PCE concentrations rebounded to a higher level.

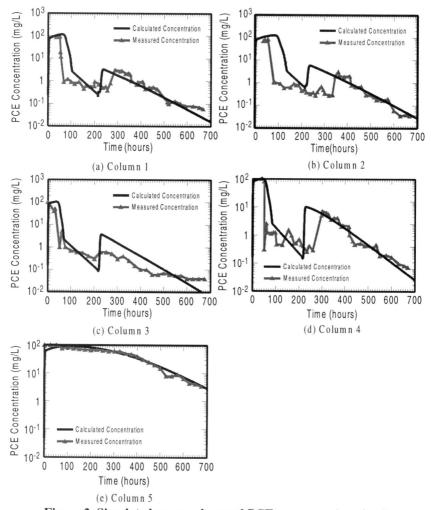

Figure 2. Simulated versus observed PCE concentrations for five column experiments.

The basic characteristics of the breakthrough curves for Cl^- were also simulated. The calibration of model parameters was based on PCE effluent concentrations. The production of Cl^- demonstrates that oxidization is occurring.

Simulation of A Field-Scale Experiment at a DOE facility

A field-scale experiment of MnO_4^- oxidation of TCE was performed at the Department of Energy Portsmouth Gaseous Diffusion Plant (PORTS) in Ohio by researchers from Oak Ridge National Laboratory (ORNL) (West et al., 1997). The field test involved injecting MnO_4^- in one horizontal well and extracting fluid from another horizontal well about 27 m up-gradient of the injection well (Figure 3).

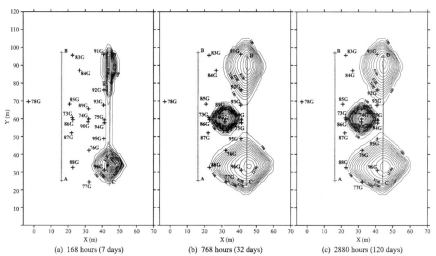

(a) 168 hours (7 days)   (b) 768 hours (32 days)   (c) 2880 hours (120 days)

Figure 3. Simulated permanganate concentration for a field scale experiment at the DOE Portsmouth site (from West et al., 1997).

In the experiment, there was preferential flow of MnO_4^- near the ends of the injection well (Figure 3). There was also difficulty in the experiment in maintaining a constant injection/withdrawal rate of 10 gpm for both the injection and extraction wells. The injection rate was reduced to 6 gpm. MnO_4^- was also added through a vertical well in the center of the treatment zone to improve the efficiency of flooding. A total of 10,740 kg of $KMnO_4$ was injected between July 26 and August 27, 1997 through the horizontal well. In addition, 1960 kg of $KMnO_4$ was delivered using the vertical well (74G injected at 2 gpm) between August 20, 1997 and August 28, 1997. There was a significant reduction in the TCE concentration in the ground-water as the experiment continued.

A two-dimensional model (51x51 nodes) was constructed to represent the treatment zone and its vicinity (Figure 3). The simulation grid was designed with relatively small node spacings inside the test region ($\Delta x = 1$m, and $\Delta y = 2$ m) and relatively large spacings outside (maximum $\Delta x = 6$ m, and maximum $\Delta y = 7.5$ m). The thickness of the simulation domain is the same as the Gallia Formation ($\Delta z = 1.5$ m). The flow from the horizontal injection well to the Gallia was not uniformly distributed along the wellbore. As noted by West et al. (1997) and Korte et al. (1997), aquifer heterogeneity near the well, clogging of the well, or a pressure drop along the well screen may have contributed the variable flooding of the zone between the wells. Without additional data, we could not determine the exact cause of this pattern of flooding. To simplify the model setup, we adjusted flow rates along the well to provide the observed pattern of MnO_4^- flooding, while keeping the hydraulic conductivity constant.

The simulation indicates that dissolved TCE concentrations were significantly reduced, where the MnO_4^- was delivered, as compared to the initial TCE concentration (Figure 4a). At time = 7 days, the contour line of 0.005 mg/L TCE (drinking water standard) is located near the horizontal injection well (Figure 4b). After 32 days, the zone of 0.005 mg/L or less TCE concentration expanded outward significantly (Figure 4c). After this time (e.g., 120 days; Figure 4d), there is relatively little enlargement of the treated zone because the zone flooded by MnO_4^- didn't expand very much.

A detailed match of simulated and measured MnO_4^- distributions is impossible because hydraulic head and information on the extent of reaction between the MnO_4^- and the oxidizable aquifer materials (mainly organic carbon content) were unavailable. The present simulation suggests that 172 mg/kg of aquifer materials has been oxidized. Further sensitivity study indicates that the amount of oxidizable aquifer materials in the aquifer may significantly affect the transformation of TCE into non-harmful products.

CONCLUSIONS

A simulation and design tool was developed to model a remediation scheme that is based on the oxidation of chlorinated compounds by MnO_4^-. The computer code can simulate coupled processes including the dissolution of NAPLs; the chemical reactions among permanganate, dissolved chlorinated compounds, and aquifer materials; and the mass transport in three dimensions. We demonstrated this code with applications to a series of column experiments (Schnarr et al, 1998), one small field experiment (Schnarr et al, 1998), and a larger-scale field demonstration of source zone flushing (West et al., 1998). The success in generally matching this variety of experiments provides an initial confirmation that the code has been able to capture the essential physics and chemistry of the processes. The computer simulations, together with column, test cell, and field experiments also suggest that the proper determination of NAPL saturation, aqueous concentration of chlorinated compounds, and the reactivity of the aquifer material with MnO_4^- are critical to a successful chemical oxidization design and field operation. This study and previous work also indicate that oxidization of chlorinated compounds is probably most efficient at low NAPL saturation. The chemical reaction between the chlorinated compounds and MnO_4^- occurs in dissolved phase, but the dissolution of NAPLs is rate-limited. Our group is presently working to devise new approaches to accelerate the rate of oxidation.

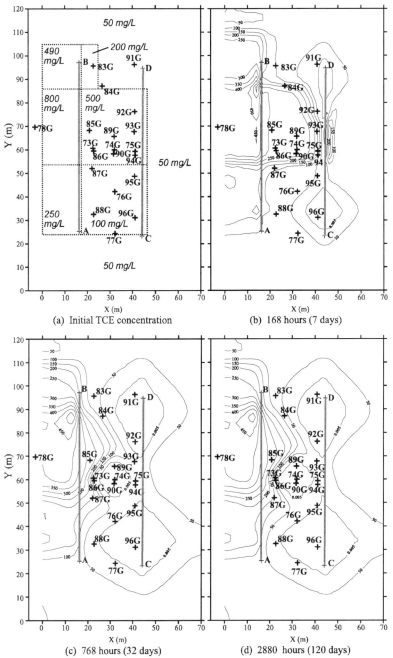

(a) Initial TCE concentration

(b) 168 hours (7 days)

(c) 768 hours (32 days)

(d) 2880 hours (120 days)

Figure 4. Simulated TCE concentration for a field scale experiment at the DOE Portsmouth site.

Clearly, there is more work to be done in the development of the modeling approach outlined in this paper. For example, the experimental work has showed that the production of $MnO_2(s)$ is significant in the oxidative treatment of chlorinated compounds with permanganate. The $MnO_2(s)$ particles are transported and deposited as colloids during the remediation process. One of our current modeling initiatives is to study the effect of $MnO_2(s)$ on the flooding efficiency with oxidative schemes.

ACKNOWLEDGEMENTS

Funding for this study was provided by Department of Energy Environmental Management Science Program under Grant No. DE-FG07-96ER14735.

REFERENCES

Korte, N., M. Muck, P. Kearl, R. Siegrist, T. Houk, R. Schlosser, J. Zutman, Field evaluation of a horizontal well recirculation system for groundwater treatment: Field demonstration at X-701B Portsmouth Gaseous Diffusion Plant, Piketon, Ohio, Oak Ridge National Laboratory, Grand Junction, Colorado, 1997.

Powers, S. E., L. M. Abriola, and W. J. Weber, Jr., An experimental investigation of nonaqueous phase liquid dissolution in saturated subsurface systems: Transient mass transfer rates, Water Resour. Res., 30, 321-332, 1994.

Schnarr, M., C. Truax, G. Farquhar, E. Hood, T. Gonullu, and B. Stickney, Laboratory and controlled field experiments using potassium permanganate to remediate trichloroethylene and perchloroethylene DNAPLs in porous media, Journal of Contaminant Hydrology, 29, 205-224, 1998.

West, O. R., S. R. Cline, W. L. Holden, F. G. Gardner, B. M. Schlosser, J. E. Thate, D. A. Pickering, T. C. Houk, A full-scale demonstration of in situ chemical oxidation through recirculation at the X-701B site, Oak Ridge National Laboratory, Oak Ridge, TN, ORNL/TM-13556, 101 p., 1997.

Yan, Y. E., and F. W. Schwartz, Oxidative degradation and kinetics of chlorinated ethylenes by potassium permanganate, Journal of Contaminant Hydrology, 37, 343-365, 1999.

SOURCE ZONE MASS REMOVAL USING PERMANGANATE: EXPECTATIONS AND POTENTIAL LIMITATIONS

N. R. Thomson, E. D. Hood, L. K. MacKinnon
(University of Waterloo, Waterloo, Ontario, Canada)

ABSTRACT: The capability of oxidation technologies to destroy certain types of contaminants *in situ* has generated a demand for unbiased performance assessments. This technical note briefly presents the findings from two trials where a source zone was actively flushed with a concentrated permanganate solution. The first trial was performed at a field site and focused on mass removal from a relatively homogeneous residual source zone while the second trial was conducted in a laboratory setting and addressed mass removal from a pooled source zone. For each of these trials extensive characterization of the pre-oxidation and post-oxidation conditions was undertaken. The findings from these investigations indicate that a substantial amount of the mass present can be removed *in situ*; however, non-ideal flushing conditions and the formation of solid phase precipitates may limit complete mass removal.

INTRODUCTION

The failure of pump-and-treat systems to remove significant DNAPL mass from the subsurface has increased the need for new remedial technologies designed to isolate, remove, or treat DNAPL sources. *In situ* chemical oxidation using permanganate is one of the promising technologies that has appeared over the last ten years with the capability of removing considerable mass from DNAPL source zones.

Since 1989, research conducted at the University of Waterloo has been directed at providing an understanding of this technology (e.g., Schnarr et al., 1998, and Hood et al., 2000). Both active flushing and passive dosing approaches are being investigated. Preliminary laboratory experiments with soil columns indicated that residual perchloroethylene (PCE) was effectively degraded by $KMnO_4$ flushing and that Cl^-, a stable end product, was a useful tracer to monitor the treatment process. Measurements of Cl^- and PCE in the effluent indicated that almost complete oxidation occurred within the soil columns. Following these column studies, two field trials were performed in a 7.5 m^3 double walled sheet pile cell at Canadian Forces Base (CFB) Borden to investigate the efficacy of this treatment strategy at the pilot field scale. In the first field trial a homogeneous PCE residual source zone (saturation ~8%) was flushed with a 10 g/L $KMnO_4$ solution for 120 days, followed by a 60 day water flush. Chloride measurements from the extraction wells indicated that 91% of the original source zone had been destroyed. Measurements in the effluent from the water flush and in core samples from the source zone indicated that PCE had been completely removed. In the second of the two field trials, ~8 L of an equal mass mixture of trichloroethylene (TCE) and PCE was released into the cell through six drive points over a period of nine days to construct a heterogeneous source zone distribution. This source zone was flushed with a solution of 10 g/L KMnO$_4$ for a period of 290 days, followed by a 120 day water flush. The chloride mass balance indicated that 62% of the source zone was destroyed. Measurements of PCE and TCE during the post-oxidation water flush confirmed that pure phase was still present, and was assumed to be in localized zones of high pure phase saturation.

This technical note presents the findings from two recent permanganate trials where either a DNAPL residual or pooled source zone was flushed with a concentrated permanganate solution. In each of the trials presented, solvent plume loading and DNAPL mass removed were determined using detailed spatial sampling of the source zone as measures of treatment effectiveness. The important findings from these two trials are presented, and issues relating to mass removal expectations and potential limitations are discussed.

FIELD TRIAL

Site Description and Methodology. This field trial was conducted in the unconfined sandy aquifer at CFB Borden from September 1995 until October 1998. A constructed source zone (1.5 x 1.0 x 0.5 m) of a homogenous multi-component DNAPL residual (12,540 g of PCE, 8,990 g of TCE, and 1,450 g of chloroform (TCM); initial volumetric saturation of 5%), was installed ~1.0 m below the water table. Initially this source zone was used to study the formation and transport of the multi-component plume (Rivett et al., 1992).

Based on initial flow modeling of the injection/extraction system, six injection and three extraction wells (Figure 1) were installed for hydraulic control of the oxidant. Various treatment units were constructed on site to recycle and treat excess oxidant solution. The oxidant recycling system was used to reduce the amount of solid permanganate required to maintain a constant permanganate concentration in the injection solution, and to avoid the cost of treating the entire effluent flow.

Four phases of fieldwork were completed at this site over the three-year duration of this field trial. The initial phase involved characterization of the solvent plume emanating from the source zone and the performance of the injection/extraction system through a tracer test. The next phase involved flushing the source zone with a concentrated permanganate solution for 485 days and regularly monitoring permanganate and chloride concentrations in the extraction wells, the injection feed, and in nine of the 98 multilevel piezometers (multilevel fence) located ~1 m down gradient from the source. In addition, periodic sets of samples were taken from numerous other multilevel pizeometers and from within the source zone using a Waterloo profiler. Following the oxidant flush, measurements of plume load and a second tracer test were conducted using the same system configuration and procedure as during the pre-oxidation tracer test. After the injection/extraction system had been off for two weeks, allowing the establishment of an aqueous phase plume under natural gradient conditions, a final round of samples was collected at the multilevel fence. The final phase of work was conducted to assess the quantity and distribution of PCE, TCE and manganese dioxide within the source zone. Twenty-two 1-m cores were collected from the source zone and sub-sampled at 5 cm intervals for solvent and MnO_2 analysis.

Pre-oxidation Solvent Mass Estimate. Frind et al. (1999) used the extensive monitoring data complied by Rivett et al. (1992) to calibrate a three-dimensional flow and transport model and estimated, as part of this effort, the amount of mass remaining in the source zone. These results were used to estimate the source mass at the beginning of the oxidation field trial (following 2400 days of ambient gradient source flushing). Their results suggested that the source contained 9,030 g PCE and 1,620 g TCE at the start of the oxidant flush, and that TCM had been essentially removed due to its high aqueous solubility.

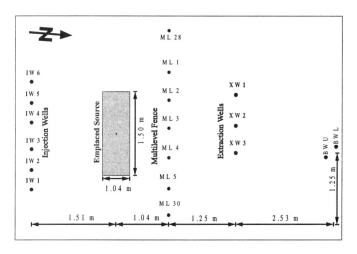

FIGURE 1. Plan view of field trial site including injection wells (IW1 to IW6), primary extraction wells (XW1 to XW3), and secondary extraction wells (BWU and BWL).

Results. The oxidant flush operated continuously for 485 days and used 892 kg of $KMnO_4$. The average oxidant concentration in the feed solution was 8 ± 3 g/L. Early monitoring results suggested that poor oxidant recovery was occurring, and the initial injection and extraction flow rates of ~400 mL/min were lowered to ~150 mL/min and two additional recovery wells were installed (Figure 1) to minimize oxidant migration below the existing extraction system. After the oxidant injection was terminated, the extraction system was run for 177 days to remove the oxidant and chloride remaining in the subsurface. During this time, the injection wells were not used and the treatment zone was flushed with ambient aquifer water.

The temporal variation of the average chloride concentration extracted and injected is presented on Figure 2. These data are difficult to interpret since the impact of oxidant recycling must be considered. The extracted chloride concentration rose after 50 days to ~150 mg/L and remained at this level for the next 100 days.   Overtime the effluent chloride decreased until it was indistinguishable from the background chloride concentration.   While the feed solution contained chloride, an increase in the chloride concentration above the background level while in the treatment zone would indicate that additional oxidation of DNAPL was occurring.   A slight lag in effluent response is evident during initial chloride breakthrough (Figure 2) but the variability of the effluent data relative to the injected concentration made a quantitative comparison impossible.   Qualitatively, however, the small differences between the injected and extracted chloride concentrations at this sampling scale suggest that after 300 days of oxidant flushing, oxidation was no longer occurring.

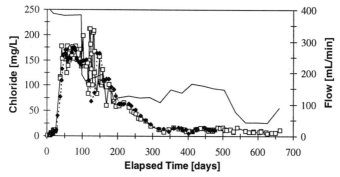

FIGURE 2. Temporal profile of the average chloride concentration extracted (□) and injected (♦). Also shown is the monthly average extraction flow rate.

While the extraction flow monitoring indicated that only low Cl⁻ concentrations remained after 300 days, a distinct chloride signature (> 250 mg/L), corresponding to the approximate location of the source, was evident at the multilevel fence up to 425 days providing a strong indication of continuing source oxidation.   However, after this time, repeated sampling of the multilevel piezometers failed to detect a chloride plume.   Based on these results, it appeared that either a small chloride plume was passing between the spacing of the piezometers installed in the multilevel fence, or that no chloride plume was present.   Source zone profiles collected after 480 days indicated the presence of a distinct chloride zone (~450 mg/L) suggestive of continued mass oxidation; however, the inability to detect this chloride plume at the multilevel fence indicated the plume was small.

A total of 339 samples from the source zone cores were analyzed for TCE and PCE.   Bulk soil concentrations of TCE ranged from non-detect to 30.9 mg/kg with a mean concentration of 0.2 mg/kg.   In comparison, PCE concentrations ranged from 0-163.1 mg/kg with a mean of 25.1 mg/kg.   The mean TCE concentration was approximately two-orders of magnitude lower than the mean PCE concentration, consistent with relatively small mass of TCE at the start of the oxidant flush and the higher TCE oxidation rate.   The higher bulk soil concentrations of TCE and PCE were erratically scattered on the down-gradient side of the source while, in contrast, the manganese dioxide was clearly present on the up-gradient edge of the source with the center being the region of highest concentration.

LABORATORY TRIAL

Physical Model Description and Methodology. The second trial involved a laboratory investigation designed to assess the effectiveness of using a permanganate solution to reduce the mass in a DNAPL pool. A two-dimensional physical model (2.5 m long by 0.45 m high by 0.15 m wide) was filled with silica sand overlying a silica flour lens simulating a two-dimensional saturated zone overlying a capillary barrier (Figure 3). A total of 51 spatially distributed sample ports were installed on one face of the model to permit periodic sampling. The sample ports were spatially distributed so that an accurate evaluation of the dissolution and oxidation processes could be undertaken near the bottom of the model where the DNAPL pool would be located. A pump at the influent end of the model and a constant head at the effluent end controlled flow through the model aquifer.

PCE was emplaced into the model aquifer over 12 hours using a series of strategically placed drive points to form a continuous PCE accumulation directly on top of the silica flour. After 35 days, the redistribution of PCE appeared to have stopped, and the final amount of PCE remaining in the physical model was ~700 g. Based on a series of separate experiments it was observed that the millimeter scale heterogeneity of the sand bed strongly influenced the flow of PCE causing it to migrate preferentially along the bedding structure forming fine layers of higher saturation.

Buffered Milli-Q water was pumped through the model aquifer to simulate water flush conditions at a flow rate of 4 mL/min. The water flush was continued for 90 days prior to the addition of $KMnO_4$, and was resumed after the oxidation phase for 150 days to evaluate the post-oxidation conditions. In each case the water flush was continued until a steady-state concentration of PCE in the effluent was reached.

The oxidant feed was prepared to a concentration of 10 g/L using laboratory grade $KMnO_4$. The progress of the oxidation was monitored by taking regular measurements of Cl^-, MnO_4^-, and pH, from the influent, effluent, and spatial sample ports. Chloride, as the conservative reaction product, was used to evaluate the mass of PCE destroyed by the oxidant. The oxidant flush was continued for 146 days.

During the post-oxidation water flush the concentration of Cl^-, PCE and Mn^{2+} in the effluent were monitored. After 140 days of the post-oxidation water flush a measurement of the spatial distribution of PCE was performed, followed by a measurement of the Cl^- spatial distribution.

The collection and analyses of nine soil cores and the excavation of the surrounding porous medium were used to evaluate the extent of oxidation, the remaining distribution of pure phase PCE, and presence of MnO_2 precipitates within the model aquifer. Each core was sub-sampled at 2 cm intervals and adjacent samples of the sand were used to evaluate the spatial distribution of PCE and recoverable Mn^{2+}.

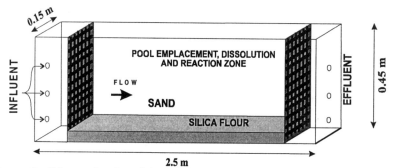

Figure 3. Schematic of model used for the DNAPL pool oxidation study.

Results. The flux-averaged effluent Cl⁻ concentration gradually rose from near zero to 450 mg/L over the course of the first 40 days (Figure 4). During the next 106 days of oxidation the concentration decreased from this maximum concentration to ~180 mg/L. Based on stoichiometric considerations, a concentration of 180 mg/L of Cl⁻ corresponds to an equivalent aqueous phase PCE concentration of 210 mg/L. This equivalent concentration of PCE was an order of magnitude higher than the steady-state pre-oxidation concentration of ~20 mg/L, evidence that the presence of the oxidant had enhanced the overall mass transfer rate from the PCE pool by an order a magnitude.

The spatial distribution of Cl⁻ indicated elevated levels (>200 mg/L) existing for first 100 days of oxidation over the lower half of the model aquifer. The maximum Cl⁻ concentration observed was near 3000 mg/L and persisted until ~60 days. After 146 days of oxidation the maximum Cl⁻ concentration observed was near 120 mg/L reflecting a reduced rate of oxidation.

The spatial variation in the MnO_4^- concentration over time also coincided with the observed trend in the Cl⁻ concentration. The MnO_4^- concentration across the upper ~25 cm was close to the influent concentration of 10 g/L during the entire oxidant flush period. Along the bottom 20 cm the MnO_4^- concentration was lowest near the silica flour and towards the effluent end. These depressed MnO_4^- concentrations suggested that the oxidant had been consumed by the reaction with the PCE to produce the elevated Cl⁻ concentrations observed in the same area. After five months of oxidant flushing the system had a uniform concentration of ~10 g/L. Along with the decrease in the Cl⁻ production, these data suggested that the reaction between the PCE and MnO_4^- had diminished considerably.

The excavation of the model aquifer provided an opportunity to visually observe the distribution of MnO_2 deposits. Along the bottom 8 cm there was a much higher quantity of MnO_2 deposits, as indicated by black coloration. At some locations the accumulation of MnO_2 had agglomerated, so that hard, rock-like deposits of MnO_2 and sand were formed. These agglomerated zones of MnO_2 deposits were heterogeneously distributed along the bottom of the sand, suggesting that the initial distribution of pure phase PCE saturation was similarly heterogeneous.

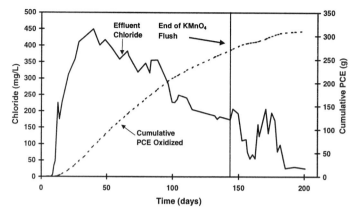

Figure 4. Flux-averaged effluent concentration over the course of the oxidant and post–oxidant flushes.

The Mn^{2+} concentration profiles for each of the soil cores were consistent with the observed distribution of MnO_2 deposits. The concentration of Mn^{2+} was, in general, highest directly above the silica flour base confirming the original distribution of PCE. Peak measurements of Mn^{2+} in each of the cores varied between 2600 to 15000 mg Mn^{2+} / kg of sand. The peak measurement of Mn^{2+} was considerably lower in cores taken near the influent end of the model.

The concentration of bulk PCE in the soil cores varied with the height above the silica flour base and the core location. The range of concentrations observed in each of the cores generally increased from the influent to the effluent end, suggesting that PCE mass removal had proceeded in that direction. The peak PCE concentrations observed in the cores varied between 0.31 and 2580 mg PCE/kg of sand, a four-order of magnitude range. The PCE distribution observed in the soil cores agreed closely with the spatial distributions of Cl^- at the end of the oxidant flush, and PCE during the post-oxidation water flush. It was also determined that the presence of PCE in the soil cores could account for the PCE mass that had not been removed during the oxidant and water flushes.

SUMMARY

These permanganate flushing trials clearly demonstrate that a substantial amount of the initial *in situ* mass can be destroyed. However, at most sites it is difficult to accurately determine the initial DNAPL mass. An alternative measure of the treatment effectiveness is related to the reduction in solvent mass flux. Table 1 summarizes the mass flux reduction and mass removal achieved for each of these trials. The mass flux estimate, if performed in a systematic fashion before and after treatment, provides a measure that is directly related to the source strength and hence the ability of the source zone to generate a plume.

Since most source zones may be a combination of residual and pooled DNAPLs, the results from these two trials provide convenient endpoints for consideration. For the residual portions of a source zone the results from the field trial indicate that nearly all of the mass can be destroyed *in situ* with an expectation of a comparable level of mass flux reduction. The remaining mass in the field trial discussed here was believed to be a result of either preferential flow pathways giving rise to diffusion limited mass removal, or desorption limitations controlled by intra-particle diffusion. For source zone regions with large DNAPL accumulations or pools, the laboratory trial findings suggest that some level of mass removal is expected, and at least in the short-term a proportionally larger reduction in the mass flux. The decline in the rate of mass removal in this laboratory trial is likely related to the small surface area to volume ratio relative to a residual DNAPL which provides less interface area for mass transfer and permits concentrated MnO_2 deposits to form at the pool interface that limit mass transfer.

TABLE 1. Summary of mass flux reduction and mass removal.

Trial	Pre-oxidation		Post-oxidation		Percent Reduction	
	Mass [g]	Flux [mg/day]	Mass [g]	Flux [mg/day]	Mass [g]	Flux [mg/day]
Field	10647	4317	0.05	239	>99%	95%
Laboratory	700	110	385	27	45%	75%

ACKNOWLEDGEMENTS
Financial support for this research has been provided by the University of Waterloo Consortium for Solvents-In-Groundwater Research Program. Additional funding was provided by a Natural Science and Engineering Research Council of the Government of Canada research grant (N. Thomson), and the former Waterloo Centre for Groundwater Research (now CRESTech).

REFERENCES

Frind, E.O., J.W. Molson, M. Schirmer, and N. Guiguer. 1999. "Dissolution and mass transfer of multiple organics under field conditions: The Borden emplaced source." *Water Resour. Res. 35*(3), 683-694.

Hood, E., N. R. Thomson, G. J. Farquhar, and D. Grossi. 2000. "Experimental determination of the kinetic rate law for the oxidation of perchloroethylene *by* potassium permanganate." *Chemosphere. 40*(12):1383-1388.

Rivett, M.O., S. Feenstra, and J.A. Cherry. 1992. "Groundwater zone transport of chlorinated solvents: a field study." In the proceedings of Modern Trends in Hydrogeology, Hamilton, Ontario, May 10.

Schnarr, M., C. Truax, G. Farquhar, E. Hood, T. Gonullu, B. Stickney. 1998. "Laboratory and Controlled Field Experiments Using Potassium Permanganate to Remediate Trichloroethylene and Perchloroethylene DNAPLs in Porous Media." *J. Contam. Hydrol. 29*(3), 205-224.

PHASE-TRANSFER-CATALYSIS ON THE OXIDATION OF TRICHLOROETHYLENE BY PERMANGANATE

Yongkoo Seol and Frank W. Schwartz
The Ohio State University, Columbus, OH, USA

Abstract: Permanganate (MnO_4^-) oxidation has been demonstrated successfully as an effective in situ process for degrading chlorinated solvents in the aqueous phase. This study evaluated the effectiveness of phase-transfer-catalysts (PTCs) in enhancing degradation rates. PTCs work by transferring MnO_4^- into the DNAPL phase where oxidation also occurs. MnO_4^- oxidation of trichloroethylene (TCE) was studied using kinetic batch experiments in conjunction with three PTCs. The influences of PTCs on the aqueous solubility of TCE were also assessed to determine the relative contributions of oxidation in the aqueous and the DNAPL phases. Columns containing sand, residually saturated with TCE, were employed to demonstrate the effect of large interfacial areas on those MnO_4^- reactions. TCE oxidation rates were estimated by measuring the chloride concentration and UV-Vis absorbance for MnO_4^- in the aqueous phase. The enhancement of TCE destruction by the PTCs was reflected in the more rapid consumption of MnO_4^- and production of chloride ions. The TCE dissolution in aqueous phases, however, was not increased with PTCs. Therefore, the selected PTCs increased the rate of TCE decomposition by catalyzing MnO_4^- oxidation in the DNAPL phase. Large interfacial area further enhanced the reaction by facilitating the PTC transfer.

INTRODUCTION

Potassium permanganate ($KMnO_4$) has been widely studied for the oxidative dechlorination of TCE in bench scale experiments (Yan, 1998). Both laboratory and mathematical modeling results suggested that the limited solubility of chlorinated solvents in the aqueous phase ultimately controls the rate of oxidative destruction. TCE oxidation occurs only in the aqueous phase and involves dissolved species, the permanganate ion (MnO_4^-) and the chlorinated ethylenes. Thus, the solubility of the contaminants indirectly controls the utilization of MnO_4^- in the reaction and keeps reaction rates relatively small.

This paper investigates the use of phase transfer catalysis to speed up the oxidation of chlorinated solvents. Phase transfer catalysis can increase the overall rate of DNAPL oxidation by enabling reactions to occur in both the aqueous and DNAPL phases. Adding a phase-transfer catalyst (PTC) transfers some of MnO_4^- to the nonaqueous phase. Thus, MnO_4^- can oxidize chlorinated compounds both in the aqueous phase and in the pure-phase solvent.

The objective of this study is to evaluate whether the catalyzed oxidation scheme can enhance the oxidation of TCE with MnO_4^-. We conducted kinetic batch and column experiments to verify the hypothesis that PTCs would oxidize TCE in the nonaqueous phase and speed up the overall rate of TCE decomposition. Our study was designed as a proof-of-concept work and concentrated on the enhancement of TCE oxidation with MnO_4^- in the presence of pure TCE and selected PTCs.

THEORETICAL BACKGROUND

Phase transfer catalysis is a technique for converting similar chemical species situated in two or more phases. Because of its polarity, an ionic species cannot normally enter nonpolar organic substances. However, once it combines with a PTC, the organic phase extracts the association of ionic species with the PTC from the aqueous phase. Common PTCs include organic-soluble cations such as quaternary ammonium or phosphonium ions, which contain both liphophilic and hydrophilic moieties. The amphiphilic catalysts distribute themselves between aqueous and organic phases, form an association with the reactive anion (e.g., MnO_4^-), and bring them into the organic phase in a form suitable for reaction.

The reaction of an organic compound with MnO_4^- in the presence of PTCs can be represented by an overall equation (Eq. 1), conceptually involving two separate steps (Starks *et al.*, 1994), a transfer step and an intrinsic displacement reaction step;

$$K_{aq}^+ MnO_{4\,aq}^- + RX_{org} \xrightarrow{[Q^+X^-]} K_{aq}^+ X_{aq}^- + RMnO_{4\,org} \tag{1}$$

where R is the organic reactant (e.g., TCE) with leaving group X (e.g., Cl⁻). The transfer step involves the extraction of MnO_4^- from the aqueous to the organic phase and the release of Cl⁻ back into the aqueous phase. Intrinsic displacement occurs when the transferred MnO_4^- is dissociated from the ion pair, oxidizes the chlorinated solvent, and disengages Cl⁻ from the organic compound. The transfer step includes a simple conversion processe (Eq. 2), taking the ion pair into the organic phase;

$$Q_{aq}^+ + MnO_{4\,aq}^- \longleftrightarrow [Q^+ MnO_4^-]_{org} \tag{2}$$

From Eq. (2), one can define the extraction constant (K_E), which quantifies the capability of the nonpolar solvent to extract an ionic species from the aqueous phase (Eq. 3).

$$K_E = \frac{[Q^+ MnO_4^-]_{org}}{[Q^+]_{aq} \cdot [MnO_4^-]_{aq}} = \frac{[Q^+ MnO_4^-]_{org}}{[MnO_4^-]_{aq}^2} \tag{3}$$

This constant is related to experimentally determined solubilities (Karaman *et al.*, 1984). K_E increases in direct proportion to the solubilities of the ion pair in the organic phase.

Phase transfers could be increased when tiny droplets of one phase end up scattered in the other. This distribution of phases maximizes the interfacial area for diffusional anion transfers. This effect could be important in ground-water settings. The relatively intimate dispersal of one phase in another is often found with the residual saturation of the nonwetting fluid in ground-water/DNAPL systems.

MATERIALS AND METHODS

Phase transfer catalyst. Three PTCs, tetra-n-ethylammonium bromide (TEA, $(C_2H_5)_4NBr$), tetra-n-butylammonium bromide (TBA, $(CH_3(CH_2)_3)_4NBr$), and pentyltriphenylphosphonium bromide (PTPP, $CH_3(CH_2)_4P(C_6H_5)_3Br$) with 99% or higher purity were obtained from Aldrich Chem. Co (Table 1). The catalysts were selected on the basis of their K_E and molecular structures. Due to the lack of available solubility data for TCE, estimated extraction constants for methylene chloride were used to guide the selection of PTCs. The concentration of selected PTCs in all experiments was kept below their aqueous phase solubility in order to avoid aggregation and precipitation of catalyst-permanganate salts.

Kinetic Batch Experiments. We conducted kinetic batch experiments to study the rate of oxidation of TCE in the presence of MnO_4^- and selected catalysts. $KMnO_4$ solution (2.5 mM) and concentrated catalyst stock solutions were prepared in a phosphate-buffered solution (0.1 M, $KH_2PO_4 : K_2HPO_4$, pH=8). 3.0-mL of $KMnO_4$ solution and 0.1-mL of concentrated catalyst solution were placed in the test tubes to provide a final catalyst concentration of 10-mole % $KMnO_4$. After the solution was homogenized with gentle shaking, we added 1.0-mL of TCE (99.5 %, Aldrich Chem. Co.) to the permanganate solution. While frequently shaken, the samples were monitored with chloride concentration and UV absorbance for TCE degradation. A 1.0-mL sample was scanned using a Varian Cary 1 UV-visible spectrophotometer at wavelengths ranging from 400 to 600 nm. The addition of PTCs to the permanganate solution did not impact the UV absorbance across the range of wavelengths of interest. Chloride concentration was measured using a Labconco Digital Chloridometer using coulometric titration of chloride ions.

Column Experiment. Column experiments were conducted in 10 cm long and 1 cm diameter stainless steel tubes. They were packed with medium sand (0.462-0.600 mm). Residual saturation with TCE was created by saturating with buffer solution, saturating with TCE upward flow, and back-flushed again with buffer solution downward flow to get rid of excess TCE. The actual experiments

Table 1. Chemical properties of selected PTC- MnO_4^- ion pairs.

Name/PTCs	Abbr.	FW^a	S_w^b	S_o^b	$logK_E^c$
tetraethylammonium	TEA	165.71	7.30×10^{-4}	0.235	1.64
tetrabutylammonium	TBA	277.92	2.10×10^{-4}	0.417	4.98
pentyltriphenylphosphonium	PTPP	413.35	2.40×10^{-4}	1.460	7.40

FW; formula weight (g/mole), S_w; solubility (M) in water, S_o; solubility (M) in methylene chloride, a; formula weight of bromide salts, b; Karaman, *et al.*, (1984), c; calculated with Eq. (3)

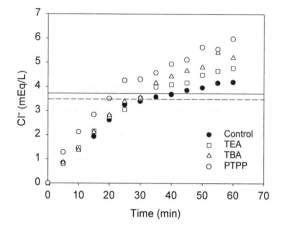

FIGURE 1. Concentration of chloride ion released from the oxidation of TCE with MnO$_4^-$ and different PTCs. Horizontal lines indicate the stoichiometrically expected chloride concentration from complete oxidation of permanganate (solid) and the experimentally measured chloride concentration (dotted).

involved two separate trials with KMnO$_4$ solution by itself and a combination of KMnO$_4$ and PTCs at a flow rate of 0.6 ml/min. The Mass of each column was measured at every step to monitor porosity, residual saturation of TCE, and TCE removal. Outflow from columns was analyzed with UV-Vis spectrometer to provide information on MnO$_4^-$ consumption.

Solubility Enhancement. The kinetics of TCE dissolution in the aqueous phase was measured in order to examine the influences of the catalysts on TCE solubility. The same kind of kinetic batch experiments was employed, except for the absence of KMnO$_4$. TCE dissolved was extracted with pentane and analyzed using a Fisons Instruments 8060 gas chromatograph equipped with a Ni[63] electron capture detector and a DB-5 capillary column.

RESULTS AND DISCUSSION

Chloride Concentration Measurements. The time variation in the concentration of chloride ion can serve as an indicator of the rate of TCE breakdown. We looked at these rates for three catalysts and a blank as a control. As Figure 1 illustrates, the extent of reaction speedup provided by the catalysts was variable. The addition of PTPP led to a marked increase in the Cl$^-$ concentration in the aqueous phase during the first 20 to 25 minutes of the experiment. These results suggest a significant enhancement in the rate of TCE degradation. PTPP, with the largest extraction constant of the three PTCs, was most efficient in transferring MnO$_4^-$ to the TCE phase. These results indicate that PTPP would be the most effective catalyst in speeding up the oxidation of TCE by MnO$_4^-$.

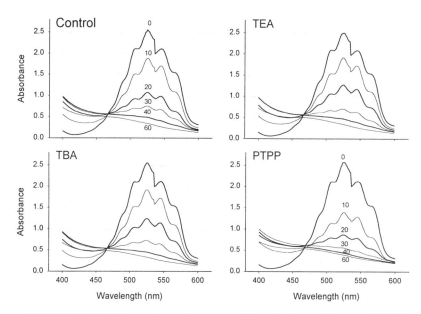

FIGURE 2. UV-Vis spectra of aqueous phase in the oxidation of TCE facilitated with MnO$_4^-$ and different PTCs. Numbers on graphs indicate the reaction time (min).

The noticeable change in reaction rates occurring at 20 to 25 minutes coincides with a concentration of Cl$^-$ of 3.5 mEq/L (Figure 1). This Cl$^-$ production represents about 93 % of the total Cl$^-$ that would be expected (i.e., 3.75 mEq/L) from the stoichiometry, given an initial permanganate concentration (2.5 mM). During the early phase of the reaction, MnO$_4^-$ would mainly participate in the oxidation. After the complete consumption of MnO$_4^-$, less efficient oxidation can be accomplished with MnO$_2$, causing TCE to breakdown more slowly. The breakpoint in reaction rates probably indicates the transition in the key oxidant from MnO$_4^-$ to MnO$_2$.

UV/Vis Absorbance Monitoring. Once catalysts were added to the system, we observed that the TCE phase, initially colorless and transparent, became purple due to the permanganate transfer. The TCE phase became clear once the reaction was completed and the MnO$_4^-$ was used up. The purple color of the TCE phase was most distinct with PTPP. We made quantitative measurements of the concentration of MnO$_4^-$ in the aqueous phase with time using UV/Vis absorbance measurements. Figure 2 displays UV/Vis spectra successively changing with time for the four different treatments. Decreases in the maximum absorbance at 526 nm indicated the loss of MnO$_4^-$ as it reacted with TCE. The rapid decrease in the maximum absorbance with PTPP confirms that PTPP is the most efficient catalyst in transferring MnO$_4^-$ into the TCE phase, and in enhancing the permanganate consumption.

The disappearance of MnO$_4^-$ with TCE oxidation can be quantified with

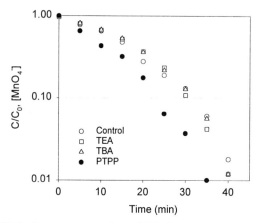

FIGURE 3. Relative concentration of MnO₄⁻ in the aqueous phase as a consequence of TCE oxidation catalyzed with different PTCs.

UV/Vis absorbance at 526 and 418 nm using the equation;

$$\frac{c_t}{c_o} = \frac{A_{526} - (\varepsilon_{MnO_2}^{526} / \varepsilon_{MnO_2}^{418}) A_{418}}{\varepsilon_{MnO_4^-}^{526} c_o} \tag{4}$$

where ε is the molar absorptivity of the manganese species (MnO_2 or MnO_4^-) at the wavelength (418 or 526 nm), c_o is the initial MnO_4^- concentration, c_t is the concentration of MnO_4^- at time t (Mata-Perez and Perez-Benito, 1985). The relationship between MnO_4^- concentration, expressed as log (c_t/c_o) with time, is depicted in Figure 3. As expected, the MnO_4^- concentration was reduced faster with PTPP than with other treatments. The functions were convex, which probably result from a shortage of MnO_4^- available to react with TCE later in the experiments. As MnO_4^- concentration decreases in the experiment, TCE concentration increases slightly because of the continuous dissolution of the pure phase. Because these two parameters vary in the course of the reaction, it cannot be considered strictly as pseudo-first-order. However, to help compare results, pseudo-first-order rate constants (k) were determined from the slopes of the log (c_t/c_o) versus time data in Table 2, using only the early straight portion of curves. The rate constants verified that the consumption of MnO_4^- was more rapid with PTPP than the other catalysts.

Column outflow was monitored in the same manner with the UV-Vis spectrometer to quantify MnO_4^- consumption (Figure 4). Due to the greater

Table 2. Pseudo-first order rate constants for permanganate consumption calculated from straight portion of relationships between log (c_t/c_o) and time.

catalyst	Control	TEA	TBA	PTPP
k (min⁻¹)	0.0475	0.0424	0.0408	0.0791
r²	0.9857	0.9926	0.9962	0.9927

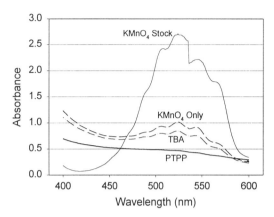

FIGURE 4. UV-Vis spectra of outflow from columns with different treatments compared with initial KMnO₄ stock solution.

interfacial area between aqueous phase and TCE at residual saturation, a large reduction of absorbance at 526 nm after 4 min of residence time was observed with MnO_4^- solution and this decline was even more significant with PTCs. PTPP consistently provided the greater consumption of MnO_4^-. The large interfacial area speeds up not only the dissolution of TCE into aqueous phase but also the transfer of MnO_4^--catalyst pair into TCE.

TCE Solubility Enhancement. In our experiments, the solubilization in the aqueous phase, if it occurred, might enhance rates of TCE destruction. Effectively, enhanced solubility could increase the rate of transfer of TCE into the aqueous phase. Accordingly, we undertook experiments to examine the potential for the solubility enhancement of TCE by PTCs (TBA and PTPP). The

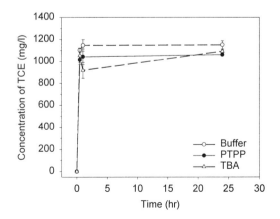

FIGURE 5. Variation in the concentration of TCE in aqueous phases as a function of time and the PTCs.

experiments involved kinetic batch experiments, looking at the TCE increase in the solution with time, as a function of the additives. The results showed no significant difference in TCE concentration for the first 30 minutes, as compared to trials without PTCs (Figure 5).

CONCLUSIONS

These preliminary experiments point to significant opportunities for enhancing the oxidation of chlorinated solvents, like TCE, by MnO_4^-. Speedup in oxidation rates was evidenced in our experiments by (1) an increase in the rate of Cl^- production, (2) a more rapid disappearance of MnO_4^- from the aqueous phase, and (3) qualitative changes in the coloration of the TCE phase. Experimental evidence discounted the possibility that the solubilization of the TCE in the aqueous phase contributed to the rate enhancement. All of the results confirm that the enhancement in TCE decomposition is due to reactions between MnO_4^- and TCE in the nonaqueous phase, in addition to the oxidation of TCE in the aqueous phase.

Work is needed to examine the affinity of PTCs for adsorption and whether this process might significantly reduce their effectiveness as the catalysts. There are potential work-arounds to this problem. For example, one could pretreat the geological medium to add other cations to exchange sites, or develop an injection technology that delivers the solution directly to the target contaminants. We are continuing work to develop the catalytic scheme for permanganate oxidation. We are interested in further studying mechanisms of catalytic reaction in nonaqueous phase with various PTCs and the impact of PTC addition on DNAPL mobility and solubility. Evaluation of phase transfer assisted solvent oxidation at larger scales, including column or tank experiments, is required before any field demonstrations.

ACKNOWLEDGEMENTS

This material is based upon work supported by the Department of Energy under Grant No. DE-FG07-96ER14735.

REFERENCES

Karaman, H., Barton, R.J., Robertson, B.E. and Lee, D.G., 1984. "Preparation and properties of quaternary ammonium and phosphonium permanganates". *J. Org. Chem*, 49: 4509-4516.

Mata-Perez, F. and Perez-Benito, J.F., 1985. "Identification of the product from the reduction of permanganate ion by trimethylamine in aqueous phosphate buffers". *Can. J. Chem*, 63: 988-992.

Starks, C.M., Liotta, C.L. and Halpern, M., 1994. *Phase-transfer catalysis; fundamentals, applications, and industrial perspectives*. Champman & Hall, New York, NY, 668 pp.

Yan, Y.E., 1998. *Abiotic remediation of ground water contaminated by chlorinated solvents*. Ph.D. Thesis, The Ohio State University, 105 pp.

EXPERIMENTAL STUDY OF OXIDATION OF POOLED NAPL

Stanley Reitsma (University of Windsor, Windsor, Ontario, Canada)
Melanie Marshall (University of Windsor, Windsor, Ontario, Canada)

ABSTRACT: Two-dimensional experiments were conducted to study important processes that occur during *in situ* oxidation of pooled non-aqueous phase liquid (NAPL). Results indicate that *in situ* oxidation of chlorinated compounds will lead to the production of various by-products that may significantly affect the over-all performance and behavior of the treatment. For potassium permanganate-NAPL systems, production of manganese dioxide and carbon dioxide are particularly important since they will change flow patterns. Carbon dioxide production may also mobilize NAPL and transport NAPL vapors. The rate of reaction affects the rate of carbon dioxide production. However, even at low reaction rates, production of carbon dioxide gas remains an issue. The behavior of the potassium permanganate flush is affected by NAPL zone permeability.

INTRODUCTION

In situ oxidation of chlorinated solvents has been researched by various authors (e.g. Jerome et al., 1997; McKay et al., 1998; Schnarr et al., 1998). *In situ* oxidation is appealing since surface treatment of extracted water may not be required, reaction enhances treatment rates (Reitsma and Dai, 2000), and cost of treatment is estimated to be competitive with other aggressive NAPL source zone treatment technologies. Several disadvantages of *in situ* oxidation have been identified as well (McKay et al., 1998; Schnarr et al., 1998) including possible plugging of the porous medium by low solubility by-products such as manganese dioxide (MnO_2) or carbon dioxide (CO_2). Since NAPL generally contains carbon compounds, production of CO_2 is likely to occur with any type of oxidant. Precipitation of low solubility products may or may not occur depending on the oxidant. Potassium permanganate produces significant quantities of MnO_2 that may have significant influence on groundwater flow and delivery of the oxidant to contaminated areas.

Experimental investigations of *in situ* oxidation have primarily been completed using one-dimensional columns where affect of by-products on aqueous flow pattern is negligible (Schnarr et al., 1998). Field pilot studies (e.g. McKay et al., 1998) show that the production of by-products significantly influences injection capacity of injection wells. However, the mechanism of plugging was not clearly been identified in these pilot studies and the influence of by-products in the vicinity of the pool in not known. Due to the lack of two-dimensional experiments and the inability to carefully monitor events during pilot studies, mobilization of NAPL during oxidation treatment was also not identified as a possible problem (or benefit).

The objective of the experimental research presented here is to examine the important processes that occur during *in situ* oxidation of pooled NAPL in porous media treated with a $KMnO_4$ solution. Experiments focussed on visual observation of processes, including oxidation, formation of MnO_2 precipitate, generation of CO_2 gas and potential DNAPL mobilization. Objectives included sampling and chemical analyses which were to provide mass transfer rate and removal efficiency data. Three experiments were completed to observe the influence of reaction rate on the oxidation process.

A two-dimensional experiment was designed to determine the important processes that occurred during the treatment of a DNAPL pool with $KMnO_4$. Experiments were conducted within a tank that allowed direct visual observation of the oxidation process. Homogeneous porous media containing a single DNAPL pool was used to establish understanding of the significant processes. Experiments were designed to observe the oxidative treatment of a DNAPL pool of defined geometry in porous media under a constant flow rate. A pretreatment water flush was designed to establish pretreatment mass transfer rates. A $KMnO_4$ flush was introduced to the system to determine oxidative treatment rates and removal efficiency followed by a post-treatment water flush to establish post-treatment mass transfer rates and effectiveness of treatment. Results were based upon visual observations of important processes that occurred throughout the course of the experiments as well as data from sample analyses. Concentration profiles and observations of reactants/products were used to characterize reactions between PCE or TCE and $KMnO_4$ and calculate mass removal rates.

MATERIALS AND METHODS

A simple porous media and DNAPL pool configuration were used to establish understanding of the significant processes that occur during *in situ* oxidation using $KMnO_4$. The experiment was designed to observe the oxidation of a single DNAPL pool of defined geometry under a constant flow rate in homogeneous porous media. DNAPL was injected into a single coarse sand lens surrounded by fine sand to produce a pool positioned in the center of the tank.

An initial water flush was completed to establish baseline mass transfer rates from pooled DNAPL prior to treatment. $KMnO_4$ solution was then introduced at a uniform flow rate at one end of the tank and effluent samples collected from the opposite end. Additional point samples were collected around the DNAPL pool to establish dissolved DNAPL concentration distributions in the vicinity of the pool. Samples taken were originally to be then be analyzed for NAPL compounds, $KMnO_4$ and Cl^-. Chloride concentrations were to be used to establish mass transfer rates during the $KMnO_4$ flush but measurement difficulties eliminated the availability of this data. The $KMnO_4$ flush was followed by an additional water flush to determine post-treatment mass transfer rates. Careful visual records of the experiments including time-lapse digital pictures were collected for analysis. Details of the experimental tank are illustrated in Figure 1.

Two quartz sands employed in the experiments were chosen to provide a flow velocity of approximately 1.0 m/day with reasonably small head drop and low operating pressure. Coarse sand was used for placement of the DNAPL pool and was chosen such that DNAPL would preferentially remain in the coarse sand due to differences in capillary characteristics between the fine and coarse sand. Because the fine-grained sand used in all experiments had a higher capillary entry pressure than the capillary pressure in the coarse sand, DNAPL injected into the coarse sand lens remained in the lens and did not enter the fine sand prior to treatment. This created a single well-defined DNAPL pool in the center of the tank that could then be studied. The porosity of the fine quartz sand was found to be 0.44 and the coarse silica sand was 0.49. The tank is shown in Figure 2.

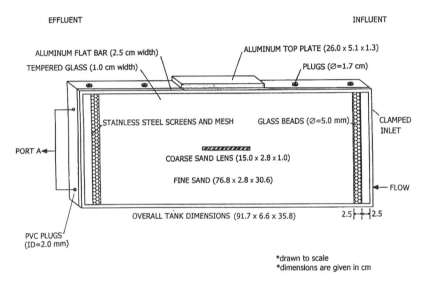

Figure 1. Experimental apparatus for two-dimensional oxidation experiments.

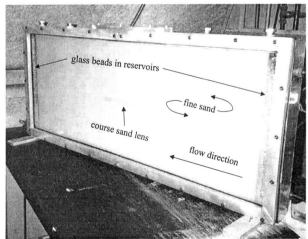

Figure 2. Oblique view of the experimental tank prior to injection of KMnO$_4$.

Solutions of $KMnO_4$ of the desired concentration were prepared by dissolving $KMnO_4$ crystals in de-ionized, de-aired water. Solutions were thoroughly mixed at a rate of 100 rpm using a Philips and Bird Stirrer Model 7790-400. Mixing was continued at the same rate during the entire course of $KMnO_4$ injection. $KMnO_4$ solutions held in 45 L glass containers were covered with aluminum foil to protect the liquid from halodecomposition.

Flow rates were maintained using a Cole-Parmer peristaltic pump equipped with a Masterflex PTFE Tubing Pump Head Model 77390-00 employing 4.0 mm OD PTFE tubing. A flow rate of 2.26 mL/min was maintained to establish an average flow velocity of 0.001 cm/s in the sand. This flow rate produced approximately 1 pore volume of flow through the tank per day or 3.0 L/day. Measurements of flow were checked at the tank effluent at least 2 times per day using a graduated cylinder. The pump speed was held constant throughout the duration of the experiment but daily flow rates fluctuated slightly between 2.17 and 2.50 mL/min.

In all three experiments, undyed DNAPL was introduced into the coarse sand lens to establish a DNAPL pool. Injection of DNAPL into the center of the coarse sand layer was accomplished using a Harvard Apparatus Pump 22 equipped with a Becton Dickinson 20 mL glass syringe (19.13 mm gauge) inserted through a port in the back of the tank. TCE or PCE was injected at a slow rate (4 to 10 mL/hr) to reduce possibility of DNAPL entry into the finer sand surrounding the course grained lens.

To study the affect of different reaction rates on treatment behavior, conditions were changed for each of the three experiments. In experiment 1, 5.0 mL of TCE was injected into the course sand lens and the concentration of injected $KMnO_4$ was 10 g/L. In experiment 2, 5.0 mL of PCE was injected into the course sand lens and the $KMnO_4$ concentration was 5 g/L. In experiment 3, 2.0 mL of PCE was injected and $KMnO_4$ concentration was 1 g/L. Due to differences in the reaction coefficient for TCE- and PCE-$KMnO_4$ systems, differences in aqueous solubility for TCE and PCE, and different injected concentrations of $KMnO_4$, the reaction rate for $KMnO_4$ and DNAPL compound in experiment 1 was approximately 200 times greater than for experiment 2 and 1000 times greater than for experiment 3.

EXPERIMENTAL RESULTS AND DISCUSSION

Experiment 1 employed a $KMnO_4$ flush of concentration 10.0 g/L to treat a DNAPL pool containing 5.0 mL of TCE. Initially, most of the flow was diverted around the pool due to reduced permeability of the coarse sand lens caused by the presence of TCE in the lens. Precipitated MnO_2 was an excellent indicator of where reaction was taking place since MnO_2 was not being transported in the aqueous phase but precipitated almost immediately after reaction. A small amount of MnO_2 was observed in the first few centimeters of the pool indicating that at early time some $KMnO_4$ solution penetrated the pool and reaction took place in the pool as can be seen in Figure 3. However, most of the reaction took place in the first millimeter of the pool, indicated by heavy build up of MnO_2 in this zone. Heavy precipitation of MnO_2 was also observed on the upper and lower bounds of the course sand lens. Figure 3 shows no MnO_2 precipitation occurred in the posterior portion of the course sand lens. The pattern of MnO_2 precipitation around the pool was indicative of transverse macro-scale mass transport from a NAPL pool (see Reitsma and Dai, 2000), particularly on the lower side of the pool.

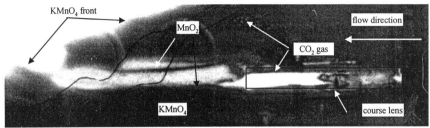

Figure 3. Close-up view of first experiment at approximately 20 hours after beginning of injection (0.8 pore volumes) with visible zones of MnO_2 and CO_2.

A distinct plume containing no visible $KMnO_4$ or MnO_2 precipitate was observed down-gradient of the pool and extending to the effluent end of the tank. Thin stringers of MnO_2 precipitate (about 1 to 2 mm in width visible in Figure 3) existed between zones with or without $KMnO_4$. The thin stringer of MnO_2 coincided with the transverse mixing zone between the dissolved TCE plume down-gradient of the DNAPL pool and surrounding $KMnO_4$ zones. Because the reaction between $KMnO_4$ and TCE is relatively rapid, these reaction zones were thin (unlike in experiment 2 and 3 where PCE was used).

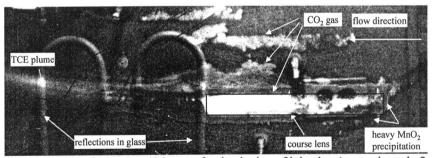

Figure 4. Experiment 1 at 48 hours after beginning of injection (approximately 2 pore volumes) with visible zones of MnO_2 and CO_2.

As the experiment continued the production of CO_2 gas was rapid and led to nearly complete de-saturation of the coarse sand lens, along with a subsequent reduction of sand permeability to the $KMnO_4$ solution (shown in Figure 4). As the gas volume increased in the pool, the gas began to travel upwards and was easily recognized as bubbles distributed in a vertical path above the coarse sand lens. As shown in Figure 4, MnO_2 formation was observed several centimeters above the coarse sand lens coinciding with zones of de-saturation where CO_2 had previously been noted. This effect suggests that the CO_2 gas transported TCE vapor from the coarse sand lens into the fine sand above the lens. The TCE vapor in the CO_2 gas then transferred into the surrounding aqueous phase where subsequent reaction and MnO_2 precipitation took place. The CO_2 gas may have significantly enhanced mass transfer by providing an additional mass transport mechanism from the TCE pool. Transport rates due to gas movement is presently being investigated.

The generation of CO_2 also caused mobilization of TCE NAPL from the course sand lens into the surrounding fine sand. Zones of MnO_2 were noted down-gradient of and below the pool that were not associated with CO_2 movement or aqueous phase transport of dissolved TCE. These 'spotty' patterns of MnO_2 precipitate were indicative of DNAPL movement in a slightly heterogeneous porous media. During the disassembly of the experiment, TCE was found in the down-gradient reservoir of the tank.

The production of CO_2 gas has been identified as a possible plugging mechanism of the porous medium and lead to poor injection performance (McKay et al., 1998). However, the subsequent movement of DNAPL compounds in the gas phase and mobilization of DNAPL have not previously been identified as important processes during *in situ* oxidation of chlorinated compounds. Experiment 1 showed that both of these processes may occur when the reaction between the DNAPL compound and the oxidant was rapid.

Experiment 2 was conducted using PCE rather than TCE (lower reaction rate and aqueous solubility) and the injected concentration of $KMnO_4$ was also reduced from 10 g/L to 5 g/L in order to determine in what systems CO_2 production may not be important. Some differences in location of the MnO_2 was observed down-gradient of the course sand lens. Production rate of CO_2 was less than in experiment 1 but eventually led to the de-saturation of the course sand lens. Mobilization of the DNAPL was again apparent. Since the results in experiment 2 were similar to that in experiment 3, details are left out for brevity.

Experiment 3 was conducted using 2.0 mL of PCE to reduce NAPL mobilization and the injected $KMnO_4$ concentration was reduced to 1.0 g/L again to reduce the CO_2 production rate. As shown in Figure 5, the reduced NAPL volume used in experiment 3 led to increased flow of oxidant through the pool when compared to experiments 1 and 2. Flow was now focused through the pool due to the higher permeability of the course sand lens. Reaction took place inside the NAPL zone as well as on its periphery.

As experiment 3 progressed, CO_2 began to de-saturate the course sand lens as shown in Figure 6 (picture taken 110 hours after the beginning of the $KMnO_4$ injection). The shape of the precipitated MnO_2 plume in experiment 3 was also much different than in the earlier experiments. As can be seen in Figure 6, heavy MnO_2 precipitation occurred throughout the NAPL zone and fanned out down-gradient of the pool. The amount of precipitation was greatest within the pool and decreased with distance from the pool. Since flow occurred in the course sand lens, reaction occurred there unlike in experiments 1 and 2. Unlike with TCE and high $KMnO_4$ concentrations, the dissolved NAPL-$KMnO_4$ reaction in experiment 3 was significantly slower than in experiment 1. Thus some $KMnO_4$ was able to pass through the entire NAPL zone and continue to react with dissolved PCE down-stream of the NAPL zone. The fan shape of MnO_2 precipitation is due to diverging streamlines down-gradient of the course sand lens. After about 300 hours (approximately 12 pore volumes of $KMnO_4$ solution), the MnO_2 plume had become twice the thickness of the course sand lens as shown in Figure 7 due to upward movement of CO_2 gas near the front of the pool. Figure 7, taken after the post-treatment water flush was complete, clearly shows the zones where MnO_2 precipitation occurred. Evidence of NAPL mobilization due to CO_2 production and degassing in the course sand lens is apparent from the patchy MnO_2 precipitation noted in Figure 7. After 12 pore volumes of $KMnO_4$ solution were injected, aqueous PCE concentrations in the course sand lens were still very near solubility indicating the PCE NAPL still existed in this lens.

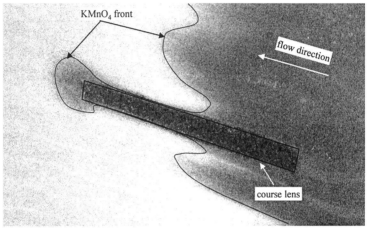

Figure 5. Front of the $KMnO_4$ flush as it passes through the course sand lens (oblique view - lens is actually horizontal) in experiment 3.

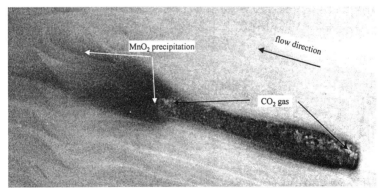

Figure 6. Experiment 3 at 110 hours after beginning of the $KMnO_4$ flush.

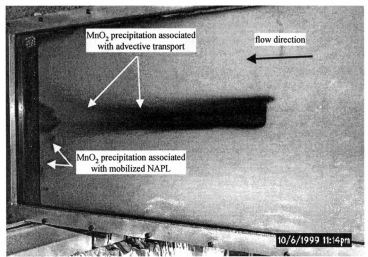

Figure 7. Experiment 3 after completion of the $KMnO_4$ flush and the post-treatment water flush.

CONCLUSIONS

In situ oxidation of chlorinated compounds will lead to the production of various by-products that may significantly affect the over-all performance and behavior of the treatment. For $KMnO_4$-NAPL systems, production of MnO_2 and CO_2 are particularly important since they will change flow patterns. CO_2 production may also mobilize NAPL and transport NAPL vapors. The rate of reaction affects the rate of CO_2 production. However, even at low reaction rates, production of CO_2 gas remains an issue. The behavior of the $KMnO_4$ flush is affected by NAPL zone permeability and this is very likely to become more apparent with greater heterogeneity.

REFERENCES

Jerome, K., 1997. "*In situ* oxidation destruction of DNAPL", In: *EPA Ground Water Currents, Developments in Innovative Ground-water Treatment*, EPA 542-N-97-004.

McKay, D., Hewitt, A., Reitsma, S., LaChance, J., and Baker, R., 1998. *In situ* oxidation of trichloroethylene using potassium permanganate: Part 2, Pilot study, *First International Conference, Remediation of Chlorinated and Recalcitrant Compounds*, Monterey, California, May 18-21.

Reitsma S., and Q. Dai, 2000. Reaction enhanced mass transfer from pooled NAPL, *Second International Conference, Remediation of Chlorinated and Recalcitrant Compounds*, Monterey, May 22-25.

Schnarr, M., Truax, J. C., Farquhar, G., Hood, E., Gonullu, T., and Stickney, B., 1998. "Laboratory and controlled field experiments using potassium permanganate to remediate trichloroethylene and perchloroethylene DNAPLs in porous media", *Journal of Contaminant Hydrology*, 29(1): 205-224.

REACTION-ENHANCED MASS TRANSFER FROM NAPL POOLS

Stanley Reitsma (University of Windsor, Windsor, Ontario, Canada)
Qunli Dai (University of Windsor, Windsor, Ontario, Canada)

ABSTRACT: The mass transfer enhancement from a single non-aqueous phase liquid (NAPL) pool due to reaction of dissolved NAPL compounds is quantified through theoretical analysis. Simulations are completed to study reaction induced mass transfer enhancement at a local-scale near the NAPL/water interface as well as at a macro-scale around the NAPL pool. For slower reaction rates, such as for potassium permanganate ($KMnO_4$) and perchloroethene (PCE), local-scale mass transfer enhancement is expected to be small. Simplified two-dimensional simulations quantifying macro-scale mass transport from a horizontal NAPL pool can be represented by mathematically equivalent falling film reactor problems. In contrast to local-scale mass transfer, due to the larger scale mass transfer from pools can be approximated assuming the oxidation reaction rate is instantaneous. Enhancement factors due to destruction of the NAPL compound(s) are mainly dependent on NAPL solubility, and oxidant concentration and to a lesser extent on reaction rate and reaction stoichiometry. Macro-scale reaction enhancement factors may be in the range of 5 to 50 for potassium permanganate and chlorinated solvents.

INTRODUCTION

In situ oxidation has had limited success as dense non-aqueous phase liquid (DNAPL) source zone cleanup technology in pilot-scale testing (Jerome et al., 1997; Schnarr et al., 1998). Mass transfer limitations at various scales may have significantly influenced the treatment efficiency. The following discussion will provide insight into the mass transfer processes that may have occurred during the *in situ* oxidation pilot studies and how these factors may influence field application.

The main objective of the following work is to theoretically quantify the influence of destruction of NAPL compounds in the aqueous phase on mass transfer rates from NAPL source zones during *in situ* oxidation. Particularly, potential increases in mass transfer rate due to reaction at both local-scale and macro-scale are explored. Local-scale mass transfer development focuses on the zone in close proximity to the NAPL-water interface and macro-scale mass transfer focuses on the transport of mass out from NAPL pools or residual zones. The theoretical development for local-scale mass transfer provides a range for expected mass transfer increases and is meant to establish an estimate of mass transfer coefficients within NAPL pools or residual zones. The development for macro-scale mass transfer extends theoretical equation development for dissolution of NAPL pools in absence of reaction. The results from macro-scale equations are intended to establish estimates of NAPL pool removal rates where reaction occurs in the aqueous phase.

LOCAL MASS TRANSFER WITH REACTION

Mass transfer of the organic contaminant from the immiscible organic phase to the aqueous phase has been represented using a single-film model (e.g. Powers et al., 1992, 1994; Imhoff et al., 1993). The single-film model is a theoretical concept that assumes a boundary layer exists near the NAPL/water interface where mass transfer is due to diffusion only. The single-film model expresses mass flux of a NAPL compound A from the NAPL to the surrounding bulk aqueous phase as:

$$N_A = k_m^* \left(C_{Ai} - C_A \right) \tag{1}$$

where N_A is the mass flux of A from the organic phase, C_{Ai} is the concentration of compound A at the interface between the organic and water phases, and C_A is the concentration of compound A in the bulk aqueous phase. Concentration of The normalized mass transfer coefficient, k_m^*, is represented by:

$$k_m^* = D_A^* / \delta \tag{2}$$

where δ is the thickness of the single film and D_A^* is the aqueous diffusion coefficient for compound A. In the case of single-component NAPL as considered here, resistance to mass transfer in the NAPL phase is nil (Treybal, 1980) and C_{Ai} is approximately equal to the effective aqueous solubility of compound A. In the case of multicomponent NAPL, both NAPL-phase and aqueous-phase resistance could be important (Holman and Javandel, 1996).

The destruction or reaction of the NAPL compound in the single film will tend to increase the mass transfer rate due to increased chemical gradients within the single film. The increase in mass transfer rate is referred to as the enhancement factor in chemical engineering literature. In all cases, the enhancement factor due to reaction is greater than or equal to one. The following development establishes expected local mass transfer enhancement due to reaction near the water/NAPL interface.

First consider an infinitely fast reaction:

$$A + bB \rightarrow P \tag{3}$$

In this reaction A is the organic compound being transferred from the water/NAPL interface to the bulk aqueous phase, B is the oxidant being transferred into the single film from the bulk aqueous phase, and b is a stoichiometric coefficient. Since the reaction is instantaneous, only A or B can exist at any location within the single film. The equation for mass flux of A from the organic phase with reaction, N_A^*, is then given by (Hatcher, 1986):

$$N_A^* = k_m^* C_{Ai} \left[1 + \frac{D_B^*}{D_A^*} \frac{C_{Bo}}{b C_{Ai}} \right] \tag{4}$$

where D_B^* are the diffusion coefficient for compound B, and C_{Bo} is the concentration of B in the bulk aqueous phase. The normalized mass transfer coefficient in Equation 4 uses the diffusion coefficient for the organic compound A since the original mass transfer coefficient to which the enhancement factor is applied is based on this value. The ratio of mass flux with to without reaction, defined as the mass transfer enhancement factor, E, is given by:

$$E = N_A^* / N_A \tag{5}$$

Using TCE and potassium permanganate for compounds A and B, respectively, and a concentration of potassium permanganate in the bulk phase of 10 g/l (63.2 mM) and bulk aqueous phase TCE concentration of 1100 mg/l (8.4 mM), the enhancement factor due to instantaneous reaction would be approximately equal to 7. In contrast, for PCE and KMnO$_4$ and an instantaneous reaction, the enhancement factor would be approximately 70. Clearly, the rate of mass transfer will be enhanced significantly if the reaction were very rapid. However, the reaction rate between permanganate and chlorinated alkenes is not expected to be sufficiently rapid to be estimated using Equation 4 as is shown below.

Again consider the reaction given by Equation 3. Assuming steady-state conditions within the single film occur rapidly and transport is diffusion-controlled, for a second-order reaction the governing equations for mass transfer of the organic and oxidant within the film are given by:

$$D_A^* \frac{d^2 C_A}{dx^2} - kC_A C_B = 0 \, , \, D_B^* \frac{d^2 C_B}{dx^2} - bkC_A C_B = 0 \tag{6}$$

where C_A and C_B are the concentrations of compound A and B, respectively, x is the location in the film, and k is a reaction coefficient. Equation 6 is non-linear and requires a iterative numerical solution. A finite difference scheme with central spatial weighting was used to solve the equations.

Two sets of simulations were completed to establish enhancement factors for the TCE- and PCE-potassium permanganate reactions. For both reactions, b is equal to two. Boundary conditions used in simulations include constant concentrations of permanganate and organic compound in the bulk aqueous phase of 10 g/l and zero, respectively, no transfer of permanganate across the interface (i.e., $\partial C_B / \partial x = 0 \, @ \, x = 0$), and the dissolved concentration of organic at the interface is assumed to be equal to the aqueous solubility of the organic compound (e.g. Imhoff et al., 1993; Powers et al., 1994). The effective diffusion coefficient of permanganate (cm^2·s^{-1}) is calculated using the Nernst Equation for dilute solutions. Because concentrations of permanganate are high, the actual effective diffusion coefficient is likely to be somewhat less than that calculated using the Nernst Equation. Other simulation input is shown in Table 1.

Table 1. Values used for calculation of mass transfer enhancement factors.

Parameter			Value
Diffusion coefficient -	TCE		10.1×10^{-6} cm^2·s^{-1}
	PCE		9.4×10^{-6} cm^2·s^{-1}
Aqueous solubility -	TCE		1100 mg·l^{-1}
	PCE		150 mg·l^{-1}
Second-order reaction coefficient - TCE[3]			0.67 mol^{-1}·s^{-1}
		PCE	0.045 mol^{-1}·s^{-1}
Ionic conductance -	K$^+$		73.5 cm^2·s·mol^{-1}
	MnO$_4^-$		61.3 cm^2·s·mol^{-1}
Effective diffusion coefficient - KMnO$_4$			16.9×10^{-6} cm^2·s^{-1}
Temperature (for use in Nernst Equation)			10 °C

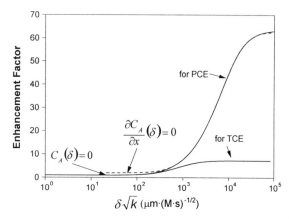

The expected range of enhancement factors for a range of single film thickness, δ, is shown in Figure 1 (based on permanganate concentration of 10 g/l and aqueous solubility of PCE and TCE). The x-axis in Figure 1 is scaled using $\delta k^{1/2}$ since results using different values of k when plotted in this manner will plot on the same curve. Results in Figure 1 show that for film thickness less than 100 μm, the enhancement factor for

FIGURE 1. Local-scale mass transfer enhancement factors for TCE- and PCE-potassium permanganate systems as a function of film thickness.

PCE- and TCE-KMnO$_4$ systems will be approximately equal to one, which indicates that very little reaction takes place within the film. For TCE, when the film thickness is greater than 200 μm, the enhancement factor increases rapidly and approaches the asymptotic value of 7, which may be calculated above using Equation 4 for instantaneous reaction. For PCE the enhancement factor does not change dramatically unless the theoretical film thickness is greater than 1000 μm. These results indicate that mass transfer enhancement will not increase greatly unless distances over which diffusion transfer processes dominate are on the order of several hundred μm.

Justification for the use of $C_A = 0$ in the bulk aqueous phase in the above analysis can be provided by looking at results where this is not the case. If no reaction were to take place in the bulk aqueous phase, representing the extreme opposite to no organic in the bulk aqueous phase, the concentration gradient for compound A at the interface between the bulk aqueous phase and the single film interface would be equal to zero (i.e. $\partial C_A / \partial x = 0 \ @ \ x = \delta$). A series of simulations were completed to solve Equation 6, now using $\partial C_A / \partial x = 0 \ @ \ x = \delta$ rather than $C_A(\delta) = 0$ such that a comparison could be made for extremes in boundary conditions at the bulk aqueous phase boundary. Figure 1 compares the results for the same PCE-KMnO$_4$ system used above for the two different boundary conditions. Results are very similar for the larger film thickness and diverge only at the lower values of $\delta \sqrt{k}$. For $\partial C_A / \partial x = 0 \ @ \ x = \delta$, the enhancement factor at low values of $\delta \sqrt{k}$ is equal to two.

Regardless of enhancement factor, mass transfer coefficients derived from dissolution experiments represent the lower limit of mass transfer coefficients during oxidation of NAPL and may be used to represent the lower limit of mass transfer from residual and pooled NAPL during oxidation treatment.

TRANSPORT FROM NAPL POOLS WITH REACTION

The transport of an organic compound from a NAPL pool occurs primarily through advection of water through the NAPL pool (Powers et al., 1998) and through dispersion to surrounding water (Sale, 1998), which is subsequently swept away through advection. For a homogeneous porous medium with uniform flow parallel to a NAPL pool and steady-state conditions, the dispersion process has been represented by Hunt (1988) using the follow governing equation:

$$v_x \frac{\partial C_A}{\partial x} = D_{TA} \frac{\partial^2 C_A}{\partial z^2} \tag{7}$$

where x and z represent the distance along the pool parallel to flow and above the pool, respectively, v_x is the aqueous phase horizontal flow velocity, and D_{TA} is the transverse dispersion coefficient for compound A. Longitudinal dispersion is assumed to be negligible in development of Equation 7 since ratio of longitudinal to transverse concentration gradients is negligible at steady-state. Boundary conditions applied by Hunt (1988) for a semi-infinite planar NAPL source located at $x \geq 0$ and $z = 0$ are: $C_A = C_{sat}$ for $x > 0$, $z = 0$, $C_A = 0$ for $x = 0$, and $C_A = 0$ for $z = \infty$. C_{sat} is the effective aqueous solubility of compound A. Mass flux from the NAPL pool, N'_A, at a particular location x is given by:

$$N'_A = C_{sat} \phi \sqrt{\frac{v_x D_{TA}}{x \pi}} \tag{8}$$

where ϕ is the porous medium porosity. Sale (1998) concludes from previously published laboratory work and from additional laboratory work that in zones in close proximity to NAPL pools, aqueous concentrations of organic compounds found in the NAPL are very near effective solubility limits. These results indicate that the boundary conditions used for Equation 7 are justified.

Hunt (1988) followed development of mass transfer relationships for film adsorption in film reactors. Using a similar approach, mass transfer from NAPL pools involving chemical reaction may be quantified using techniques developed for analogous mass transfer processes occurring in a film reactor where a well-mixed gas contacts a downward flowing liquid film. A compound in the gas, such as CO_2, diffuses into the film and reacts with a compound within the aqueous phase. Figure 2a depicts the variation in concentration in the falling film resulting from the adsorption with reaction process. Figure 2b compares the mathematically equivalent problem of mass transfer with reaction from a NAPL pool in a uniform flow field to the falling film reactor shown in Figure 2a.

Since the two systems shown in Figure 2 are mathematically equivalent, equation development for film reactors is adopted to analyze mass transfer from NAPL pools. For a semi-infinite planar NAPL source located at $x \geq 0$ and $z = 0$, the following boundary conditions may be applied: $C_A = C'_{Ai}$ for $x > 0$, $z = 0$, $C_A = 0$ for $x = 0$, $C_A = 0$ for $z = \infty$, $C_B = C'_{Bo}$ for $x = 0$, $z \geq 0$, and $C_B = C'_{Bo}$ for $x \geq 0$, $z = \infty$. C'_{Ai} is the aqueous solubility of compound A at the interface between the NAPL pool and surrounding water, and C'_{Bo} is the concentration of compound B in the surrounding water prior to any reaction.

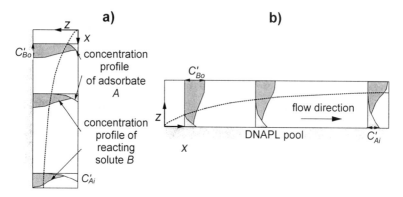

FIGURE 2. Development of concentration profiles in a) film reactor and b) mathematically equivalent problem for mass transfer from a NAPL pool in a uniform flow field.

The mass transfer rate of compound A from a single NAPL pool with instantaneous irreversible reaction, N'^*_A, irrespective of reaction mechanisms is (modified after Danckwerts (1970)):

$$N'^*_A = E'_i C'_{Ai} \phi \sqrt{\frac{D_{TA} v_x}{\pi x}}$$
(9)

where E'_i is the enhancement factor for an instantaneous reaction. The enhancement factor for an instantaneous reaction is:

$$E'_i = \left[\text{erf} \left(\frac{\beta *}{\sqrt{D_{TA}}} \right) \right]^{-1}$$
(10)

The term $\beta *$ may be solved by iterative solution of:

$$\exp\left(\frac{\beta *^2}{D_{TB}} \right) \text{erfc}\left(\frac{\beta *}{\sqrt{D_{TB}}} \right) = \left(\frac{C'_{Bo}}{z C'_{Ai}} \right) \left(\frac{D_{TB}}{D_{TA}} \right)^{1/2} \exp\left(\frac{\beta *^2}{D_{TA}} \right) \text{erf}\left(\frac{\beta *}{\sqrt{D_{TA}}} \right)$$
(11)

where D_{TB} is the transverse dispersion coefficient for compound B. When Ei $\gg$ 1 then:

$$E'_i = \frac{\sqrt{D^*_A}}{\sqrt{D^*_B}} \left(1 + \frac{D^*_B}{D^*_A} \frac{C'_{Bo}}{b C'_{Ai}} \right)$$
(12)

Based on a number of numerical simulations for film reactors, Brian et al. (1961) provides a functional equation that may be used to represent the enhancement factor for mass transfer from a NAPL pool, E', given by:

$$E' = \frac{M^{1/2} \eta *}{\tanh\left(M^{1/2} \eta * \right)}, \quad M = \frac{\pi k C'_{Bo} x}{4 v_x}, \quad \eta * = \left[\frac{E'_i - E'}{E'_i - 1} \right]^{1/2}$$
(13)

The variable M is a dimensionless number relating reaction rate and exposure time. Equation 13 requires iterative solution.

Equation 14 is used to compare expected mass transfer enhancement factors for a NAPL pool consisting of either TCE or PCE and initial

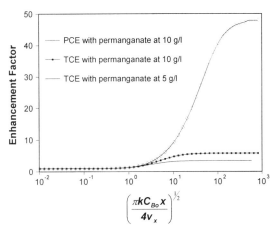

FIGURE 3. Functional expression for enhancement factor (adopted from Brian et al. 1961) due to chemical reaction as a function of the dimensionless exposure time for a second-order reaction.

concentrations of potassium permanganate of 5 g/l or 10 g/l. Results are shown in Figure 3 where the enhancement factor for the pool is plotted versus the dimensionless number, $M^{1/2}$. The maximum expected enhancement factor, which may well be the most important result, may be directly calculated using Equation 10 and 11 or Equation 12. From Equation 12, it is clear that the maximum enhancement factor is dependent mainly on the solubility of the DNAPL compound, the stoichiometry of the reaction, the concentration of the dissolved reactant, and to a lesser extent the transverse dispersion coefficients of the reactant and NAPL compounds.

CONCLUSIONS

At the local-scale and relatively slow reaction rates, very little reaction is expected to take place within the single film near the DNAPL/water interface. If very little reaction takes place here, no significant increases in concentration gradient will be achieved in the film. On the other hand, if the diffusion distance is large or reaction rate is rapid, local-scale mass transfer may be increased greatly.

For macro-scale mass transfer from NAPL pools, mass transfer rate enhancement can be approximated using the approach from Brian et al. (1961) developed for falling film reactors. For the cases explored in this work, the mass transfer enhancement factor may generally be approximated using the maximum enhancement factor that would result if the reaction rate were instantaneous. However, the analysis does not include pool interference, reaction within the pool, or local mass transfer limitations within the pool and these will require further study before the above conclusion can be made more general. For a PCE pool and injected permanganate concentration of 10 g/l, mass transfer rates could well be 45 times greater then without permanganate. For TCE and similar permanganate concentration, the enhancement of mass transfer rate is about 5 times that with only dissolution.

REFERENCES

Brian, P. L. T., Hurley, J. F., and Hasseltine, E. H., 1961. "Penetration theory of gas adsorption accompanied by a second-order chemical reaction", *Association of International Chemical Engineering Journal, 7*: 226-239.

Danckwerts, P. V., 1970. *Gas-Liquid Reactions*, McGraw Hill, N. Y.

Hatcher, W. J., 1986. "Reaction and mass transport in two-phase reactors", In *Handbook of Heat and Mass Transfer, Volume 2: Mass Transfer and Reactor Design*, N. P. Cheremisinoff (ed.), Gulf Publishing Company, Houston.

Holman, H.-Y. N., Javandel, I., 1996. "Evaluation of transient dissolution of slightly water-soluble compounds from a light nonaqueous phase liquid pool", *Water Resources Research, 32*: 915-923.

Hunt, J. R., Sitar, N., and Udell, K. S., 1988. "Nonaqueous phase liquid transport and cleanup 1. Analysis of mechanisms", *Water Resources Research, 24*(8), 1247-1258.

Imhoff, P. T., Jaffe, P. R., and Pinder, G. F., 1993. "An experimental study of complete dissolution of a nonaqueous phase liquid in saturated porous media", *Water Resources Research, 30*(2): 307-320.

Jerome, K., 1997. "*In situ* oxidation destruction of DNAPL", In: *EPA Ground Water Currents, Developments in Innovative Ground-water Treatment*, EPA 542-N-97-004.

Powers, S. E., Nambi, I. M., and Curry, Jr., G. W., 1998. "Non-aqueous phase liquid dissolution in heterogeneous systems: Mechanisms and local equilibrium modeling approach", *Water Resources Research, 32*(12): 3293-3302.

Powers, S. E., Abriola, L. M., and Weber, Jr., W. J., 1992. "An experimental investigation of NAPL dissolution in saturated subsurface systems: Steady state mass transfer rates", *Water Resources Research, 28*(10): 2691-2706.

Powers, S. E., Abriola, L. M., and Weber, Jr., W. J., 1994. "An experimental investigation of nonaqueous phase liquid dissolution in saturated subsurface systems: Transient mass transfer rates", *Water Resources Research, 30*(2): 321-332.

Sale, T., 1998. *Interphase Mass Transfer From Single Component DNAPLs*, Doctoral Thesis, Colorado State University.

Schnarr, M., Truax, J. C., Farquhar, G., Hood, E., Gonullu, T., and Stickney, B., 1998. "Laboratory and controlled field experiments using potassium permanganate to remediate trichloroethylene and perchloroethylene DNAPLs in porous media", *Journal of Contaminant Hydrology, 29*(1): 205-224.

Treybal, R. E., 1980. *Mass-Transfer Operations*, 3rd edn., McGraw-Hill Book, New York, NY. 621 pp.

EFFICIENCY PROBLEMS RELATED TO PERMANGANATE OXIDATION SCHEMES

X. David Li and Franklin W. Schwartz
The Ohio State University, Columbus, Ohio, USA

ABSTRACT: This study investigates problems that potentially could impact a permanganate oxidation scheme, such as competitive permanganate utilization in reactions with aquifer materials, MnO_2 precipitation around zones of high DNAPL saturation, and permeability-related flow bypassing. A series of batch experiments were conducted to examine whether typical aquifer materials (glacial deposits, alluvium and carbonate-rich sand) reacted with permanganate. The results show a dramatic consumption of the oxidant and a significant change in the concentrations of some chemical species, which may be of environmental concern. 1-D column and 2-D flow tank experiments have been conducted to examine mass removal rates and related flushing efficiencies. The results indicate that mass removal rates are also greatly influenced by the MnO_2 precipitation and flow bypassing. It is anticipated that in actual field settings, the issue of flushing efficiency needs to be considered in the design.

INTRODUCTION

Oxidation schemes for the *in situ* destruction of chlorinated solvents, using potassium permanganate, are receiving considerable attention. Potassium permanganate is a powerful oxidant that has been used for long time in wastewater treatment. Batch scale experiments have demonstrated the rapid mineralization of contaminants with final products that are environmentally safe (Yan and Schwartz, 1999). Researchers at the University of Waterloo (Schnarr et al., 1998) and Oak Ridge National Laboratory have conducted experiments to investigate permanganate-oxidation schemes in field and laboratory tests. The results suggested that permanganate oxidation is a promising technology for the remediation of chlorinated solvents. This approach also has been implemented by consultants at several sites in Florida and California.

Indications from these field studies and our work are that permanganate oxidation schemes have inherent problems that could severely limit their applicability. As a powerful oxidant, permanganate is not only capable of oxidizing chlorinated ethylenes, but also inorganic compounds in ground water and solid aquifer materials. For example, Drescher and others (1998) reported the significant consumption of permanganate by humic acid in a system consisting of sand, TCE, permanganate and humic acid. With aquifer materials consuming some of the permanganate, the overall efficiency of the cleanup is reduced and costs are increased. Reaction also can release metals (e.g. Cr^{3+}) to the aqueous phase at concentration of regulatory concern. Given the present, relatively limited understanding of the oxidation reactions, there is a need for studies to understand these problems and to improve the efficiency of permanganate delivery. Another potential problem is that key reaction products, manganese dioxide and carbon dioxide, can cause plugging and flow diversion. There is little research on the impact of these problems on mass removal rate and on possible approaches to inhibit precipitation. This study specifically addresses gaps in knowledge that bear on the realistic use of permanganate-oxidation schemes in DNAPL clean ups. Here, we present preliminary data on the likely metals that can be released due to permanganate-aquifer interaction. We also discuss the results of various column and flow tank experiments that elucidate problems of flooding inefficiencies, and pore plugging related to reaction products.

MATERIALS AND METHODS

The study involves two types of experiments. We utilize batch studies to investigate the consumption of MnO_4^- by aquifer materials and the release of potentially hazardous metals. Column (1-D) and flow tank studies (2-D) are used to examine the impact of oxidation products on DNAPL removal efficiencies.

Permanganate Consumption/Metal Release. Batch experiments with natural aquifer materials are used to examine the interaction of potassium permanganate with aquifer material and the extent to which metals are released. Samples were collected from various aquifer materials, such as, alluvium, glacial till, glacial outwash, and carbonate sand. Special efforts were made to collect samples at depth, away from soil horizons.

Samples were placed in aluminum pans to air dry for two to three days. Experiments were conducted at room temperature in 25 mL vials. 15 gram of each sample was mixed with 15 mL of 0.5 g/L potassium permanganate solution. A control was prepared with the same potassium permanganate solution, but without any aquifer material. After the permanganate solution was added to the aquifer solids, the vials were covered with aluminum foil and black cloth, and kept in the dark to avoid photo-induced degradation of potassium permanganate. Sample vials were shaken by hand once a day to ensure mixing during the experiment. At predetermined time steps, 5 mL of solution was removed from the vial and filtered through a 0.22 μm glass fiber membrane to remove any suspended solids immediately prior to the concentration measurement. Permanganate concentrations were measured with a Varian Cary 1 UV–visible spectrophotometer at a wavelength 525 nm.

The study of metals released by the oxidation of aquifer materials followed a similar procedure. The controls in this case were mixtures of 15 mL Milli-Q deionized water and aquifer materials. By comparing the differences between metals released from the control experiments and oxidized materials, the effect of MnO_4^- treatment could be established. After ten days, when the permanganate had mostly been consumed, solutions of both the controls and samples were separated from large solid particles and filtered to remove suspended solids. The samples were analyzed by a Perkin-Elmer Sciex ELAN 6000 Inductively Coupled Plasma Mass Spectrometer (ICP-MS). As a first step, a semi-quantitative screening analysis was performed to identify the elements with most dramatic change in concentration. A quantitative analysis was then carried out with specific external standards. 12 elements were quantitatively analyzed. The results were then converted to the pore water concentration, based on the porosity of the media.

1-D Column Experiment. The 1-D column experiment was conducted using a glass column with Teflon end fittings and packed with medium silica sand (US Silica, Ottawa, IL). Details of the experimental set up are summarized in Table 1. Once the column was packed, it was positioned upside down. 1 mL of TCE was added to the column evenly across the opening to create a zone of residual DNAPL saturation. The column was returned to an upright position and fluid was pumped upward through the column using an Ismatec tubing bed pump (Cole Parmer Instrument Co. Vernon Hills, IL). Effluent samples were collected periodically for chemical analysis. MnO_4^- concentration was measured with a UV-visible spectrophotometer as mentioned earlier. TCE was measured with a Fisons Instruments 8060 gas chromatograph equipped with a Ni^{63} electron capture detector and a DB-5 capillary column (J&W Scientific, Rancho Cordova, CA). Cl^- ion concentration was measured with a Buchler Digital Chlorodometer. Samples were taken at 12 hour intervals over a two week period. The experiment was concluded when the effluent TCE concentration fell below 5 μg/L. After the experiment finished, several porous medium samples were taken along the column. The samples were treated with thiosulfate to dissolve the MnO_2. The resulting solution was analyzed by ICP-MS to quantify the distribution of MnO_2 along the column.

Table 1. Experimental design for the 1-D column and 2-D flow tank experiments

Column /tank	Dimension (mm)	Medium	Q (mL/min)	$KMnO_4$ concentration	Porosity
1 – D	605(L)x50 (ID)	Silica sand	1.2	1 g/L	0.385
2 – D	305(L)x50(H)x3 (T)	Glass beads	0.5	0.2 g/L	0.42

2-D Flow Tank Experiments. A thin, small 2-D glass flow tank with Teflon end fittings was constructed for the experiment (Table 1). The flow tank was filled with transparent borosilicate glass beads with a mean grain size of 1 mm. Two PTFE tubes with inside diameter of 1.32 mm were installed in the two ends of the column to function as recharge and discharge wells. Both tubes had ends open at a depth of 0.5 cm from the bottom of column. Once the column was packed, it was saturated with Milli-Q water. A Spectroflow 400 solvent delivery system (Kratos analytic, NJ) was used to maintain inflow to the tank, the Ismatec pump removed fluid at the outlet.

With both pumps running at the same steady rate, flow was horizontal along the length of the tank. 1 mL of the TCE was added to the tank from the top to form a zone of residual DNAPL saturation across the vertical depth of the tank and a small DNAPL pool at the bottom. The tank was flushed for about two weeks with $KMnO_4$. Effluent samples were taken three times a day and were analyzed as before. Images of the column were taken with a digital camera to monitor the development of the growing zone of MnO_2 precipitation.

RESULTS

Results of batch experiments to examine the consumption of MnO_4^- by natural materials are plotted in Figure 1. After 10 days, the MnO_4^- concentration dropped to near zero for every sample of natural aquifer material, irrespective of the type of sample. There are two exceptions, with virtually no change in the MnO_4^- concentration with time. These two samples are the silica sand and the glass beads, which were used later for the column experiments. The decrease in the MnO_4^- concentration appears follow a first order kinetic rate law. Because permanganate is likely conservative in the aqueous phase, it is unlikely concentration reductions were caused by sorption onto the solids.

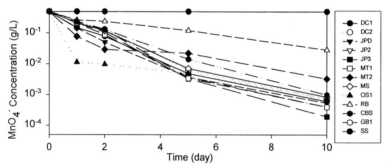

FIGURE 1. Utilization of MnO_4^- through oxidation of various aquifer materials.

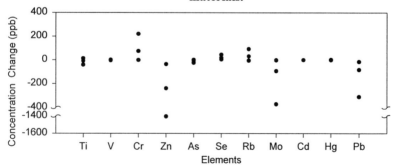

FIGURE 2. Average, maximum and minimum concentrations for the elements in pore water after treatment with 0.5 g/L $KMnO_4$ for ten days.

The decreasing MnO_4^- concentrations with time implies that the aquifer materials have being oxidized. Results of the ICP-MS analysis of the pore water MnO_4^- treatment (Figures 2, 3) confirmed this conclusion. In spite of the variability in the types of sample, there are common features among the results. Of the 12 elements that were quantitatively analyzed, chromium, selenium and rubidium generally increased in concentration; and vanadium, zinc, cadmium and lead, decreased. The remaining elements (titanium, arsenic, molybdenum, cesium and mercury) generally showed no significant change. While the pattern of change is relatively consistent, the magnitude of the change is variable. In some cases, the concentration could change several hundred fold.

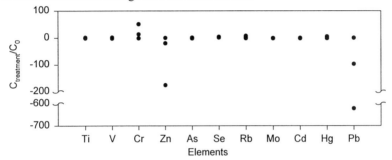

FIGURE 3.Average, maximum and minimum ratio of concentrations of elements in pore water before and after treatment with 0.5 g/L $KMnO_4$ for ten days.

The 1-D column experiment confirms the capability of MnO_4^- to oxidize TCE present in the column. Mass balance calculations, based on the measured TCE and chloride concentration in the effluent from the column, indicated the nearly complete removal of the pure TCE in a relatively short time (Figure 4). Intergrating the TCE removal curve shows that after 365 hours of permanganate flushing, 96.9% of the initial TCE was removed from the column. The rate of TCE removal was highest when MnO_4^- first fully saturated the column, and decreased with time dramatically. 90% of the TCE was removed in the first 115 hours of the experiment, while another 250 hours was required to remove the last 7%. Flushing by MnO_4^- was halted when Cl^- measurements suggested that the oxidation reaction had stopped. Flushing continued with Milli-Q water being pumped through the column. Interestingly, TCE concentrations rebounded to about 130 μg/L. Apparently, the Cl^- measurements were not sensitive at the lower level of detection to indicate that small quantities of TCE remained in the column. It is estimated that another couple hundred hours would be needed for the TCE concentration drop to below 5 μg/L.

The distribution of manganese dioxide precipitation in the column after the experiment is presented in Figure 5. The majority of the Mn was presented at or very close to the DNAPL zone. These precipitates tended to plug the column and toward the end of the experiment flushing was increasingly difficult.

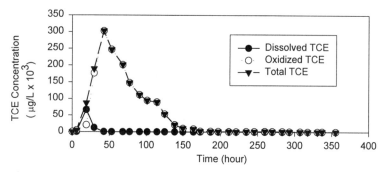

FIGURE 4. TCE concentration with time in the effluent of the 1 – D column and calculated removal based on Cl⁻ stoichiometry.

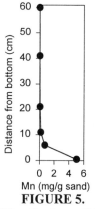

FIGURE 5. Distribution of Mn along the column.

The results from the 2-D flow tank experiment (Figure 6) were similar to the 1-D experiment, except that removal rates were noticeably smaller. After 313 hours of flushing with MnO_4^-, only 34.9% of the initial TCE was removed from the column. Half of the TCE was removed in the first 25% of the elapsed time. Note that in Figure 6 at the end of the experiment TCE concentration rebounded, once the injection of MnO_4^- had stopped.

The visual observation system provided a useful way to monitor the growth of the zone of MnO_2 precipitation. Precipitation started once MnO_4^- came in contact with residual DNAPL. There was a tendency for the MnO_4^- flood to bypass the zone with the highest DNAPL saturation, moving instead through a much less saturated zone in the upper portion of the tank. With time, the precipitation of MnO_2 reduced the permeability. MnO_2 rapidly formed a precipitation rind above the DNAPL pool. Greater injection pressure was required to maintain the flow close to the end of the experiment. The experiment came to a halt when MnO_2 plugged the tank nearly completely and MnO_4^- could no longer be injected. Carefully examination of the tank after the experiment indicated the presence of tiny CO_2 bubbles, produced from the oxidation reaction. The gas bubbles likely played a role in reducing the permeability and the flow in the system. Flushing with MnO_4^- appears effective in removing residual DNAPL. However, much of the original volume of the pool of DNAPL at the bottom of the tank was evident at the end of the experiment.

SUMMARY AND DISCUSSION

In the batch experiments, the gradual disappearance of MnO_4^- and the change in chemistry of the aqueous phase pointed to significant interactions between the porous medium and the MnO_4^-. Given the variability in mineralogy

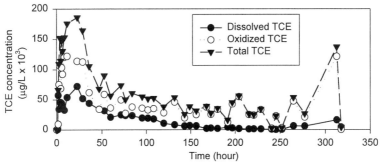

FIGURE 6. Effluent TCE concentration change with time in the 2 – D column experiment.

and grain coatings, it is not possible to propose reaction mechanisms. We are now in the processes of elucidating the MnO_4^-- solid reactions in more detail. The permanganate likely oxidizes the solid metal oxide. For example, chromium, Cr (III), which often exists as Cr_2O_3 solid, was probably oxidized to Cr (VI), a dissolved species of chromium. Any metals sorbed to organic matters would be added to solution as MnO_4^- oxidizes the organic matter. Clearly, MnO_4^- is a powerful oxidant that is capable of oxidizing solids and organic contaminants indiscriminately. The heavy metal loading is of regulatory concern, although it is not clear how mobile these constituents might be. The ability for most natural aquifer materials to consume MnO_4^- reduces concern that the presence of MnO_4^- in solution would emerge as a major problem of contaminant at most sites.

Understanding how the chemical composition of the aqueous phase will change is more complicated because the oxidation of TCE is capable of shifting the equilibra of dissolved-precipitated phase, sorbed-solution phase and complexations by changing pH, E_h and the concentration of Cl⁻. For example, selenium tends to much more mobile in the oxic rather than reducing condition (Alloway, 1995) as the E_h goes up with the treatment. Mobilization of several metals (Hg, Cd, Pb) was found to be very sensitive to the concentration of Cl⁻. A given element will likely respond to the permanganate treatment differently. Our results suggest caution in the application of oxidation schemes to environmentally sensitive areas and aquifers containing high concentrations of metal oxides at lower oxidation states.

Results from both the 1-D and 2-D experiments indicate that early in the flushing process, the mass removal rate is greatest. As the treatment proceeds, the rate decreases dramatically. Running the experiments to achieve complete DNAPL removal will produces a long tail in the later part of the treatment, which points to inefficiencies in the flushing. This tailing effect has also been noted by researchers studying DNAPL dissolution. Towards the end of their experiments when a large portion of the contaminant was dissolved, small droplets remained in hydraulically stagnant zones in the medium. Preferential flow paths appear to develop even with the homogeneous media. With MnO_4^- oxidation schemes, the tendency for preferential flow paths to develop will be further promoted as MnO_2

precipitates in the zones of higher DNAPL saturation or CO_2 bubbles are trapped in the porous medium. The rapid oxidation and relatively slow mass transfer rate from the DNAPL to the aqueous phase means that MnO_2 would tend to precipitate at or immediately adjacent to the DNAPL, which results in a zone of more concentrated precipitates around the zones of greatest saturation. Flow would tend to bypass these zones and follow a more permeable flow path, causing the DNAPL oxidation process to become diffusion controlled.

These effects are most obvious in the 2-D tank experiment where DNAPL is present in zones of high residual saturation or pools. The pooled DNAPL persisted because the interfacial area is relatively small and MnO_4^- moved around but not through these zones. In zones of residual saturation, DNAPL oxidization was more efficient because MnO_4^- could move through the zones. With the opportunities for MnO_4^- to move around zones of DNAPL saturation in the two dimensional system, there was a much smaller overall destruction of TCE in the 2-D tank as compared to the 1-D column. These results suggest that with 3-D flow conditions, evident at actual contaminated sites, the efficiency of removal would probably be smaller. It appears that MnO_4^- oxidation is more effective in cleaning up residually saturated DNAPLs rather than pool of DNAPL. Clearly, in actual field settings, the issue of flushing efficiency will be of concern in the design. Experiments are underway to find an effective additive that minimizes the precipitation effects.

ACKNOWLEGEMENTS: This paper is based on research supplied by the Department of Energy under Grant No. DE-FG07-96ER14735.

REFERENCES

Alloway, B.J. (Eds.). 1995. *Heavy metals in soils* (2nd edition). Blackie academic & Professional, London, pp. 106-351.

Drescher, E., A. R. Gavaskar, B. M. Sass, L. J. Cumming, M. J. Drescher and T. K. J. Williamson. 1998."Batch and column testing to evaluate chemical oxidation of DNAPL source zones." In G. B. Wickramanayake and R. E. Hinchee (Eds.), *Physical, Chemical, and Thermal Technologies, Remediation of Chlorinated and Recalcitrant Compounds*, pp. 425-432. Battelle Press, OH.

Schnarr, M., C. Truax, G. Farguhar, E. Hood, T. Gonullu, and B. Stickney. 1998. "Labortory and controlled field experiments using potassium permanganate to remediate trichloroethylene and perchloroethylene DNAPLs in porous media." *Journal of Contaminant Hydrology.* 29(3): 205-224.

Yan, Y.E. and F.W. Schwartz. 1999. "Oxidative degradation and kinetics of chlorinated ethylenes by potassium permanganate." *Journal Of Contaminant Hydrology.* 37(3-4): 343-365.

IN SITU DESTRUCTION OF CHLORINATED SOLVENTS WITH KMnO$_4$ OXIDIZES CHROMIUM

Jane Chambers, Alan Leavitt, and Caryl Walti (Northgate Environmental Management, Inc., Oakland, California, USA)
Cindy G. Schreier (PRIMA Environmental, Sacramento, California, USA)
Jeff Melby & Lucas Goldstein (LFR Levine·Fricke, Emeryville, California, USA)

ABSTRACT: In-situ chemical oxidation is an attractive option for aggressively remediating chlorinated volatile organic compounds (VOCs) in soil and ground-water. This technology has the potential to rapidly destroy chemicals of concern, restoring groundwater quality much faster than can be accomplished through typical pump-and-treat remedies. The authors completed a pilot-scale evaluation of in-situ oxidation using potassium permanganate (KMnO$_4$) to remediate chlorinated solvents in soil and groundwater at a former electronics manufacturing facility in Sunnyvale, California. Our study evaluated the soil's natural capacity to reduce chromium to its more stable trivalent form.

INTRODUCTION AND BACKGROUND

The authors performed a chemical oxidation treatability study and pilot test for a VOC release site in Sunnyvale, California. The property was formerly used for electronics manufacturing activities that resulted in the release of trichloroethene (TCE) to soil and groundwater. There has been significant degradation of TCE to cis-1,2-dichloroethene (DCE) in site groundwater. Site cleanup activities included removal of waste handling facilities and associated chemical-affected soils; soil-gas extraction; and groundwater extraction. However, after implementing those cleanup measures, the total VOC concentrations in ground-water at the site are asymptotic and remain as high as 8,000 micrograms/liter (µg/L). The residual VOC plume is partially located under the existing buildings at the Site, as shown in Figure 1. Groundwater at the site is approximately 9 feet (2.7 meters) below the ground surface (bgs). Sediments in this region are a complex sequence of alluvial deposits, consisting of heterogeneous deposits of relatively fine-grained materials (i.e., sandy silts and clays) with localized sand stringers. The ground-water-bearing zones have been classified as A-zone (approximately 9 to 25 feet [2.7 to 7.6 meters] bgs) and B-zone (approximately 25 to 45 feet [7.6 to 13.7 meters] bgs). The sediments have a typical organic content of 800 to 3,500 mg/kg.

KMnO$_4$ was selected as the oxidant for this evaluation because it effectively degrades DCE and TCE (ESTCP, 1999, and LaChance et al.), is more persistent than other oxidants (thus is good for areas of limited access, such as under buildings where in can flow further from point of injection than other oxidants without breaking down), is relatively safe to handle in the field, and does not generate large quantities of heat or gases (concern for building occupants as well as its effect on soil permeability).

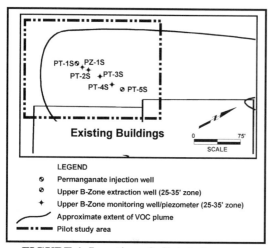

PT-1S⊙ ₊PZ-1S
PT-2S ₊PT-3S
PT-4S⁺ ⊘ PT-5S

Existing Buildings   0 _____ 75'
                                SCALE

LEGEND
⊙   Permanganate injection well
⊘   Upper B-Zone extraction well (25-35' zone)
✛   Upper B-Zone monitoring well/piezometer (25-35' zone)
╱   Approximate extent of VOC plume
▬ ▪▪ ▬   Pilot study area

FIGURE 1. Location of pilot study area.

Comprehensive bench-scale tests were performed prior to implementing the pilot-scale test to estimate the dose requirements needed, to assess the effect of $KMnO_4$ on soil permeability, and to observe geochemical reactions. Results of the treatability study are reported separately in these proceedings (Chambers et al., 2000). Column and batch studies indicated that as much as 2,510 µg/L of hexavalent chromium were present after the introduction of a $KMnO_4$ solution. The detected hexavalent chromium was caused by oxidation of naturally occurring chromium in soils and from chromium impurities in the $KMnO_4$. This finding alerted the authors to the need to evaluate hexavalent chromium attenuation during $KMnO_4$ treatment. The federal and California maximum contaminant levels (MCLs) for total chromium in drinking water are 100 and 50 µg/L, respectively. Naturally occurring trivalent chromium concentrations in California soils range from 23 to 1,579 milligrams/kilogram (mg/kg) (Kearney Foundation, 1996). Site soils contain an average of about 76 mg/kg of trivalent chromium. $KMnO_4$ also contains chromium at concentrations up to approximately 60 mg/kg.

PILOT STUDY

The pilot study was conducted from October 1999 through January 2000. $KMnO_4$ solution was injected over a one-week period and pilot study area wells were monitored during the injection and over the next three months.

Pre-Pilot Study Activities. Before injection, extensive cone penetration testing (CPT) was performed in the source area to evaluate the complex lithology, and to further define the distribution of VOCs in site groundwater. Wells were installed to provide baseline chemical data, to monitor chemical changes, and to evaluate geochemical reactions during and after injection. Baseline groundwater samples were analyzed for VOCs, general minerals, and various field parameters (pH, temperature, oxidation reduction potential [ORP] dissolved oxygen [DO], conductivity and turbidity).

The injection well, PT-1S, was installed on the upgradient margin of the source area. PT-1S was screened between 25 and 35 feet (7.6 and 13.7 meters) bgs. An extraction well, PT-5S, was installed approximately 60 feet (18.3 meters) northeast and hydraulically downgradient from the injection well to enhance the existing hydraulic gradient. Three monitoring wells (PT-2S through PT-4S) were installed between the injection and extraction wells at 15-foot (4.6-meter) intervals.

Pilot Study Injection and Monitoring. A solution of 40 to 50 grams/liter (g/L) $KMnO_4$ was injected into the upper B-zone sediments at injection well PT-1S intermittently over a one-week period. Approximately 4,000 gallons (15,140 liters) were injected at a rate of 1 to 2 gallons per minute (3.8 to 7.6 liters/minute) and a maximum wellhead pressure of 10 pounds per square inch (0.70 kilograms/centimeter^2).

An additional monitoring well, PZ-1, was installed during the week of injection, when it became apparent that a monitoring well was needed closer to the injection well to monitor the geochemical effects of $KMnO_4$.

During the week of injection, field parameters were monitored at least once each day at pilot study and perimeter wells. Following injection, field monitoring and low flow groundwater sampling were performed at key pilot study wells on a weekly basis for four weeks and monthly thereafter. Field parameters were measured using a Hydrolab MiniSonde® multi-parameter instrument with a flow-through cell and included pH, temperature, ORP, DO, conductivity, and turbidity. The presence of $KMnO_4$ was monitored in the field by measurement of percent absorbance. Selected groundwater samples were analyzed for VOCs, hexavalent chromium, total chromium (sum of trivalent and hexavalent chromium), $KMnO_4$, manganese, and chloride.

Results. Pilot study monitoring data indicate several interesting results. Although $KMnO_4$ concentrations decreased substantially following injection, low $KMnO_4$ concentrations persisted in site groundwater longer than anticipated. $KMnO_4$ was detected in site groundwater at the injection well (PT-1S) and in the nearest downgradient well (PZ-1) for 9 weeks following injection. Figure 2 shows $KMnO_4$ concentrations in these wells during and after injection.

The authors estimate that approximately two thirds of the injected oxidant short-circuited upward into the A-zone, which in this location is more permeable. Because of the injection pressure and the tendency of permanganate to increase the permeability of sediments, as was noted in the treatability study (Chambers et al., 2000), injection well design (location, screened intervals, and construction) is an important factor in final remedial design.

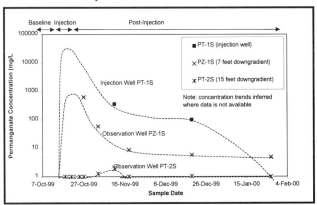

FIGURE 2. $KMnO_4$ concentrations in B-zone pilot study wells.

KMnO$_4$ flowed in the more permeable A-zone farther than was observed in the B-zone: 50 feet (15 meters) in the A-zone versus 30 feet (9.1 meters) in the B-zone, as measured downgradient from the injection well. Figures 3a and 3b illustrate the VOCs (sum of TCE and DCE) in the B-zone source area before and after KMnO$_4$ injection, respectively. The crescent-shaped area on Figure 3b shows where most of the VOC destruction occurred in the B-zone source area. In this area, concentrations of VOCs were above 5,000 µg/L before the injection of KMnO$_4$. Within a few days after injection, concentrations decreased by approximately 60 to 70%. KMnO$_4$ was only observed at low concentrations in well PT-2S (Figure 2) yet VOC concentrations in both PT-2S and PT-3S decreased significantly, suggesting that the leading edge of KMnO$_4$ was at PT-3S, located 30 feet (9.1 meters) downgradient from the injection well. Because substantial decreases in VOC concentrations were observed rapidly in both of those wells, it appears that thin, sand stringers provided a preferential pathway for KMnO$_4$ to flow radially from the injection well.

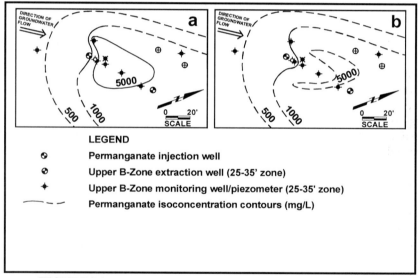

FIGURES 3a and 3b. Pre- and post-injection extent of VOCs in B-zone groundwater.

Figure 4 illustrates the VOC destruction and rebound measured in monitoring wells PT-2S and PT-3S. [Note: Baseline data were not available for PZ-1S because it was installed after the start of injection, therefore PZ-1 data are not presented on Figure 4.] Concentrations of VOCs at well PT-2S decreased from 8,000 µg/L to 3,200 µg/L shortly after injection began. Over the next three months, the VOC concentration increased to 4,300 µg/L. Likewise, VOC concentrations at well PT-3S declined dramatically immediately following injection (from 7,200 µg/L to 2,300 µg/L) and remained at this lower concentration over the next three months. Rebound of VOC concentrations was not significant in wells PT-2S and

PT-3S suggesting that a significant mass of VOCs adsorbed to soil was destroyed by KMnO$_4$.

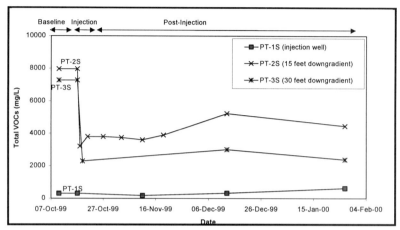

FIGURE 4. Trends in VOC concentrations in B-zone groundwater.

In addition to evaluating oxidation of VOCs, another objective of the pilot study was to monitor geochemical reactions, particularly the fate of hexavalent chromium. Neither trivalent nor hexavalent chromium were detected in site groundwater before injection. However, chromium is naturally present in site soils (approximate concentration of 70 mg/kg), and is also an impurity of KMnO$_4$ (approximate concentration of 30 to 60 mg/kg). As previously discussed, earlier treatability studies for the site indicated that hexavalent chromium could be produced as a result of KMnO$_4$ treatment (Eary and Rai, 1987, and Fendorf and Zasoski, 1992). It is possible that chromium is oxidized not only by KMnO$_4$, but also by MnO$_2$, a by-product of the KMnO$_4$ treatment. MnO$_2$ has the potential to migrate significantly downgradient from the treatment zone. Therefore, as part of this pilot study, site groundwater within and downgradient from the treatment zone was monitored for the presence of chromium.

Laboratory analyses of groundwater samples collected during and after injection indicated that matrix interference with KMnO$_4$ precluded reliable analysis for hexavalent chromium. Therefore, samples were analyzed for total dissolved chromium, and the authors conservatively assumed that all dissolved chromium detected was in the hexavalent form. This is a reasonable assumption considering that the strong oxidizing conditions of the permanganate would favor the formation of hexavalent chromium, and that trivalent chromium is relatively insoluble in water.

Maximum dissolved chromium concentrations of approximately 2,000 µg/L were detected in shallow groundwater within 5 feet (1.5 meters) of the injection well approximately one week after injecting KMnO$_4$ at the site. As shown in Figure 5, chromium concentrations rapidly attenuated within two to three weeks to concentrations on the order of 100 µg/L. Three months after initiating the pilot test, dissolved chromium concentrations were less than 5 to 10 µg/L in all groundwater samples. These data demonstrate the significant capacity for site soils to convert hexavalent chromium back to trivalent chromium. This natural attenuation process has been reviewed by the U.S. EPA (Palmer and Puls, 1994) and others. It is likely that hexavalent chromium was reduced by naturally occurring organic matter (e.g., humic and fulvic acids), ferrous iron, and other substances. For example, total organic carbon (TOC) measurements for soil samples from the site ranged from 800 to 3,500 mg/kg. TOC concentrations of this magnitude would provide sufficient reducing capacity for soils to naturally attenuate chromium.

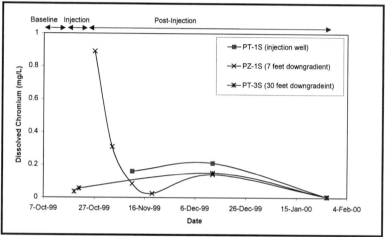

FIGURE 5. Trends in chromium concentrations in B-zone groundwater.

CONCLUSIONS

This pilot study demonstrated that $KMnO_4$ can effectively remediate 1,2-DCE and TCE in saturated-zone sediments and groundwater at the site. $KMnO_4$ was detected up to 50 feet (15 meters) downgradient from the injection well, in the permeable sediments, and as far as 30 feet (9.1 meters) downgradient, in the more fine-grained B-zone sediments interval.

Total VOC concentrations decreased by approximately 60 to 70% in groundwater samples collected from monitoring wells located 15 and 30 feet (4.6 and 9.1 meters) downgradient from the injection well. Post-injection monitoring data, collected three months after $KMnO_4$ treatment, showed little or no evidence of rebound. These data suggest that VOCs adsorbed to the sediments were oxidized in the vicinity of the downgradient monitoring wells.

Site soils exerted a substantial demand for $KMnO_4$ that was several orders of magnitude higher than that required simply to oxidize the VOCs dissolved in groundwater. In addition to the adsorbed VOCs, site soil contains organic carbon and inorganic constituents (including chromium) that compete to consume the $KMnO_4$. However, the concentrated dose of $KMnO_4$ used for the pilot study was very persistent compared to other oxidants that survive for only a few hours or days (e.g., ozone and peroxide). After an initial decline in concentration, residual $KMnO_4$ was observed for approximately 2 months in the injection area. The presence of this oxidant provides continued treatment of VOCs as they slowly diffuse from fine-grained sediments. The overall cost effectiveness of this technology will depend on several factors, including site geology, chemicals of concern, and the size of the remediation area.

It is essential to understand the geochemistry before implementing full-scale treatment. In this case study, chromium was naturally present in site soils in the stable chromium form. Additionally, chromium is a common impurity of $KMnO_4$. With the high concentrations of $KMnO_4$ needed to overcome the soil demand, a significant amount of chromium could be introduced to the subsurface during full-scale injection. Although chromium concentrations temporarily increased in groundwater, site soils had enough reducing capacity to decrease chromium to nondetectable levels within three months after treatment. This may not be the case at other sites, where sediments have less organic carbon, ferrous iron, or other constituents capable of reducing hexavalent chromium. Also, the formation of hexavalent chromium could be a significant concern if groundwater downgradient from the treatment zone is used as a water supply. Considering these factors, the authors recommend that geochemical evaluations and treatability testing be performed before using this in-situ oxidation technology on a pilot-scale and full-scale basis.

REFERENCES

Chambers, J. D., A. L. Leavitt, C. L. Walti, C. G. Schreier, J. T. Melby, and L. Goldstein. 2000. "Treatability Study – Fate of Chromium During Oxidation of Chlorinated Solvents." The Second International Conference on Remediation of Chlorinated and Recalcitrant Compounds. (Monterey, California, May 22-25, 2000).

Eary, L. E., and D. Rai. 1987. "Kinetics of Chromium(III) Oxidation to Chromium(IV) by Reaction with Manganese Oxide." *Environ. Sci. Technol.* 21(12): 1187-1193.

Environmental Security Technology Certification Program. 1999. *Technology Status Review In Situ Oxidation.*

Fendorf, S. E., and R. J. Zasoski. 1992. " Chromium(III) Oxidation by δ-MnO₂. 1. Characterization". *Environ. Sci. Technol.* 26(1): 79-85.

Kearney Foundation. 1996. *Background Concentrations of Trace and Major Elements in California Soils, Special Report.* Kearney Foundation of Soil Science, Division of Agriculture and Natural Resources, University of California.

LaChance, J. C., S. R. Reitsma, D. McKay, and R. Baker. "n.d.". "In Situ Oxidation of Trichloroethene using Potassium Permanganate Part 1: Theory and Design."

Palmer, C. D., and R. W. Puls. 1994. " Natural Attenuation of Hexavalent Chromium in Ground Water and Soils." *EPA Ground Water Issue.* U.S. EPA, EPA/540/S-94/505.

TREATABILITY STUDY--FATE OF CHROMIUM DURING OXIDATION OF CHLORINATED SOLVENTS

Jane Chambers, Alan Leavitt, Caryl Walti
(Northgate Environmental Management, Oakland, California, USA)
Cindy G. Schreier (PRIMA Environmental, Sacramento, California, USA)
Jeff Melby (Levine Fricke, Emeryville, California, USA)

ABSTRACT: A comprehensive laboratory study was performed to evaluate potassium permanganate ($KMnO_4$) for *in situ* remediation of chlorinated volatile organic compounds (VOCs) in groundwater at a site in Sunnyvale, California. In addition to determining the dose requirements needed to oxidize the chemicals of concern, estimating the soil demand of $KMnO_4$, and assessing the effect of $KMnO_4$ on soil permeability, the study evaluated the effect of adding a strong oxidizing agent on the speciation of naturally occurring chromium. Evaluating the effect of $KMnO_4$ addition on chromium speciation is important because chromium is an impurity in $KMnO_4$ and because both $KMnO_4$ and manganese dioxide (a by-product of permanganate oxidation) can oxidize trivalent chromium to toxic hexavalent chromium.

INTRODUCTION

Injection of potassium permanganate solution is an emerging technology for the *in situ* oxidation of chlorinated solvents in soil and groundwater. Several recent laboratory tests and field applications have been conducted to evaluate the ability of $KMnO_4$ to enhance removal of trichloroethylene (TCE) and perchloroethylene (PCE) dense, non-aqueous phase liquids (DNAPLs). [1-4]

The oxidation of cis-dichloroethylene (cis-DCE, $C_2H_2Cl_2$) and TCE (C_2HCl_3) by $KMnO_4$ is believed to occur according to the reactions:

$$3C_2H_2Cl_2 + 8KMnO_4 \rightarrow 6CO_2 + 6KCl + 2KOH + 8MnO_2(s) + 2H_2O$$
$$C_2HCl_3 + 2KMnO_4 \rightarrow 2CO_2 + 2KCl + 2MnO_2(s) + HCl$$

Based on this stoichiometry, 4.3 g of $KMnO_4$ is needed to completely oxidize 1 g of cis-DCE, while 2.4 g of $KMnO_4$ is needed to completely oxidize 1 g of TCE. However, $KMnO_4$ is a non-selective oxidizing agent and will also be consumed by natural organic matter and other oxidizable species. $KMnO_4$ demand in soil and groundwater was therefore determined.

Manganese dioxide (MnO_2), a by-product of $KMnO_4$ oxidation, can potentially affect the permeability of the aquifer material. In addition, it may be possible for MnO_2 to oxidize soil chromium(III) [Cr(III)] to toxic, water soluble Cr(VI). Total Cr has a California Maximum Contaminant Level (MCL) of 0.05 mg/L. The effects of MnO_2 will depend primarily on the total mass of $KMnO_4$ injected and the area over which it is dispersed.

Because of the mobility and toxicity of Cr(VI), the presence of Cr impurities in solid $KMnO_4$, and the potential for both $KMnO_4$ and MnO_2 to oxidize soil Cr(III), the formation and attenuation of Cr(VI) during treatment of site soils by $KMnO_4$ was investigated as part of this study.

MATERIALS AND METHODS

Sample Description. Groundwater from the site in Sunnyvale, California contains up to 8 mg/L of cis-DCE and TCE. Soil at this location also contains approximately 70 mg/kg of Cr(III), which is typical of Northern California and other regions of the United States with soils partly derived from chromium-bearing minerals. Clean site groundwater, cis-DCE/TCE impacted groundwater, and clean and impacted sands, silts and clays were collected and delivered to PRIMA Environmental for testing.

Determination of $KMnO_4$ Groundwater Demand. Concentrations of cis-DCE (1,800 µg/L) and TCE (130 µg/L) in site groundwater were used to calculate the "stoichiometric" dose requirement of 8,000 µg $KMnO_4$ / L site water. To evaluate the potential reaction of $KMnO_4$ with other groundwater constituents, samples of site groundwater were allowed to react with $KMnO_4$ at concentrations greater than the stoichiometric dose for cis-DCE and TCE. Site water (100mL) was added to three clear glass serum bottles equipped with cork-style Mininert™ stoppers. $KMnO_4$ solution was then added to two of the bottles so that the initial concentrations of $KMnO_4$ in the bottles were 16,000 µg/L (2 x stoichiometry) and 40,000 µg/L (5 x stoichiometry). A control containing no $KMnO_4$ was also prepared. The solutions were mixed for 5.75 hours, then analyzed for volatile organic compounds, pH and $KMnO_4$.

Determination of $KMnO_4$ Soil Demand. The $KMnO_4$ demand of each soil type (sandy, silty, clayey) was estimated by adding 15 mL of 10 g/L $KMnO_4$ solution to 15 g of field-moist soil from the site, mixing, then periodically measuring residual $KMnO_4$ via UV-visible spectroscopy (560 nm).

Effect of $KMnO_4$ Treatment on Soil Permeability. Permeability tests (one before $KMnO_4$ treatment, one after) were performed on sandy soil using three $KMnO_4$ concentrations: 1g/L, 5 g/L, and 25 g/L. A standard permeability test (ASTM D 5084) using clean water was first conducted to obtain initial permeability and porosity information. A modified permeability test using $KMnO_4$ rather than water as the permeant was then conducted. Because only two cores were prepared (due to sample limitations), one core was flushed with 1 g/L $KMnO_4$ and the other with 5 g/L. At the end of this portion of the test, the core receiving 1 g/L $KMnO_4$ was then flushed with 25 g/L $KMnO_4$. The cores were flushed with $KMnO_4$ until 15 pore volumes of $KMnO_4$ solution had been put through. Effluent from the sand columns was collected approximately after each pore volume and analyzed for $KMnO_4$. Effluent from the sand column exposed to 5 g/L $KMnO_4$ was also analyzed for manganese dioxide and Total Cr. The sand column flushed with 5 g/L $KMnO_4$ was subsequently flushed with about 9 pore volumes of clean site water, and the effluent collected and analyzed for Cr(VI).

Oxidation of Soil Chromium. For each soil a batch test was conducted in which equal amounts of soil and $KMnO_4$ solution were mixed. Once the $KMnO_4$ had completely reacted, the mixtures were centrifuged and filtered. The filtrates were analyzed colorimetrically for Cr(VI).

Reduction of Cr(VI) After $KMnO_4$ Treatment. Ferrous iron, ascorbic acid, and molasses were evaluated as possible reducing agents for Cr(VI). Eight replicates containing sand and 0.5 g/L $KMnO_4$ were prepared. The replicates were mixed for approximately 3 days, until the purple color of $KMnO_4$ was gone. The mixtures were then centrifuged and each aqueous phase was analyzed for Cr(VI). After analysis of Cr(VI), the aqueous phase of Replicates 1-4 was decanted and replaced with reductant solution (see Table 1). After mixing for 24 hours, the mixtures were again centrifuged and the aqueous phase analyzed for Cr(VI). The supernatant was then decanted and replaced with clean site water. Cr(VI) in the aqueous phase was then analyzed periodically. Replicates 5-8, which still contained the original solution, were analyzed periodically for Cr(VI) in order to determine whether the Cr(VI) concentration changed over time.

TABLE 1. Reductants Used to Reduce Cr(VI).

Replicate Number	Reductant*
1	Ferrous Iron (400 mg/L $FeSO_4 \cdot 7H_2O$)
2	Ascorbic Acid (128 mg/L ascorbic acid)
3	Molasses (2 mL Brer Rabbitt, full flavor molasses / L water)
4 (Control)	None (clean site water added)

*All solutions made up in clean site water.

Natural Attenuation of Cr(VI) Under Simulated Subsurface Conditions. A column was filled with sand then flushed with clean site water that had been purged with nitrogen gas until the dissolved oxygen (DO) concentration was less than 1 g/L. (Low DO was used to simulate conditions at the site.) The column was then flushed with site water spiked with 2,500 µg/L Cr(VI). Breakthrough of Cr(VI) was monitored. Once the effluent concentration was near the influent concentration, the column was sealed to prevent contact with air. After standing undisturbed for 7 days, the column was drained and the pore water analyzed for Cr(VI).

RESULTS AND DISCUSSION

Determination of $KMnO_4$ Groundwater Demand. The results of the groundwater test are shown in Table 2. The 5X stoichiometric dose completely oxidized the cis-DCE and TCE in the sample.

TABLE 2. Cis-DCE and TCE in Groundwater After
Treatment with KMnO$_4$.

Test	Cis-DCE, mg/L	TCE, mg/L
0X Stoichiometric Dose (0 mg/L KMnO$_4$--Control)	1.9	0.18
2X Stoichiometric Dose (16.1 mg/L KMnO$_4$)	0.026	0.013
5X Stoichiometric Dose (40.3 mg/L KMnO$_4$)	< 0.001	< 0.001

Determination of KMnO$_4$ Soil Demand. The results of the time course are shown in Figure 1. All three soils exerted a strong initial KMnO$_4$ demand, with the silt and clay materials consuming about twice as much KMnO$_4$ (4271 and 4376 mg/kg, respectively) as the sandy material (2,191 mg/kg). After 24 hours, the rate of KMnO$_4$ consumption slowed significantly. For example, after 14 days, the sand had only consumed an additional 2296 mg/kg KMnO$_4$, the silt an additional 3335 mg/kg and the clay 2787 mg/kg. This slower rate was linear for all three materials.

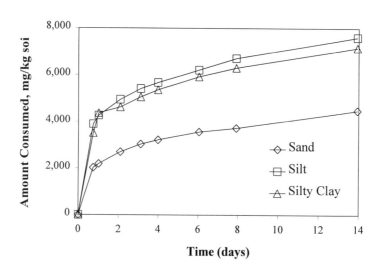

FIGURE 1.   KMnO$_4$ Consumed by Site Soils.

The initial and later rates are given in Table 3. The results probably overestimate the KMnO$_4$ demand of the soils in the field, especially for silt and clay, because the soil and KMnO$_4$ solutions in the laboratory were well-mixed.

TABLE 3. Rate of KMnO$_4$ Consumption in Site Soils.

Soil	Initial Rate (0-1 days), mg/kg/day	Later Rate (1-14 days), mg/kg/day
Sand	2288 (r^2 = 0.962)	162 (r^2 = 0.924)
Silt	4447 (r^2 = 0.966)	244 (r^2 = 0.932)
Clay	4419 (r^2 = 0.994)	218 (r^2 = 0.956)

Effect of KMnO$_4$ Treatment on Soil Permeability. Results of the permeability tests (Table 4) indicate that KMnO$_4$ treatment does not reduce soil permeability. In fact, the hydraulic conductivity increased by almost one order of magnitude after application of 25 g/L KMnO$_4$ solution.

TABLE 4. Effect of KMnO$_4$ on Permeability of Site Soils.

Soil Type	Hydraulic Conductivity, cm/s					
	1 g/L		5 g/L		25 g/L	
	Before* KMnO$_4$	After** KMnO$_4$	Before KMnO$_4$	After** KMnO$_4$	Before KMnO$_4$	After** KMnO$_4$
Sand	3.2x10^{-5}	3.4x10^{-5}	2.2x10^{-5}	2.5x10^{-5}	3.2x10^{-5}	1.1x10^{-4}

* Permeant was water.
** Permeability measured after column flushed with approximately 15 pore volumes KMnO$_4$ solution.

Permeameter Effluent Tests for KMnO$_4$ and Cr. Effluent from the sand permeability columns was analyzed for KMnO$_4$ and other parameters. The concentration of KMnO$_4$ is shown in Figure 2. KMnO$_4$ breakthrough occurred after the first pore volume, and the effluent concentration nearly reached the influent concentration by the third pore volume. Flowrate through the columns varied, but given that 15 pore volumes were put through in five days and assuming that the pore volume was 75 mL, the average residence time within the columns can be roughly estimated at 8 hours. Assuming that the rate of KMnO$_4$ consumption is about 162 mg/kg/day (see above) and the mass of sand in each column was only 0.5 kg, it is not surprising that the effluent concentration quickly approached the influent concentration. The amount of manganese dioxide suspended in the effluent varied from about 37 mg/L to 75 mg/L, but show no obvious trend.

Cr was detected in effluent from the 5 g/L Sand column. The Cr concentration ranged from 1.79 mg/L to 2.51 mg/L during flushing with KMnO$_4$, but decreased to 0.459 mg/L after flushing with about 2 pore volumes of clean site water (Figure 3). After flushing with five to six pore volumes the Cr concentration decreased to below the MCL of 0.05 mg/L. It must be noted that total Cr rather than Cr(VI) was measured in effluent that contained KMnO$_4$ because the purple color of KMnO$_4$ interferes with the Cr(VI) colorimetric method. Since the site water used to prepare the KMnO$_4$ solutions contained < 0.005 mg/L Cr, it was assumed that any Cr in the effluent was due to treatment. Furthermore, because KMnO$_4$ and MnO$_2$ are known to oxidize Cr(III) to Cr(VI), any chromium measured in the effluent was assumed to be present as Cr(VI).)

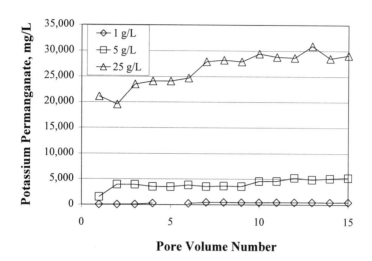

FIGURE 2. KMnO₄ in Effluent from Sand Permeability test.

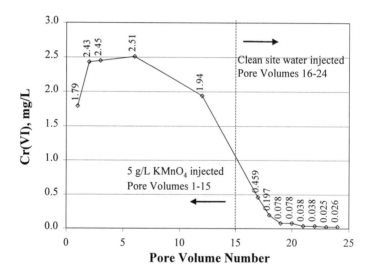

FIGURE 3. Chromium Concentration in Effluent from
Sand Permeability Tests.

Oxidation of Soil Chromium. The results of the batch tests in which KMnO₄ solution was mixed with an equal mass of soil are shown in Table 5. The concentration of Cr(VI) was approximately 0.8 mg/L for all three soils exposed to KMnO₄, and non-detectable in untreated samples. (A sand control was not run in this experiment.)

TABLE 5. Aqueous Cr in Batch Tests Containing Soil Exposed to KMnO₄ Solution.

	Cr(VI), mg/L (colorimetric, diphenylcarbazide)	Cr(VI), mg/L (ion chromatography)
Sand (0.5 g/L KMnO₄)	0.81	0.69
Silt (1 g/L KMnO₄)	0.84	n.a.*
Silt (0 g/L KMnO₄)	< 0.05	n.a.
Clay (1 g/L KMnO₄)	0.81	n.a.
Clay (0 g/L KMnO₄)	< 0.05	n.a.

* n.a. = not analyzed.

Some of the Cr was apparently due to impurities in the solid KMnO₄. Analysis indicated that the KMnO₄ contained 63 mg Cr/kg KMnO₄, which was presumably present as Cr(VI) due to the oxidizing nature of KMnO₄. However, this was not enough to account for most of the chromium observed in the various tests. For example, a 0.5 g/L KMnO₄ solution would contribute 0.003 mg/L Cr, yet the concentration of Cr(VI) in the sand batch test was 0.81 mg/L. Similarly, a 1 g/L KMnO₄ solution would contribute 0.006 mg/L Cr, while a 5 g/L solution would contribute 0.031 mg/L Cr. Analysis of the soils for Total Cr yielded values ranging from 86 mg/kg to 93 mg/kg These concentrations are more than enough to account for all of the Cr(VI) observed. However, it must be emphasized that not all of this soil chromium is likely to be available to react with KMnO₄ or with MnO₂.

Effect of Reductants on Cr(VI). One possible means of controlling the Cr(VI) produced during KMnO₄ treatment would be to add a reducing agent after injection of KMnO₄. This would reduce Cr(VI) as well as residual KMnO₄, and possibly even solid MnO₂. Three common Cr(VI) reducing agents were tested: ferrous iron, ascorbic acid (Vitamin C) and molasses. The results are shown in Table 6. Comparison of the three reductants indicates that ferrous iron was most effective, while treatment with molasses was not significantly different than treatment with clean site water. In all cases, though, the concentration of Cr(VI) was less than the Cr MCL of 0.05 mg/L within 24 hours after addition of the reductant solution.

Evaluation of the data in Table yields two important points. First, the amount of Cr(VI) generated by mixing sand with KMnO₄ solution was reproducible, indicating that the formation of Cr(VI) is not an artifact. Second, the concentration of Cr(VI) decreases with time, even if no reductant is added, suggesting that natural attenuation in the field is likely.

It is interesting to note that of the replicates not treated with a reductant (Replicates 5-8), the concentration of Cr(VI) decreased almost twice as much in Replicate 5 than in Replicates 6-8. The reason for this difference is unclear since the only apparent difference between Replicate 5 and Replicates 6-8 is the number of times they were sampled. Possibly, the oxygen exposure and the soil to liquid ratio affected the rate of Cr(VI) attenuation.

TABLE 6. Effect of Reducing Agents on Aqueous Cr(VI) Concentration.

Cumulative Time (days)	3	4	5	12	13	26
Replicate	Cr(VI), mg/L					
	After KMnO$_4$	After Reductant	After Clean Water			
1-Fe(II)	0.234	< 0.01	<0.01	< 0.01	n.a.**	< 0.01
2-Ascorbic Acid	0.225	0.028	<0.01	< 0.01	n.a.	< 0.01
3-Molasses	0.242	0.040	0.012	< 0.01	n.a.	< 0.01
4-Clean Site Water	0.221	0.042	0.013	0.020	n.a.	< 0.01
5-None*	0.240	0.230	0.214	0.081	n.a.	< 0.01
6-None*	0.248	n.a.	n.a.	n.a.	0.155	< 0.01
7-None*	0.237	n.a.	n.a.	n.a.	0.153	< 0.01
8-None*	0.240	n.a.	n.a.	n.a.	0.163	0.013

* KMnO$_4$ solution not replaced.
** n.a. = not analyzed.

Natural Attenuation of Cr(VI) Under Simulated Subsurface Conditions. The initial Cr(VI) concentration in soil pore water was 1.15 mg/L. After seven days, the concentration in pore water exposed to the soil for seven days was 0.68 mg/L, a decrease of 41%. This is consistent with the reductant tests described above in which 32-66% of Cr(VI) disappeared in Replicates 5-8 within 12-13 days.

CONCLUSIONS

The results of this laboratory investigation demonstrate that KMnO$_4$ solution can effectively treat site groundwater containing cis-DCE and TCE. The amount of KMnO$_4$ needed to completely degrade the contaminants in site water was negligible (40.3 mg KMnO$_4$/L) compared to the amount expected to react with the site soils (162 mg/kg/day for sand). The formation of MnO$_2$ during KMnO$_4$ treatment is not expected to adversely affect the groundwater flow since column tests showed no decrease in permeability when sand cores were flushed with up to 15 pore volumes of 25 g/L KMnO$_4$ solution.

Although some Cr(VI) was formed during laboratory testing (presumably due to Cr impurities in the solid KMnO$_4$, as well as oxidation of soil chromium by KMnO$_4$ and MnO$_2$), it disappeared when left in contact with the site soils, indicating that these soil have a high capacity to naturally attenuate chromium.

A field pilot test was performed based on the results of this study and is described elsewhere in these proceedings [5]. The authors recommend that treatability studies be conducted to evaluate geochemical reactions associated with the introduction of strong oxidants such as KMnO$_4$ into the subsurface. Testing is important because the oxidant may contain impurities (KMnO$_4$ contains chromium) and native soils may contain chemicals that could be mobilized as a result of the oxidant.

REFERENCES

1. D. McKay et al. .1998. "In Situ Oxidation of Trichloroethylene Using Potassium Permanganate: Part 2. Pilot Study." In G. B Wickramanayake and R.E. Hinchee (Eds.), *Physical, Chemical and Thermal Technologies—Remediation of Chlorinated and Recalcitrant Compounds* pp. 377-382, Battelle Press, Columbus, OH.
2. J. C. LaChance et al. 1998. "In Situ Oxidation of Trichloroethylene Using Potassium Permanganate: Part 1. Theory and Design." In G. B Wickramanayake and R.E. Hinchee (Eds.), *Physical, Chemical and Thermal Technologies—Remediation of Chlorinated and Recalcitrant Compounds* pp. 397-402, Battelle Press, Columbus, OH.
3. Y.E. Yan and F.W. Schwartz. "1998. Oxidation of Chlorinated Solvents by Permanganate." In G. B Wickramanayake and R.E. Hinchee (Eds.), *Physical, Chemical and Thermal Technologies—Remediation of Chlorinated and Recalcitrant Compounds* pp. 403-408, Battelle Press, Columbus, OH.
4. E. Drescher et al. 1998. "Batch and Column Testing to Evaluate Chemical Oxidation of DNAPL Source Zones." In G. B Wickramanayake and R.E. Hinchee (Eds.), *Physical, Chemical and Thermal Technologies—Remediation of Chlorinated and Recalcitrant Compounds* pp. 425-432, Battelle Press, Columbus, OH.
5. J. Chambers et al. "In Situ Destruction of Chlorinated Compounds with KMnO4 Oxidizes Chromium." Proceedings of the Second International Conference on Remediation of Chlorinated and Recalcitrant Compounds.These proceedings. Battelle Press, Columbus, OH. Project publication date: 2000.

PERMANGANATE TRANSPORT AND MATRIX
INTERACTIONS IN SILTY CLAY SOILS

Amanda M. Struse (IT Corporation, Englewood, Colorado)
Robert L. Siegrist (Colorado School of Mines, Golden, Colorado)

ABSTRACT: The transport of permanganate in low permeability media (LPM) and its ability to degrade trichloroethylene (TCE) *in situ* were studied using intact soil cores and a diffusion transport cell. With a set of uncontaminated cores, the effective diffusion coefficient of a solute tracer was measured and the tortuosity of the LPM was determined. Then potassium permanganate (5000 mg/L $KMnO_4$) was introduced to the influent well of the system and diffusive transport of bromide and permanganate were observed over a period of 30 to 60 days. The transport cells were then disassembled and the cores were dissected for morphological and chemical analyses. Concentration gradients of several matrix constituents were determined. The experiment was then repeated after 2 μL of pure phase TCE were delivered into the center of each of two intact cores. Permanganate transport through the contaminated cores and the extent of TCE degradation were determined. The research demonstrated that permanganate can migrate by diffusion at rates that can be predicted based on the effective diffusion coefficient and matrix interactions of the permanganate. Through physical and chemical characterization, it was determined that there was little physical effect on the soil pore structure due to the $KMnO_4$ treatment and appreciable soil organic matter remained even after extended exposure to permanganate. During diffusive transport of permanganate, TCE originally present in the silty clay soil was degraded to non-detectable levels.

INTRODUCTION

Contamination of soil and ground water with petrochemicals and chlorinated solvents is common in many industrial settings throughout the United States. The remediation of these sites presents a major challenge, especially in low permeability media (LPM) (Siegrist and Lowe, 1996). In bulk deposits of LPM or zones of LPM within otherwise permeable deposits, mass transfer due to advection is often negligible and diffusion dominates the transport of ground water solutes. As a result, *in situ* technologies that depend upon mass recovery of contaminants, such as pump-and-treat and soil vapor extraction, are often inefficient and expensive. For *in situ* technologies that rely on mass destruction, like chemical oxidation, there are challenges with the delivery of treatment agents throughout the contaminated zone. There is increasing interest in *in situ* treatment of organics in LPM using oxidants like potassium permanganate ($KMnO_4$) that might be able to migrate by diffusion and chemically degrade organic contaminants of concern (COCs) within the LPM. The success of *in situ* chemical oxidation of organics within LPM relies heavily on the ability to accurately predict the rate and extent of mass transport of the oxidant in the subsurface.

Research was completed at the Colorado School of Mines (CSM) in collaboration with Oak Ridge National Laboratory (ORNL) to experimentally measure diffusive transport properties for $KMnO_4$ in intact cores of silty clay soil, to explore the permanganate interactions with the soil matrix, as well as determine the ability of $KMnO_4$ to degrade TCE_{DNAPL} *in situ*. The laboratory and modeling research are highlighted herein and further details may be found in Struse (1999). Information on a companion field study may be found in Siegrist *et al.* (1999).

EXPLORATORY MODELING

Methods. As a simulation of field conditions, a mass transport model was used to illustrate the rate of diffusion of MnO_4^- from an *in situ* reactive barrier such as that created by soil fracturing with an oxidative particle mixture (OPM) containing $KMnO_4$ (Siegrist *et al.*, 1999). The model was used to explore the factors affecting the mass transport of MnO_4^- in the subsurface and to design the appropriate lab experiments. A large portion of the modeling was completed using an existing transport model, TRANS1D (Dawson, 1997). TRANS1D is a spreadsheet-based program capable of predicting one-dimensional groundwater solute transport, including first-order degradation. In this work, TRANS1D modeled the MnO_4^- fracture as a continuous source of constant mass while the MnO_4^- concentration in the pore water of the fracture remained higher than the solubility limit (40,000 mg/L). After the fracture pore water concentration begins to drop below solubility, TRANS1D no longer accurately predicts the MnO_4^- mass transfer because the fracture no longer acts as a constant source. Then, diffusive transport of MnO_4^- was modeled using Trans1D in combination with a mass balance based on a form of Fick's Law (assuming a linear concentration gradient) combined with pseudo first-order decay.

A sensitivity analysis was completed by varying several model parameters in order to assess the magnitude of their impact on MnO_4^- migration and the time required for oxidant depletion in the OPM fracture. The parameters examined include the initial mass of permanganate present, the half-life of permanganate in the soil matrix, the porosity, and the effective diffusion coefficient. First, one parameter was altered, it's affect on the system assessed and then it was returned to its original value and another parameter was then altered. The mass was first halved to 0.60 g/cm^2 and then doubled to 2.40 g/cm^2. The half-life was varied between 15 and 120 days and the porosity was varied between 0.3 and 0.5 v/v. Finally, the effective diffusion coefficient was varied by an order of magnitude (7.6E-06 vs. 7.6E-07 cm^2/s).

Results. The model predicts a reactive zone (defined as the region where the MnO_4^- pore water concentration is >1.0 mg/L) of ~15 cm after one month which expands to more than 60 cm in 12 months. These model predictions were similar to the observations made during a companion field study involving soil fracturing of silty clay soils with permanganate OPM (Siegrist *et al.*, 1999). From the sensitivity analysis it was shown that the initial mass of $KMnO_4$ had no effect on the rate at which permanganate migrated from the OPM-filled fracture, only the length of time during which the fracture was active. The thickness of the entire reactive zone was increased only by 20 cm when the half-life was increased from 14 to 120 days and only 8 cm when the porosity was varied between 0.3 and 0.5 v/v. However, the extent of MnO_4^- migration increases by nearly 26 cm when the effective diffusion coefficient is increased by an order of magnitude.

EXPERIMENTAL STUDIES

Materials and Methods. A one dimensional transport cell was constructed to measure the diffusive transport and matrix interactions of $KMnO_4$ within intact cores of a low permeability media (Figure 1). Initially the effective diffusion coefficient of a conservative solute was determined using an uncontaminated, relatively intact soil core. Then $KMnO_4$ was added to the influent well of the transport cell and diffusion of permanganate to the effluent well was measured. *In situ* chemical oxidation of TCE_{DNAPL} was then assessed by spiking an uncontaminated core with pure phase TCE before adding the $KMnO_4$ to the system. At the end of each test, the transport cell was disassembled and the cores were dissected for morphologic and chemical analysis. To aid interpretation of the results, companion soil cores were characterized for properties such bulk density, porosity, and total organic carbon (TOC).

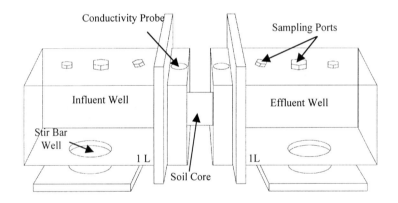

FIGURE 1. Diffusion transport cell apparatus.

Intact soil cores were collected from an uncontaminated location within a land treatment unit at a DOE site in Ohio. Comprised of unconsolidated Quaternary age fluvial deposits, the site has a surficial deposit of 2 to 8 m of low-permeability clays and silts ($K_{sat} < 10^{-5}$ cm/s) known as the Minford member. These sediments are reportedly comprised of 70 to 95% silt (quartz and feldspars) and clay particles (illite, quartz, kaolinite, and smectite) with pH of ~6.0, cation exchange capacity of ~17.5 meq/100g, and total organic carbon (TOC) of 500 to 1500 mg/kg (Siegrist *et al.*, 1999). Intact cores were collected from the site at 2.0 to 2.5 m bgs by ORNL using a hydraulic probe truck and thin-tube samplers. The cores (each 4 cm diam. by 30 to 60 cm long) were then shipped to CSM in foam-padded containers. Some of the soil cores were used for characterization of soil properties including porosity (0.4 v/v), moisture content (26% dry wt.%), and dry bulk density (1.58 g/cm^3), and total organic carbon content (700 to 900 mg/kg). The oxidant demand of the silty clay soil was measured during 14-day batch tests and found to be ~2.8 and 10.8 mg-MnO$_4^-$/g soil at oxidant doses of 500 and 5000 mg/L KMnO$_4$, respectively.

In the laboratory, core segments (3.8 cm diam. by 2.54 cm length) were then taken from the intact cores received from the field and they were carefully inserted into an acrylic tube that was attached in the transport cell between the influent and effluent wells (Figure 1). Once the transport cell was fully assembled, the influent cell was filled with a simulated ground water and the soil core was upflow saturated over several days. With the core saturated and the influent and effluent wells filled with simulated ground water, the effective diffusion coefficient (D$_{eff}$) for Br$^-$ was measured and the tortuosity (τ_a) of the untreated silty clay media was determined. The diffusion test began with the introduction of 100 mg/L Br$^-$ (as KBr salt) into the influent well of the transport cell (Figure 1). Then, the Br- concentration was monitored in the effluent well until a relatively constant change in concentration of Br- was observed in the effluent well. The mass flux density (J) was then determined according to equation 1.

$$J = \left\{ \frac{\Delta C}{t} \right\} \frac{V}{A} \tag{1}$$

where, J   =   mass flux density (ML^{-2}T^{-1}),
ΔC =   average change in solute concentration over time t (ML^{-3}),
V   =   volume of solution (L^3),
A   =   area through which solute is diffusing (L^2), and
t   =   duration of effluent monitoring (T)

Once the mass flux density was evaluated, it was possible to determine D$_{eff}$ from equation 2 assuming a linear concentration gradient through the soil core, after which τ_a could be calculated from equation 3.

$$J = -D_{eff} \frac{\Delta C}{\Delta x} \tag{2}$$

$$D_{eff} = D_o \tau_a \tag{3}$$

where, J   =   mass flux density (ML^{-2}T^{-1}),
D$_{eff}$ =   effective diffusion coefficient (L^{2}T^{-1}),
ΔC =   change in C over the length of the core (ML^{-3}),
Δx =   distance over which mass transfer occurred (L),
D$_o$ =   bulk diffusion coefficient (L^{2}T^{-1}), and
τ_a =   apparent tortuosity (LL^{-1}).

The apparent tortuosity includes the effects of actual tortuosity, water-filled porosity, anion exclusion, and reduced "mobility" due to the increased viscosity of water adjacent to the clay mineral surfaces relative to the bulk water

(Shackelford, 1991). It is essentially impossible to sort out the effects of these factors in soil diffusion studies and therefore, they were reported collectively as the "apparent tortuosity", τ_a. Although the water-filled porosity was determined separately during the experimental work, it was not factored out so that τ_a as measured in this research could be compared to literature values.

Two additional experiments were completed with intact cores that were contaminated by injecting 2 μL of pure phase TCE into the center of the core midway between the influent and effluent wells. This mass of TCE was equivalent to a average TCE concentration within the soil core of 60 mg/kg (assuming uniform distribution of the TCE throughout the core). The same methods were used as with the uncontaminated cores, except for the chemical and physical characterization at the end of the experiment. Measurements of matrix interactions were not possible, since it was desired to extract the entire core and determine the extent of TCE destruction that had occurred at the time MnO_4^- was detected in the effluent well. Also, due to time constraints, the initial Br- mass flux was not established before $KMnO_4$ was added to the influent well.

EXPERIMENTAL RESULTS AND DISCUSSION

Permanganate Transport and Matrix Effects. During the pre-$KMnO_4$ period, the Br$^-$ concentration in the effluent wells for both cores rose at a relatively constant rate during the initial 35 days (Figure 2). For the uncontaminated cores 1 and 2, the D_{eff} averaged 1.33E-06 cm^2/s which is about 23% greater than the estimated value of 1.69E-06 (at n=0.4 v/v) (Table 1). The calculated τ_a values are in the range of that reported by Shackleford (1991) for silty clay loam (0.19-0.30).

The cells then ran for about 4 more days when $KMnO_4$ (as 5,000 mg/L $KMnO_4$) was added to the influent wells. In both cases, soon after $KMnO_4$ was introduced to the influent well, there was an increase in the Br$^-$ mass flux density into the effluent cell (Table 1 and Figure 2). The transport cells were then allowed to operate for about 40 additional days after the $KMnO_4$ was introduced to the influent well. The monitoring of Br$^-$ concentration continued and the D_{eff} and τ_a were calculated for this time period (see Table 1). The tortuosity of the matrices appears to have decreased (indicated by values of τ_a closer to 1) as a result of the addition of $KMnO_4$ to the system (see Table 1). However, based on the physical and chemical analysis of the treated cores, there are no apparent changes to the soil pore system that should reduce τ_a, thereby increasing the D_{eff}. It is speculated that the accelerated mass transport of bromide through the soil core was due to co-diffusion with potassium cations.

Following the diffusion measurements, the transport cells were disassembled and the intact cores were visually inspected and then dissected for micromorphologic and chemical examination. Visual examination revealed that the MnO_4^- concentration gradient along the length of the core was distinct and visibly more concentrated at the end of the core nearest the source.

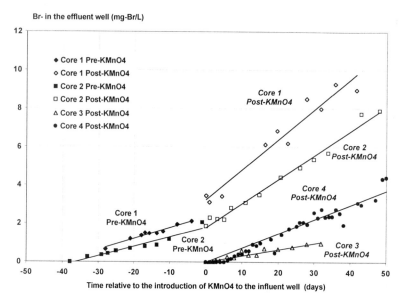

**FIGURE 1. Bromide tracer concentrations in the effluent well
of the intact core transport cell.**

TABLE 1.   Experimental parameters and results.

Parameter (units)	Core 1	Core 2	Core 3	Core 4
	Pre-KMnO₄			
J (mg/cm²s)	5.42E-08	5.03E-08	-	-
D_{eff} (cm²/s)	1.38E-06	1.28E-06	-	-
τ_a	0.24	0.22	-	-
	Post KMnO₄			
J (mg/cm²s)	1.60E-07	1.16E-07	3.36E-08	8.80E-08
D_{eff} (cm²/s)	4.06E-06	2.95E-06	8.64E-07	1.93E-06
τ_a	0.71	0.52	0.15	0.34

Note: Cores 1 and 2 were uncontaminated while cores 3 and 4 were spiked with
pure phase TCE.

Portions of the treated core were also viewed under a scanning electron
microscope (SEM). The SEM analysis revealed only subtle structural differences
between the permanganate treated cores as compared to background cores from
the site. SEM elemental analysis revealed that there were no elevated levels of
Mn indicating little deposition of manganese oxide particles.

The soil cores were divided into four equal-length segments that were
characterized chemically. The concentrations of selected matrix constituents were
determined at 0.64 and 1.90 cm from the KMnO₄ influent well (Figure 1) yielding
the results shown in Table 2. In general, extractable cation concentrations
increased with distance away from the MnO₄⁻ source and anion concentrations

decreased. Due to the much higher K^+ concentration in the soil pore water after $KMnO_4$ addition, Ca^{2+} and Mg^{2+} may have been displaced from the silty clay soil matrix through cation exchange and then migrating through the core by diffusion. As expected, the greatest manganese oxide concentrations occurred near the influent well where the greatest amount of MnO_4^- would have reacted with the soil matrix. Initially, it was hypothesized that the deposition of manganese oxides, as a by-product of the oxidation reaction, might increase the apparent tortuosity of the matrix. Based on chemical analysis and calculations made for δ-MnO_2, the volume of manganese oxide present in cores 1 and 2 was very small and averaged only 0.084% of the total pore space (assuming n=0.40 v/v). This is consistent with the observations made during the SEM analyses.

The TOC content within cores 1 and 2 was determined before and after treatment with $KMnO_4$. Using the TOC content determined for the untreated core as a reference, it was calculated that the TOC was decreased through permanganate oxidation by roughly 28% and 1.7% at 0.64 cm and 1.90 cm from the influent well, respectively. These results indicate that although the natural organic matter initially present in both cores was decreased, substantial organic material was apparently resistant to mineralization by MnO_4^-, even after 40 days of exposure. This TOC data compares well with comparable field data from the site where the soil cores were obtained. At that site, *in situ* treatment using $KMnO_4$ was demonstrated and appreciable soil TOC remained even after 10 months of exposure to high permanganate loadings (Siegrist *et al.*, 1999).

TABLE 2. Matrix constituent concentrations.

Core -	Extractable constituents								
cm from influent	K^+ mg/g	Ca^{2+} mg/g	Mg^{2+} mg/g	Mn^{2+} mg/g	Fe^{2+} mg/g	Br^- mg/L	Cl^- mg/L	MnO_4^- mg/L	MnO_2 mg/g
1 0.64	2.06	0.35	0.22	0.08	0.00	71.0	59.8	1886	2.15
1.91	1.54	1.30	0.54	0.15	0.00	45.9	50.2	840	1.14
2 0.64	2.9	0.54	0.26	0.28	0.11	23.0	69	1727	2.0
1.91	1.2	0.72	0.36	0.03	0.05	8.6	40	169	1.5

TCE Destruction and Effects. The contaminated soil cores seem to have behaved in the same general manner as the uncontaminated cores except that the time until MnO_4^- breakthrough in the effluent well of the TCE cores was approximately 17 days longer than the uncontaminated cores. The delay in MnO_4^- breakthrough was expected since the initial MnO_4^- concentration gradient would be diminished by the oxidation of TCE_{DNAPL} injected within the core.

Cores 3 and 4 were allowed to operate for roughly 35 days after KBr and $KMnO_4$ were added to the system simultaneously at time zero. Values for the mass flux density, D_{eff}, and τ_a were again calculated. As seen in Table 1 and Figure 2, the Br^- mass flux densities were similar to cores 1 and 2 in the pre-$KMnO_4$ period. This suggests that the Br^- sweep, which was observed after the introduction of $KMnO_4$ to the uncontaminated systems, is diminished in the TCE system. It is possible that the acceleration of bromide that was observed after the addition of $KMnO_4$ in the two uncontaminated cores is tempered in the contaminated system by the generation of chloride anions released by the oxidation of TCE. The value of τ_a in the contaminated cores had significant variance, but averaged to 0.24, again matching cores 1 and 2 (pre-KMnO$_4$) and indicating no effects from the deposition of MnO$_2$.

After approximately 30 days, permanganate had reached ~5 mg/L in the effluent well. The contaminated cores were then removed from the transport cells and a soil extraction was made using reagent grade hexane and deionized water (1:2 v/v) to determine the concentrations of residual soil TCE and other chlorinated organics. Gas chromatographic (GC) analysis of the organic phase showed no detectable TCE levels within the soil core. Other chlorinated organics, including tetrachloroethylene (PCE) and 1,1,1-trichloroethane (TCA) were found in the core but at relatively low levels.

ACKNOWLEDGEMENTS

This research was sponsored in part through the Subsurface Contaminants Focus Area of the DOE Office of Science and Technology and the Office of Environmental Restoration at the DOE Portsmouth Gaseous Diffusion Plant. Oak Ridge National Laboratory staff are acknowledged for their assistance in obtaining intact soil cores for the experimental work conducted. Dr. Helen Dawson is acknowledged for her advice on the modeling and Rich Harnish provided assistance with the experimental analyses.

REFERENCES

Dawson, H.E. 1997. *Screening-Level Tools for Modeling Fate and Transport of NAPLs and Trace Organic Chemicals in Soil and Groundwater.* Colorado School of Mines, Special Programs and Continuing Education, Golden, CO.

Shackelford, C.D. and D.E. Daniel. 1991. "Diffusion in Saturated Soil. I: Background." *J. Geotechnical Engineering.* Vol. 117, No. 3, March.

Siegrist, R.L., K.S. Lowe, L.C. Murdoch, T.L. Case, and D.A. Pickering 1999. "In Situ Oxidation by Fracture Emplaced Reactive Solids." *J. Environmental Engineering.* ASCE. 125(5):429-440.

Siegrist, R.L. and K.S. Lowe (Eds.). 1996. *In Situ Remediation of DNAPL Compounds in Low Permeability Media: Fate/Transport, In Situ Control Technologies, and Risk Reduction.* ORNL/TM-13305. Oak Ridge National Lab.

Siegrist, R.L., M.A. Urynowicz, and O.R. West. 1999. "An Overview of In Situ Chemical Oxidation Technology Features and Applications." Proc. Conf. on *Abiotic In-Situ Technologies for Groundwater Remediation.* August 31-September 2, 1999, Dallas Texas. U.S. EPA CERI, Cincinnati, Ohio.

Struse, A.M. 1999. "Mass Transport of Potassium Permanganate in Low Permeability Media and Matrix Interactions." M.S. Thesis, Colorado School of Mines, Golden, CO.

CHEMICAL DEGRADATION OF TCE DNAPL BY PERMANGANATE

Michael A. Urynowicz, P.E. (Colorado School of Mines, Golden, Colorado)
Robert L. Siegrist, Ph.D. P.E. (Colorado School of Mines, Golden, Colorado)

ABSTRACT: Experiments were completed to determine the rate and extent of degradation of dense nonaqueous phase trichloroethene (TCE_{DNAPL}) by chemical oxidation with permanganate (MnO_4^-). Controlled experiments were completed using special micro-reaction/extraction (MRE) vessels. The MRE vessels contained equal volumes of an aqueous phase (buffered deionized water) and an organic solvent-phase (hexane), with the aqueous phase in the bottom half of the vessel and the organic solvent phase residing in the top half. A Teflon coupon of known surface area was placed at the bottom of each vessel to support a droplet of TCE_{DNAPL} within the aqueous phase. Permanganate was added to the aqueous phase at time zero and after a prescribed reaction period, the bulk aqueous and hexane phases were sampled independently for TCE. Then an in-vessel extraction was carried out and the hexane phase was sampled a second time. The mass of TCE_{DNAPL} remaining and that destroyed were determined by mass balance. Experiments were conducted under static and mixed conditions with and without pH buffering and ionic strength adjustment. The presence of MnO_4^- in the bulk aqueous phase can markedly increase the rate and extent of degradation of the TCE_{DNAPL}. However, at higher MnO_4^- concentrations (e.g., 2.5 wt.% $KMnO_4$), the initial significant increase in the dissolution and destruction rate decreases with continued reaction time due to interfacial resistance caused by the local deposition of manganese oxide solids produced as a result of chemical oxidation. At high interfacial area to volume ratios (e.g., with dispersed ganglia), TCE_{DNAPL} may be completely degraded before resistance becomes dominant; however, at lower ratios (e.g., pools) and with higher MnO_4^- concentrations, mass transfer resistances may develop sufficiently fast that a stabilized residual evolves before complete TCE_{DNAPL} degradation is achieved.

INTRODUCTION

In situ chemical oxidation has emerged as a remediation technology that can achieve rapid and extensive destruction of TCE and other chlorinated ethenes (Siegrist *et al.*, 1999a; EPA, 1998). Oxidants that have been evaluated and applied for *in situ* remediation include hydrogen peroxide and Fenton's reagent, ozone, and permanganate. There has been increasing interest in the application of permanganate for *in situ* chemical oxidation because it is capable of oxidizing a wide range of common organic contaminants under a range of environmental conditions and is not subject to nonproductive autodecomposition or transport-limitations that constrain the spatial effectiveness of free radical oxidation processes. During the past five years, *in situ* chemical oxidation systems have been engineered to include potassium ($KMnO_4$) and sodium permanganate ($NaMnO_4$) solutions at concentrations ranging from 100 to 40,000 mg/L delivered into contaminated ground water using vertical and horizontal wells operated as injection wells or as injection/recovery/recirculation well networks (West et al., 1998; Schnarr et al., 1998). For treatment of contaminated soil, $KMnO_4$ in concentrated solution (5000 to 40,000 mg/L) or solid form (~50% by wt.) has been delivered and dispersed using probe injection, deep soil mixing, or hydraulic fracturing techniques (Siegrist et al., 1999b). Amendments to alter system pH or provide a catalyst have not been necessary.

Most controlled environmental studies with MnO_4^- have focused on chemical oxidation of soil and ground water without nonaqueous phase liquids (NAPLs) present (Vella and Veronda; 1993; Gates et al., 1995; Yan and Schwartz, 1999; Huang et al., 1999). Field demonstrations and pilot tests have explored the effects of chemical oxidation on the dissolution/destruction of NAPLs in soil and ground water environments (Gates et al., 1995; West et al., 1998; Schnarr et al., 1998; Drescher et al., 1999). However controlled experimentation has been lacking and questions regarding process effectiveness for DNAPL sites have not been fully resolved. To enhance the current base of understanding, a series of controlled experiments have been completed at the Colorado School of Mines (CSM) to evaluate the interface mass transfer and oxidative destruction of TCE_{DNAPL} in the presence of different concentrations of $KMnO_4$. This paper provides highlights of one facet of the research while additional details may be found in Urynowicz (2000) and in forthcoming publications.

BACKGROUND

Compared to hydrogen peroxide and Fenton's Reagent, permanganate oxidation of soil and ground water has more recently been studied for in situ treatment of chlorinated solvents (e.g., TCE, PCE) and petrochemicals (e.g., phenols, phenanthrene, pyrene and phenols) (Vella and Veronda; 1993; Gates et al., 1995; Schnarr et al., 1998; West et al., 1998; Siegrist et al., 1999a,b; Yan and Schwartz, 1999). The stoichiometry and kinetics of permanganate oxidation at contaminated sites can be quite complex as there are numerous reactions that manganese can participate in due to its multiple valence states and mineral forms. For degradation of chlorinated ethenes such as TCE, the oxidation involves direct electron transfer rather than free radical processes that characterize oxidation by Fenton's reagent or ozone. The stoichiometric reaction for the complete mineralization of TCE is given by equation 1:

$$2KMnO_4 + C_2HCl_3 \rightarrow 2CO_2 + 2MnO_2 + 2K^+ + H^+ + 3Cl^- \quad (1)$$

The rate of oxidative degradation of TCE is 1st-order with respect to the TCE and MnO_4^- concentrations or 2nd-order overall as given by equation 2:

$$\frac{d[TCE]}{dt} = -k_2 [TCE][MnO_4^-] \quad (2)$$

where: $[TCE]$ = concentration of TCE being oxidized (mol L^{-1}),
$[MnO_4^-]$ = the concentration of permanganate in water (mol L^{-1}), and
k_2 = second-order reaction rate constant (mol^{-1} L s^{-1}).

The 2nd- order reaction rate constants (k_2) reported for TCE degradation are in the range of 0.6 to 0.9 mol^{-1} L s^{-1} and the reaction rate appears independent of pH (over a range of 4-8) and ionic strength (up to 1.57 M Cl⁻) (Yan and Schwartz, 1999; Huang *et al.*, 1999; Urynowicz, 2000). The rate of reaction is temperature dependent as described by the Arrhenius relationship with an activation energy on the order of 35 to 70 kJ/mol (Yan and Schwartz, 1999; Huang *et al.*, 1999). The rate of disappearance of MnO_4^- is determined not just by target organic chemicals (e.g., TCE), but also by other reductants in the system including natural organic matter (NOM) and inorganic reductants (e.g., Fe^{2+} or Mn^{2+}). It is important to note that if the MnO_4^- is rapidly and extensively consumed by NOM or other nontarget demands, the MnO_4^- concentration may be depleted sufficiently that the rate and extent of reaction with the target chemical (e.g., TCE) may be adversely diminished.

The efficiency of permanganate oxidation for treatment of DNAPLs in ground water is based on a conceptual model that attributes an increased rate of DNAPL mass removal to chemical reaction within the stagnant film boundary layer (δ) (Figure 1). Chemical oxidation can yield increased oxidant and reductant concentration gradients resulting in an increase in the dissolution rate and oxidant mass flux towards the interface. This conceptual model is based on mass transfer via diffusion through a stagnant film boundary layer of fluid within which oxidation is occurring. In a system without oxidant present, the mass transfer may be described by equation 3:

$$J = - K_o (Cs - C) \tag{3}$$

where, J = mass flux of solute per unit area per unit time ($M L^{-2} T^{-1}$),
K_o = mass transfer coefficient ($L T^{-1}$),
C = solute concentration ($M L^{-3}$), and
C_S = aqueous phase concentration that corresponds to a condition of thermodynamic equilibrium with the DNAPL ($M L^{-3}$).

FIGURE 1. Schematic of interface mass transfer effects of permanganate oxidation.

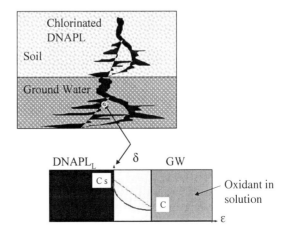

The stagnant film model described by equation 3 can be used to describe interface mass transfer where homogeneous chemical reaction occurs irreversibly in the bulk aqueous phase and the concentration gradient across the boundary layer is no longer linear. As a result of homogeneous chemical reaction, the interface mass transfer increases according to the relationship given in equation 4 (Cussler 1997).

$$\frac{K}{K^o} = \left[\frac{Dk_1}{\left(K^o\right)^2} \right]^{\frac{1}{2}} \coth\left\{ \left[\frac{Dk_1}{\left(K^o\right)^2} \right]^{\frac{1}{2}} \right\} \tag{4}$$

where, K = mass transfer coefficient with chemical oxidation (cm s^{-1}),

D = diffusion coefficient (cm^2 s^{-1}), and

k_1 = first- or pseudo-first order reaction rate constant (s^{-1}).

Equation 4 only accounts for increased interface mass transfer as a result of chemical oxidation. It does not account for any changes in interfacial resistance that may occur as a result of chemical oxidation. Permanganate oxidation of a DNAPL can yield manganese oxide solids that may deposit at the interface and result in a reduced interface mass transfer rate and DNAPL degradation rate. This is a complex process that is not fully understood, but it is speculated that the rate and extent of change in interfacial resistance development is dependent on environmental conditions including the concentration of MnO_4^- in the bulk aqueous phase. Moreover, the effect on the rate and extent of degradation of the DNAPL initially in the system will be the net result of the oxidation-induced degradation that occurs before interface mass transfer becomes limiting. At high interfacial area to volume ratios (e.g., with dispersed ganglia), TCE_{DNAPL} may be completely degraded before resistance becomes dominant; however, at lower ratios (e.g., concentrated pools), mass transfer resistances may develop sufficiently fast that a stabilized residual evolves before complete degradation is achieved.

EXPERIMENTAL METHODS

Batch experiments were performed in series using closed, zero headspace MRE vessels (Figure 2). The MRE vessels were 40-mL glass vials with Teflon-lined solid caps. Each vessel contained three separate liquid phases; an organic solvent light nonaqueous phase (hexane), an aqueous phase (buffered de-ionized water with or without $KMnO_4$) and TCE_{DNAPL}. The bulk hexane and aqueous phase liquids were present in equal volumes (20 mL). A Teflon coupon of known surface area (diameter = 2.87 mm ± 0.02) was used to support a droplet of TCE_{DNAPL} within the aqueous-phase liquid. Experiments were conducted at ambient temperature (~20° C) under static and mixed conditions at various MnO_4^- concentrations. Tests were completed with and without pH buffering and ionic strength adjustment. Each experiment was run in triplicate and included a control with no permanganate in the aqueous phase. At the onset of the experimentation, a series of control experiments were made to verify the integrity of MRE vessels over the experimental duration. Results without oxidant showed that the mass of TCE within the vessels was maintained over the course of the experiments and exhibited no downward trend with time. The arithmetic mean (3.0 µl) and relative error (8.9%) were also well within the limits of analytical error. As a result, losses due to volatilization within experimental time frames were considered negligible.

FIGURE 2. Schematic of 40-mL micro-reaction\extraction vessel. Organic solvent phase = 20 mL hexane; aqueous phase = 20 mL buffered deionized water; initial TCE_{DNAPL} = 3.0 µl; pH ~ 6.9; I = 0.3 M. Diameter of Teflon coupon = 2.87 mm ± 0.02.

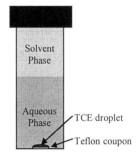

At time zero, a 3-µL droplet of TCE_{DNAPL} was deposited on the surface of the Teflon coupon and a predetermined volume of concentrated $KMnO_4$ solution was delivered into the aqueous phase of a set of triplicate MRE vessels. At the end of a prescribed reaction period, the bulk aqueous and hexane phases in a given vessel were sampled independently for TCE, and then an in-vial extraction was performed. For this in-vial extraction, the MRE vessel was shaken vigorously for approximately 30 seconds so that the residual TCE within vessel would partition into the hexane phase. Then the bulk hexane phase was sampled again for TCE (TCE_{TOTAL}). Following extraction, the aqueous phase was sampled for MnO_4^-. Then sodium thiosulfate was added to the aqueous phase to reduce any residual MnO_4^- and this remaining volume was stored for Cl⁻ analysis.

Since the mass of TCE in the bulk hexane phase (TCE_{HEX}), the mass of TCE in the bulk aqueous phase (TCE_{AQ}) and the total mass (TCE_{TOTAL}) were determined from the measured concentrations and known volumes of each phase, the mass of TCE_{DNAPL} at the end of the experiment could be calculated by mass balance (equation 5). Also, since the mass of TCE_{DNAPL} injected into the reaction vial at the start of the experiment was known (TCE_{IN}), the amount of TCE oxidized (TCE_{OX}) could also be determined (equation 6).

$$TCE_{DNAPL} = TCE_{TOTAL} - (TCE_{AQ} + TCE_{HEX}) \qquad (5)$$

$$TCE_{OX} = TCE_{IN} - TCE_{TOTAL} \qquad (6)$$

The aqueous phase used in the experiments was buffered de-ionized water (pH ~ 6.9, I ~ 0.3M). The organic solvent phase used was HPLC grade hexane (Fischer). The oxidant was technical grade $KMnO_4$ (Carus Chemical Company) and the TCE was analytical grade (Fisher). The other chemical reagents used were also of analytical grade (Fischer). Samples were analyzed for TCE using a liquid-liquid hexane extraction followed by gas chromatography (GC) with a

Hewlett Packard (HP) Model 6890 Series GC with 7683 Series autoinjector and equipped with an HP 624 (30 m x 0.53 mm x 3.0 μm film thickness) column and an electron capture detector (ECD). Chloride was analyzed by ion chromatography (IC) using a Dionex DX-300 ion chromatograph equipped with an Ionpac As-14 column. pH was determined with an Orion pH electrode and an Orion model 420A pH meter (Orion Research Incorporated). Permanganate ion was determined spectrophotometrically at 525 nm using a Hach DR/2010 spectrophotometer. Prior to permanganate analysis, samples were filtered through a 0.45 μm glass fiber filter. Secondary dilution standards were prepared from a stock standard solution in order to provide five-point calibration for TCE, Cl^- and MnO_4^- analysis.

RESULTS AND DISCUSSION

Degradation of TCE_{DNAPL} in ground water includes two key processes: interface mass transfer and oxidative destructive in the stagnant film or bulk aqueous phase. Based on the magnitude of the 2nd-order rate constant measured for the oxidation of TCE by MnO_4^- ($k_2 = 0.89$ L $mol^{-1}s^{-1}$), MnO_4^- was predicted to increase the rate of dissolution of the TCE_{DNAPL} depending on the initial MnO_4^- concentration. For the control and the lowest oxidant concentration ($[MnO_4^-]o = 0$ and 0.0016 mol dm^{-3}, respectively) the oxidant mass in the aqueous phase was stoichiometrically insufficient to react with all of the TCE present at time zero. At the higher two concentrations ($[MnO_4^-]o = 0.0158$ and 0.158 mol dm^{-3}) there was substantially more permanganate than required to oxidize all the TCE.

Estimates of the degradation rate effects were made using equations 3 and 4 based on the following input parameter values: $K° = 0.000271$ cm s^{-1} (the value determined for the $[MnO_4^-]_o = 0$ mol dm^{-3} experiment under static conditions), $k_1 = [MnO_4^-]_o \times k_2$ (0.89 ± 0.07 mol^{-1} dm^3 s^{-1}), and $D = 0.94 \times 10^{-5}$ cm^2 s^{-1}. At the lowest concentration of permanganate ($[MnO_4^-]_o = 0.0016$ mol dm^{-3}), the oxidant was expected to have only a slight impact on the rate of dissolution of the TCE_{DNAPL} ($K/K° = 1.06$). However at the highest concentration ($[MnO_4^-]_o = 0.1583$ mol dm^{-3}) chemical reaction with MnO_4^- was expected to significantly increase the TCE_{DNAPL} dissolution rate ($K/K° = 4.2$). These predictions are generally consistent with the results observed for the early reaction periods studied under the conditions of this experimentation, as illustrated by the data shown in Figure 3. However at later time points with higher MnO_4^- concentrations, the development of an interfacial skin at the TCE_{DNAPL}-water interface reduced the rate of degradation of the TCE_{DNAPL}. It appears that under these concentration and reaction conditions, chemical reaction may have a significant affect on the rate of mass transfer and degradation of the TCE_{DNAPL} by reducing the rate of interface mass transfer and protecting the DNAPL from oxidative destruction. Modeling of the interface mass transfer and degradation as affected by environmental and reaction conditions, as well as investigation of the interfacial film formation and its effects are in progress at CSM.

Dissolved TCE concentrations in the aqueous phase were also monitored with time. The mass of TCE_{AQ} in the MRE's with $[MnO_4^-]_o = 0$ mol dm^{-3} increased steadily with time. By the end of an experiment more than 25% (> 0.75

μl) of the original TCE_{DNAPL} was in the aqueous phase. In comparison, the TCE_{AQ} mass for the $[MnO_4^-]_0 = 0.0158$ mol dm^{-3} experiment never exceeded 1% of the total mass and the majority of the aqueous phase samples collected during the $[MnO_4^-]_0 = 0.1583$ mol dm^{-3} experiment were non-detect for TCE.

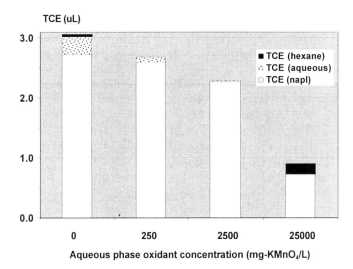

FIGURE 3. Degradation of TCE_{DNAPL} during 3 hrs. of reaction as affected by increasing concentrations of permanganate in the aqueous phase.

ACKNOWLEDGEMENTS

This research was sponsored in part through the Subsurface Contaminants Focus Area of the DOE Office of Science and Technology. Oak Ridge National Laboratory staff are acknowledged for their advice on conduct of the experimental work. Rich Harnish of CSM also provided valuable assistance with the experimental analyses.

REFERENCES

Cussler, E. L. 1997. *Diffusion.* Cambridge University Press. New York, NY.

Drescher, E.A., R. Gavaskar, B.M. Sass, L.J. Cumming, M.J. Drescher, and T. Williamson. 1999. "Batch and Column Testing to Evaluate Chemical Oxidation of DNAPL Source Areas." Proc. First International Conference on Remediation of Chlorinated and Recalcitrant Compounds. Monterey, CA. pp. 425 –432.

Gates, D.D., R.L. Siegrist and S.R. Cline. 1995. "Chemical Oxidation of Contaminants in Clay or Sandy Soil." Proc. ASCE National Conference on Environmental Engineering. American Society of Civil Engineers, Pittsburgh, PA.

Huang, K., G.E., Hoag, P. Cheda, B.A. Woody, and G.M. Dobbs. 1999. "Kinetic Study of Oxidation of Trichloroethylene by Potassium Permanganate." *Environmental Engineering Science.* 16(4):265-274.

Schnarr, M., C. Truax, G. Farquhar, E. Hood, T. Gonullu, and B. Stickney. 1998. "Laboratory and Controlled Field Experiments using Potassium Permanganate to Remediate Trichloroethylene and Perchloroethylene DNAPLs in Porous Media." *Journal of Contaminant Hydrology*, 29, pp. 205-224.

Siegrist, R.L., M.A. Urynowicz, and O.R. West, 1999a. "An Overview of In Situ Chemical Oxidation Technology Features and Applications." Proceeding of the Conf. on *Abiotic In-Situ Technologies for Groundwater Remediation.* August 31 – September 2, 1999. Dallas, Texas. U.S. EPA, Cincinnati, Ohio.

Siegrist, R.L., K.S. Lowe, L.C. Murdoch, T.L. Case, and D.A. Pickering. 1999b. "In Situ Oxidation by Fracture Emplaced Reactive Solids." *J. Environmental Engineering.* Vol.125, No.5, pp. 429-440.

Urynowicz, M.A. 2000. "Reaction Kinetics and Mass Transfer During *In Situ* Oxidation of Dissolved and DNAPL Trichloroethylene by Permanganate." Ph.D. Dissertation, Environmental Science & Engineering, Colorado School of Mines.

U.S. Environmental Protection Agency (EPA). 1998. *In Situ Remediation Technology: In Situ Chemical Oxidation.* EPA 542-R-96-005. Office of Waste and Emergency Response. Washington, D.C.

Vella, P.A. and B. Veronda, 1993. "Oxidation of Trichloroethylene; A Comparison of Potassium Permanganate and Fenton's Reagent." Proc. 3rd Intern. Symposium on Chemical Oxidation-Technology for the Nineties, Nashville, TN.

West, O..R., S.R. Cline, R.L. Siegrist, T.C. Houk, W.L. Holden, F.G. Gardner, and R.M. Schlosser. 1998. "A Field-Scale Test of In Situ Chemical Oxidation Through Recirculation." Proc. Spectrum '98 Intern. Conf. On Nuclear and Hazardous Waste Management. Denver, Sept. 13 – 18, pp. 1051-57.

Yan, Y.E. and F.W. Schwartz. 1999. Oxidative Degradation and Kinetics of Chlorinated Ethylenes by Potassium Permanganate. *Journal of Contaminant Hydrology* 37, pp. 343-365.

NUMERICAL SIMULATION OF *IN SITU* CHEMICAL OXIDATION

Eric D. Hood[1], and Neil R. Thomson
(University of Waterloo, Waterloo, Ontario, Canada)

ABSTRACT: A three-dimensional flow and transport numerical model was developed to simulate *in situ* chemical oxidation, an emerging DNAPL mass destruction technology which can use permanganate as the flushing reagent. This model incorporates representative kinetic formulations for dissolution and the oxidation reaction. The impact of the injected oxidant concentration on the amount of DNAPL removed was assessed for a two-dimensional synthetic heterogeneous aquifer after a finite volume PCE release. The results from this analysis suggest that marginal increases in mass removal may be expected above a permanganate concentration of 15 g/L, and indicate that the oxidant flux into the source zone may be a factor limiting mass removal.

INTRODUCTION

In situ chemical oxidation (ISCO) is an emerging remediation technology that has focused on mass removal from dense non-aqueous phase liquid (DNAPL) source zones. One proposed ISCO approach involves flushing the zone of DNAPL contamination with a reactive oxidant solution that degrades the contaminant in the subsurface. The advantage of this technology is that the rapid chemical reaction that occurs between the oxidant and the contaminant enhances the rate of dissolution from the non-aqueous to the aqueous phase. This mass transfer enhancement coupled with complete mineralization makes ISCO an attractive technology to remove DNAPLs from source zones.

A limited number of research studies (see Hood (2000) for a current listing), have examined the use of the oxidant potassium permanganate ($KMnO_4$) as a flushing reagent. Permanganate is widely utilized as a reagent in drinking water treatment; however, few laboratory or field evaluations carefully characterizing the effectiveness of using permanganate as part of an *in situ* treatment strategy have appeared in the literature. As a consequence, little performance data or design guidance is available to assess the relative effectiveness of various flushing alternatives.

Most of the reported ISCO field demonstrations using permanganate have employed forced-gradient schemes to flush contaminated soil (Hood et al., 1997; West et al., 1997; Schnarr et al., 1998). In many cases the conceptual design of the flushing system consists of a series of oxidant injection wells on the up-gradient side of the source zone with extraction wells located down-gradient to contain the reaction products and residual oxidant. The interaction of non-uniform groundwater flow with the interdependent kinetics of the oxidation reaction and non-aqueous phase

---

[1] *Now with Geosyntec Consultants, Guelph, Ontario, Canada.*

dissolution suggests that deploying ISCO systems requires sophisticated modeling tools for conceptual process design and analysis.

Economic and effective design of an oxidant flushing treatment system is controlled by a variety of interrelated factors with the concentration of the injected permanganate solution being one of the major design parameters. Previous field studies have chosen the injected oxidant concentration either arbitrarily or based on results from column studies. The complex kinetics make it difficult to extrapolate column study results to mass removal expectations at field sites. As part of the ongoing investigation into the ISCO technology at the University of Waterloo, a three-dimensional flow and mass transport model was developed to simulate the relevant transport processes and chemical reactions that occur during a permanganate oxidant flush. In this technical note, this model was used to address the impact of the injected oxidant concentration on the removal of DNAPL mass.

MODEL FORMULATION

The developed three-dimensional numerical model incorporates transient multi-phase flow and multi-component reactive transport. The transport component of the model includes rate-limited mass transfer from a multi-component non-wetting phase, an oxidation reaction rate formulation based on the reported rate data for chlorinated alkenes (Yan and Schwartz, 1999; Hood et al., 2000), sorption, and the standard advective-dispersive processes. The multi-phase flow equations are coupled to the transport equations by the water-phase velocity which appears in both the advective transport term and the dissolution mass transfer rate coefficient, and through the phase saturations.

The transport species consist of an organic contaminant(s) (of which an arbitrary number can be accommodated within a homogenous non-aqueous phase mixture), an oxidant which reacts with the organic contaminant in the aqueous phase, and a conservative reaction product. The dissolution of the non-aqueous phase organic contaminant(s) is controlled by the kinetic mass transfer between the aqueous and non-aqueous phases. Since the oxidation rate is dependent on the aqueous concentration of the oxidant as well as the concentration of the organic species, transport of both oxidant and organic species was necessary to appropriately simulate the ISCO process. Transport of a generic reaction product (chloride in this case of permanganate) was included since most applications of ISCO involve extensive monitoring of the reaction product. The model also includes an additional kinetic term describing the oxidation reaction between the oxidant and the porous media.

The aqueous phase transport equation is given by

$$\frac{\partial(\phi S_w C_\alpha)}{\partial t} + \frac{\partial(\rho_b C_{s\alpha})}{\partial t} - \frac{\partial}{\partial x_i}\left[\phi S_w D_{\alpha ij}\frac{\partial C_\alpha}{\partial x_j}\right] + \frac{\partial}{\partial x_i}(q_i C_\alpha) \pm \gamma_\alpha = 0$$
$$\alpha = 1,...,n_c+2 \quad (1)$$

where C_α is the mass concentration of component α, $C_{s\alpha}$ is the sorbed concentration of component α, ρ_b is the soil bulk density, $D_{\alpha ij}$ the hydrodynamic dispersion tensor for component α, q_i is the Darcy flux, ϕ is the medium porosity, S_w is the water phase saturation, γ_α represents either sources or sinks of component α, and x_i are the Cartesian coordinate directions. The non-aqueous phase is assumed to consist of a homogeneous mixture with n_c constituents. The total number of transport components is given by (n_c+2), where the $(n_c+1)th$ component is the oxidant species and the $(n_c+2)th$ component is the conservative reaction product. The source and sink terms represent oxidation reactions, mass transfer, and well boundary conditions.

Dissolution was described using the well-known stagnant film model,

$$\gamma_{\alpha,\, n \to w} = -\gamma_{\alpha,\, w \to n} = \phi S_w \lambda_\alpha (C_\alpha^{sat} - C_\alpha) \qquad \alpha = 1, \ldots n_c \qquad (2)$$

where C_α^{sat} is the effective aqueous solubility calculated using Raoult's Law, and λ_α is the lumped dissolution mass transfer rate coefficient that was determined empirically using the correlation reported by Powers et al. (1994). This empirical correlation relates the mass transfer rate coefficient to the Reynold's number, the normalized grain diameter, the uniformity index, and the non-aqueous phase saturation.

The kinetic reaction terms describing the degradation of the organic contaminant and permanganate, and the simultaneous production of chloride are:

$$\gamma_\alpha \equiv -\phi S_w k_\alpha C_\alpha C_{n_c+1} \qquad \alpha = 1, \ldots n_c \qquad (3)$$

$$\gamma_{n_c+1} \equiv -\phi S_w \sum_{\alpha=1}^{n_c} (M_{f_{\alpha,n_c+1}} k_\alpha C_\alpha C_{n_c+1}) \qquad (4)$$

$$\gamma_{n_c+2} \equiv \phi S_w \sum_{\alpha=1}^{n_c} (M_{f_{\alpha,\, n_c+2}} k_\alpha C_\alpha C_{n_c+1}) \qquad (5)$$

where k_α is the second-order reaction rate constant, and $M_{f\alpha,\, nc+1}$ and $M_{f\alpha,\, nc+2}$ are the stoichiometric mass ratios of permanganate decomposed or chloride produced relative to a unit mass of oxidized solvent.

The governing equations were approximated utilizing a control volume numerical formulation that employed a flexible system for assigning boundary conditions. To reduce the computational burden, a sequential approach was used to simulate the flow and transport components within the model. The resulting set of highly non-linear equations were solved using a full Jacobian iterative approach in conjunction with a pre-conditioned conjugate-gradient solver.

METHODOLOGY

To investigate the impact of oxidant concentration on the mass removal rate from a DNAPL source zone the model was applied to a two-dimensional (20 m long by 10 m deep) synthetic scenario consisting of a sandy aquifer overlying an aquitard (Figure 1). The domain was discretized into 20,000 control volumes (100 x 200) that were 0.2 m in the horizontal direction and 0.1 m in the vertical direction. The left and right sides of the domain were assigned constant water phase heads equivalent to a hydraulic gradient of 0.005 m/m across the domain. This hydraulic gradient was held constant throughout these simulations. A random spatially correlated permeability field (Figure 1) was generated using the algorithm developed Robin et al. (1993) based on the Fast Fourier Transform spectral technique. The permeability field was generated using geostatistics representative of the well-characterized Borden aquifer (mean of 6.9 x 10^{-12} m^2, variance of 1.0, and a correlation length of 2.0 m and 0.2 m in the horizontal and vertical directions respectively).

The general simulation approach consisted of the formation of the DNAPL source zone, water flushing the source to establish a solvent plume, oxidant flushing for 450 days, and a second water flush for 470 days.

The multi-phase flow component of the model was used to simulate the two-phase flow of a 405 kg release of a homogeneous DNAPL assigned the physical and chemical properties of perchloroethylene (PCE). The DNAPL was introduced into the domain at the location shown on Figure 1 at a constant flow rate for ~6 days and allowed to redistribute for an additional 30 days. The resulting pure phase spatial distribution (source zone) was comprised of pools and residual regions as shown in Figure 2. After DNAPL redistribution, the source zone was water flushed for 350 days to establish an realistic aqueous phase plume. The pure phase and aqueous phase distribution at 350 days were identical for each of the oxidant flushing simulations considered in this technical note.

For the oxidation portion of these simulations the oxidant and reaction product were assigned reaction stoichiometry and rates representative of PCE, permanganate and chloride. For the purpose of this investigation, the oxidation reaction between the oxidant and the porous media was assumed to be negligible. The oxidant was introduced along the entire left-hand side of the domain for 450 days. The relative impact of oxidant concentration on DNAPL mass removal performance was examined by performing simulations using injected permanganate concentrations of 5, 10, and 15 g/L. The effectiveness of each of these alternatives was defined as the mass of DNAPL removed by the oxidant flush in excess of the mass of DNAPL removed by an equivalent water flush (the simulation with no oxidant addition).

RESULTS AND DISCUSSION

The temporal variations of the aqueous phase mass for PCE, oxidant, and chloride over the 1270 day simulation period are shown in Figure 3(a to c) while Figure 3(d) provides the corresponding changes in DNAPL mass. A summary of the DNAPL mass removed in each simulation is presented in Table 1. In the simulation without oxidant addition, the aqueous phase PCE reaches a maximum mass in

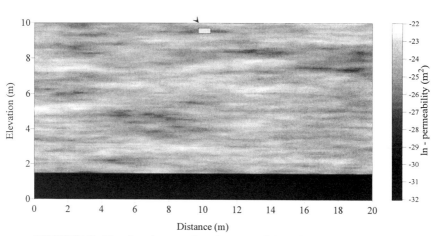

FIGURE 1. Randomly correlated permeability field based on the geostatistical properties of the Borden aquifer.

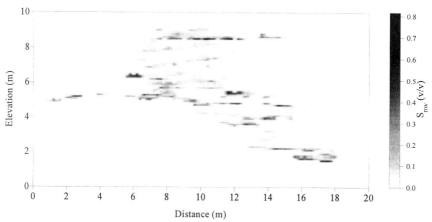

FIGURE 2. Spatial distribution of non-wetting phase saturation prior to the initial water flush.

solution after 150 days and gradually decreases over the duration of the simulation as DNAPL mass is depleted from the domain. Over 1270 days of water flushing only 20 kg of pure phase PCE was removed which is consistent with the low solubility of PCE in water.

The results from the various oxidation simulations produced results that were generally similar given the difference in oxidant concentrations. The aqueous phase PCE mass rapidly dropped following the introduction of the oxidant. At the same time, the mass of chloride resulting from mineralization of PCE by the oxidant increased to a maximum as the oxidant migrated into the source zone. As time progressed, the chloride mass in the domain for all three oxidant concentrations decreased, reflecting a decrease in the contact time between the oxidant and PCE as DNAPL was depleted on the up-gradient side of the source, and the impact of decreasing DNAPL saturations on the mass transfer coefficient. In each of the simulations, the rate of DNAPL removal (Figure 3(d)) and the mass of chloride produced was proportional to the oxidant concentration.

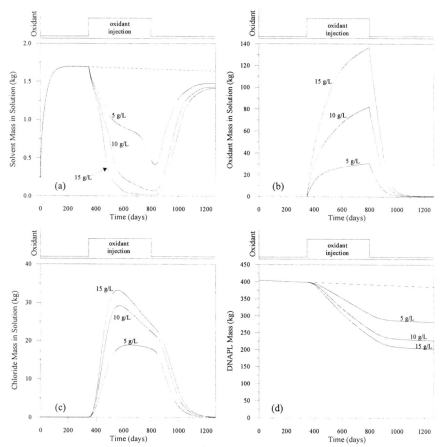

FIGURE 3. Time-series data of simulated (a) solvent mass in solution, (b) oxidant mass in solution, (c) chloride mass in solution and (d) DNAPL mass for oxidant concentrations of 5, 10, and 15 g/L (solid lines) and without oxidant (dashed line). The duration of oxidant application is also shown.

The variation in the oxidant mass in the domain (Figure 3(b)) indicated that the rate of increase of oxidant mass in solution decreased over time. In fact, the mass of oxidant in the domain approached the upper limit (i.e., the pore volume x input oxidant concentration) slowly as oxidant was depleted by the reaction. This effect was less observable for the high oxidant concentration since the higher oxidant concentration could pass through portions of the DNAPL zone without being completely depleted.

The effect of increasing oxidant concentration on DNAPL mass removal is evident in Table 1. As the oxidant concentration increased the mass of DNAPL removed over a 450 day oxidant flush increased asymptotically, which suggests that the rate of mass removal was limited by the kinetics of the dissolution process rather than those of the oxidation reaction.

TABLE 1. DNAPL mass removal for each design alternative.

$KMnO_4$ (g/L)	DNAPL Remaining (kg)	Excess % mass removed
0	385	0%
5	282	26%
10	227	39%

SUMMARY

Increasing the injected oxidant concentration provided only a marginal increase in mass removal efficiency. This result has several important consequences for field application of oxidation technologies. It suggests that the expected performance of an oxidant flushing system is not limited by reaction kinetics but rather by the ability to maintain an adequate oxidant flux into the source zone. This requirement further emphasizes the need to design an effective flushing system and means of assessing the hydraulic performance of these systems. Increasing the Darcy flux of the oxidant solution through the source zone would likely provide the benefit of increased oxidant flux while also increasing the mass transfer rate coefficient.

Additional simulations are being conducted to further investigate the findings discussed in this technical note and of other design parameters on the effectiveness of oxidant flushing.

ACKNOWLEDGMENTS

Financial support for this research has been provided by the University of Waterloo Consortium for Solvents-In-Groundwater Research Program, and Natural Science and Engineering Research Council of Canada.

REFERENCES

Hood, E. D. 2000. *In Situ Chemical Oxidation Bibliography* [Online]. Available: http://www.civil.uwaterloo.ca/groundwater/oxlitrev.html [2000, February 15].

Hood, E. D., N. R. Thomson, D. Grossi and G. J. Farquhar. 2000. "Experimental Determination of the Kinetic Rate Law for Oxidation of Perchloroethylene by

Potassium Permanganate." *Chemosphere.* *40*(12):1383-1388.

Hood, E. D., N. R. Thomson, and G. J. Farquhar. 1997. *"In Situ* Oxidation: an Innovative Treatment Strategy to Remediate Trichloroethylene and Perchloroethylene DNAPLs in Porous Media." In the proceedings of *the 6th Symposium on Groundwater and Soil Contamination*, Montreal, Quebec.

Powers, S. E., L. M. Abriola, and W. J. Weber, Jr. 1994. "An Experimental Investigation of Nonaqueous Phase Liquid Dissolution in Saturated Subsurface Systems: Transient Mass Transfer Rates." *Water Resources Research.* *30*(2): 321-332.

Robin, M. J. L., A. L. Gutjahr, E. A. Sudicky, and J. L. Wilson. 1993. "Cross-Correlated Random Field Generation With the Direct Fourier Transform Method." *Water Resources Research.* *29*(7): 2385-2397.

Schnarr, M. J., C. L. Truax, G. J. Farquhar, E. D. Hood, T. Gonullu and B. Stickney. 1998. "Laboratory and Controlled Field Experiments using Potassium Permanganate to Remediate Trichloroethylene and Perchloroethylene DNAPLs in Porous Media." *Journal of Contaminant Hydrology* [Online]. *29*(3): 205-224. Available: http://www.elsevier.nl/cas/tree/store/conhyd/sub/1998/29/3/608.pdf [2000 February. 15].

Yan, Y. E, and F. Schwartz. 1999. "Oxidative Degradation and Kinetics of Chlorinated Ethylenes by Potassium Permanganate." *Journal of Contaminant Hydrology.* *37*(3): 343-365.

West, O. R., S. R. Cline, W. L. Holden, F. G. Gardner, B. M. Schlosser, J. E. Thate, D. A. Pickering, and T. C. Houk. 1997. "A Full-scale Demonstration of *In Situ* Chemical Oxidation Through Recirculation at the X-701B: Field Operations and TCE Degradation." Environmental Sciences Division, Oak Ridge National Laboratory, Oak Ridge TN.

DESIGN CONSIDERATIONS FOR *IN-SITU* CHEMICAL OXIDATION

Daniel W. Oberle and David L. Schroder
SECOR International, Inc.

ABSTRACT: *In-situ* chemical oxidation has become a well-recognized technology for remediating contaminated soils and ground water. The two most common oxidants used for process applications are hydrogen peroxide and potassium permanganate. While chemical oxidation has proven to be an effective remedial option for a variety of chlorinated and recalcitrant compounds, many applications exist where *in-situ* chemical oxidation may be ineffective or uneconomical. Studies were conducted to evaluate the applicability of various oxidizing agents on multiple contaminant types. Testing further evaluated oxidation performance under an assortment of soil and ground water conditions. The results showed that oxidizers are consumed in a variety of secondary oxidation and degradation reactions. Bench-scale testing was used to design full-scale, *in-situ* chemical oxidation systems for several sites contaminated with various chlorinated solvents. Complete contaminant destruction was achieved in highly permeable soil conditions, but the effectiveness was less dramatic under low permeability conditions due to contact limitations. Treatment efficiencies were greatly influenced by the method of introducing the oxidizing agent into the subsurface. The results of both the bench-scale testing and field applications demonstrate that the site geology, geochemistry and the method of chemical application are important factors to consider when designing an *in-situ* chemical oxidation system.

INTRODUCTION

In-situ chemical oxidation involves the injection of oxidizing agents into the subsurface to convert contaminants into inert by-products. It is attractive because the desired oxidation reactions typically occur within a time-frame of several hours. This allows remediation activities to be completed in a matter of days or weeks instead of years. The technology shows particular promise for remediating recalcitrant compounds, such as chlorinated solvents and methyl tertiary butyl ether (MTBE). Application of this innovative technology has spread rapidly during the last decade due to the ability to readily address subsurface contamination under site conditions where conventional remedial technologies have failed. Regulatory acceptance of the process has also increased, thereby facilitating permitting issues associated with its implementation.

To ensure a successful project, several design steps need to be completed prior to proceeding with field implementation of a chemical oxidation system. The consultant should possess a basic knowledge of oxidation reaction mechanisms in order to determine the oxidizing agents best suited for the specific contaminants of concern. An understanding of relative reaction rates and the life span of reactants is required to ensure adequate contact time for the desired reactions. The chemical demand associated with pH adjustment needs to be evaluated since many of the oxidation reactions are pH-dependant. In addition, geochemical characteristics of the site must be identified to help predict how naturally-occurring mineral and organic fractions within the soil and ground water will affect the process. Finally, the consultant needs to be familiar with federal, state, and local regulatory requirements associated with the performance of *in-situ* chemical oxidation. Environmental consultants, working through each of these steps, will be able to determine the cost effectiveness and success potential for an *in-situ* chemical oxidation program prior to initiating full-scale activities.

PROPER OXIDANT SELECTION

The two most common oxidizing agents used for *in-situ* chemical oxidation are potassium permanganate and hydrogen peroxide. Hydrogen peroxide is typically implemented using one of two application techniques. The first method involves the injection of highly concentrated solutions of hydrogen peroxide into the subsurface to promote contaminant volatilization; utilizing the heat generated during the exothermic dissociation of the oxidant. The second technique (Fenton's reaction) oxidizes contaminants in place by combining hydrogen peroxide with an iron catalyst under reduced pH conditions to generate the powerful hydroxyl radical. Potassium permanganate oxidation creates little heat or gas, therefore contaminant treatment occurs primarily through oxidation. All approaches have unique applications, and selection of the proper oxidant is based on several factors.

A chemical's oxidation potential is often used to determine relative effectiveness for oxidizing organic constituents. The hydroxyl radical has an oxidation potential of 2.8 volts, compared to hydrogen peroxide and potassium permanganate which have oxidation potentials of 1.8 and 1.7 volts, respectively. These values suggest Fenton's reaction provides the strongest option for destroying organic compounds. However, the oxidation potential only provides a partial view of an oxidizing agent's capabilities. Specific mechanisms and reaction pathways also play a very important role in the reaction kinetics.

The oxidation mechanisms for hydrogen peroxide and potassium permanganate are quite different. Fenton's reagent is capable of oxidizing a wide range of compounds while potassium permanganate is more selective. However, potassium permanganate often provides more rapid destruction for specific compounds when compared to Fenton's. The reaction process and potential by-products for both oxidants differ considerably. Therefore, it is important to gain an understanding of the reaction mechanisms in order to predict how the oxidation reactions will occur.

Potassium Permanganate. Potassium permanganate is an oxidizing agent with a unique affinity for organic compounds containing carbon-carbon double bonds, aldehyde groups or hydroxyl groups. The permanganate ion is strongly attracted to the negative charge associated with electrons in the π-cloud of carbon-carbon double bonds found in chlorinated alkenes such as tetrachloroethylene (PCE), trichloroethylene (TCE), dichloroethylene (DCE), and vinyl chloride (VC). The permanganate ion borrows electron density from the π-bond in these compounds, disturbing the carbon-carbon double bond to form a bridged oxygen compound known as the hypomanganate diester. This intermediate product is unstable and further reacts by a number of mechanisms including hydroxylation, hydrolysis or cleavage. Under normal subsurface pH and temperature conditions, the primary oxidation reaction for alkenes involves spontaneous cleavage of the carbon-carbon bond. Once this double bond is broken, the highly unstable carbonyl groups are immediately converted to carbon dioxide through either hydrolysis or further oxidation by the permanganate ion.

Fenton's Reagent. The second reaction mechanism implemented for chemical oxidation involves the use of hydrogen peroxide to generate hydroxyl radicals. The process was first discovered by H.J.H. Fenton in 1876. Fenton's reagent consists of a low-pH solution of hydrogen peroxide containing dissolved iron. It is typically prepared using a solution of hydrogen peroxide, sulfuric acid, and ferrous sulfate heptahydrate. The dissolved iron acts as a catalyst for generating the hydroxyl radical, resulting in free-radical oxidation. In order to keep the iron in the ferrous state a low pH needs to be maintained. The reaction can be performed successfully at a pH range between 5 and 7, but optimal conditions are observed between 3 and 5. Obtaining optimal subsurface pH conditions is often limited by the soil buffering capacity.

Hydrogen Peroxide. Hydrogen peroxide performs poorly as an oxidizing agent for many organic compounds when used at ambient pressure and temperature conditions. Specifically, it exhibits little effectiveness for the oxidation of chlorinated solvents and other recalcitrant compounds such as MTBE. Many consultants and contractors continue to use hydrogen peroxide alone for the remediation of chlorinated solvents and MTBE, often demonstrating successful applications. However, it is important to evaluate the mechanisms responsible for the observed contaminant reductions.

Volatilization often is the primary mechanism of contaminant removal when *in-situ* oxidation is attempted using elevated concentrations of hydrogen peroxide. Concentrations as low as 11 percent can cause ground water boiling, but levels between 35 and 50 percent are frequently used for remediation. The addition of high-strength peroxide produces heat and gaseous vapor, generated during dissociation of the oxidant. The rate of hydrogen peroxide decomposition doubles with every $10^{\circ}C$ rise in temperature. Subsurface heating occurs rapidly, with 1,200 BTUs of energy and up to six cubic feet of oxygen gas released by each pound of hydrogen peroxide. Volatile organic compound (VOC) contamination is volatilized and driven from the soils by the resulting pressure gradient. This technique can be successful for sites contaminated with non-flammable VOCs if vapor recovery and treatment are not required by regulatory agencies. However, when used for flammable compounds, the mixture of VOCs and oxygen gas in a heated environment can create an explosion hazard.

DESIGN CONSIDERATIONS

Before installing a full-scale, *in-situ* chemical oxidation system, it is important to obtain essential data to ensure proper chemical addition ratios and reaction times are achieved. This information includes data on the reaction kinetics, pH conditions, and naturally-occurring interferences within the subsurface. The interferences and required pH adjustments will determine how much chemical is needed to complete the reaction, while reaction kinetics provide information on how long the reaction must proceed.

Evaluation of Reaction Kinetics. An understanding of reaction times provides design information needed to ensure adequate oxidant/contaminant contact time. Testing has demonstrated that the destruction of chlorinated solvents by Fenton's oxidation occurs quickly, typically within one to three hours. However, information regarding specific reaction mechanisms is currently in the research stage. Bench-scale oxidation studies were performed to investigate the reaction rates and mechanisms for the previously discussed oxidants to provide answers to these questions.

Reaction kinetic studies were first performed to evaluate oxidation of benzene, toluene, ethyl benzene, xylenes (BTEX), and MTBE. Testing was performed using hydrogen peroxide, Fenton's reagent, and potassium permanganate. The Fenton's reagent sample was prepared using three percent hydrogen peroxide solution and a 30 mg/l dissolved ferrous iron solution at a pH of 3.0. The permanganate sample was prepared using a two-percent solution of potassium permanganate at a pH of 7.0. The effects of high concentrations of hydrogen peroxide were evaluated with a 35 percent solution. The samples were prepared in 40 ml vials and spiked with a chemical standard, creating a solution with 7.9 mg/l benzene, 6.7 mg/l toluene, 7.2 mg/l ethyl benzene, 23.5 mg/l xylenes, and 7.2 mg/l MTBE. The samples were allowed to react at 60°F for 24-hours before analysis by gas chromatography (GC). The analysis results were below detection limits (5 ug/l per constituent) for BTEX and MTBE in the Fenton's reagent sample. No reduction was observed for the 35-percent hydrogen peroxide and potassium permanganate samples. These results indicate Fenton's oxidation can be very effective for BTEX and MTBE, but that potassium permanganate or hydrogen peroxide by itself are poor oxidants for these compounds.

A second test was performed to evaluate the oxidation performance on chlorinated solvents in Fenton's reagent, hydrogen peroxide, and potassium permanganate. The testing was performed in 40 ml vials containing solutions of 600 ug/l PCE, 700 ug/l TCE, 1,800 ug/l 1,2-DCE, and 250 ug/l 1,1,1-trichloroethane (TCA). The oxidant samples were prepared in the same manner as the previous test and allowed to react for 24 hours before analysis by gas chromatography. The results were below detection limits (10 ug/l per constituent) for all compounds in the Fenton's reagent sample. No reduction occurred in the

35 percent hydrogen peroxide sample. The potassium permanganate sample was below detection limits for PCE, TCE, and 1,2-DCE, but no reduction was observed for TCA. This testing demonstrates that Fenton's oxidation is effective for remediation of both chlorinated alkenes and alkanes, but potassium permanganate only appears effective on the chlorinated alkenes. Hydrogen peroxide by itself was an ineffective oxidizing agent for these chlorinated compounds.

A final test was completed to evaluate the rate of oxidation for each of the previously tested contaminants. Potassium permanganate and Fenton's reagent samples were prepared in the same manner as the earlier tests. The samples were spiked with BTEX, MTBE, TCE, PCE, DCE, VC, and TCA. VOC analyses were performed while oxidation proceeded for the first three hours of the reaction. The results of the analyses showed that the Fenton's reagent effectively oxidized all the contaminants within three hours, with calculated first order reaction rate constants ranging from 0.03 to 0.07 min^{-1}. As expected, the MTBE and BTEX compounds in the potassium permanganate samples did not react. However, PCE was completely oxidized within two hours, TCE was oxidized within one hour and the DCE and VC were each oxidized within 15 minutes. The calculated first order reaction rate constants for these chlorinated alkenes in solution at 60°F was 0.03 min^{-1} for PCE, 0.07 min^{-1} for TCE, 0.6 min^{-1} for DCE and 1.7 min^{-1} for VC.

Effects of pH on Oxidation Reactions. The demand for pH adjustment during chemical oxidation should be evaluated, since many of the reactions involved are pH dependant. Maintaining the proper pH conditions for Fenton's oxidation is crucial to the availability of the ferrous iron catalyst. The buffer capacity of soil will determine whether pH adjustment for chemical oxidation can be effective and economical. To evaluate the effects of pH, testing was performed using potassium permanganate and Fenton's reagent under acidic, neutral, and basic conditions. Water samples were spiked with known concentrations of MTBE, PCE, TCE, 1,2-DCE, and TCA. Fenton's reagent samples were prepared using three percent hydrogen peroxide adjusted to a pH of 3.0, 7.0, and 11.0 with sulfuric acid or sodium hydroxide. Potassium permanganate samples were prepared at a concentration of two percent and were also adjusted to pH values of 3.0, 7.0, and 11.0. Samples of the test solutions were analyzed by gas chromatography after 2-1/2 hours of reaction. The results of the analyses showed that the pH had little to no effect on the oxidation rate of chlorinated alkenes using potassium permanganate. However, the pH had a significant effect on the performance of Fenton's oxidation. Concentrations were below detection limits for the pH 3.0 sample. The oxidation efficiencies observed with the pH 7.0 solution were approximately 35 percent for MTBE and 70 percent removal for the chlorinated compounds. No destruction was observed in the test solution adjusted to a pH of 11.0.

Organic and Inorganic Reaction Interferences. Site-specific soil and ground water conditions can affect chemical oxidation performance through direct competition with contaminants for the oxidant. The primary interference with Fenton's oxidation is carbonate and bicarbonate which influence pH conditions and compete with contaminants for the hydroxyl radical. Elevated soil organics will react with Fenton's reagent and potassium permanganate. Iron also interferes with both chemicals, but the degree is dependant upon the oxidation state of the iron and the pH conditions encountered.

Concentrations of calcium carbonate often represent the greatest barrier for Fenton's applications. Elevated conditions require large amounts of acid to reduce and maintain the pH of the soil/water system while oxidation occurs. Organic acids react with the carbonate at slower rates when compared to mineral acids. However, the organic acids are also susceptible to oxidation by Fenton's reagent. Therefore, the use of organic acids for pH adjustment may be counter-productive due to these reaction interferences. Tests were performed to evaluate the potential for interference using a mixed soil/water system containing low concentrations of PCE, TCE, and TCA. The samples were adjusted to a pH of 4.0 using citric, acetic, and sulfuric acid. The sulfuric acid sample achieved between 80 and 90 percent reduction of the compounds before carbonate buffering caused the reaction to stop. The citric acid sample formed a chelate with the dissolved iron which immediately limited the effectiveness of the reaction, resulting in less than 50 percent removal of contaminants. The acetic acid sample achieved approximately 70 percent reduction of contaminants, but the GC analyses showed that the acetic acid was also oxidized by Fenton's oxidation, thus increasing the overall hydrogen peroxide demand.

REGULATORY ISSUES AFFECTING *IN-SITU* CHEMICAL OXIDATION

Congress enacted various environmental laws throughout the 1970s and 1980s to protect the environment from uncontrolled releases of hazardous chemicals. Although Congress intended to focus the laws on unintentional or uncontrolled releases of hazardous chemicals, the laws may also restrict chemical applications during remedial treatment. For example, the public drinking water program and underground injection control program of the Safe Drinking Water Act (SDWA) may prohibit the injection of safe chemicals into the groundwater for treatment of contamination. Similarly, the treatment prohibitions of the Resource Conservation and Recovery Act (RCRA) may forbid on-site treatment of contamination. Finally, federal regulatory agencies might question whether the use of oxidizing chemicals for *in-situ* remediation constitutes a reportable release under the Comprehensive Emergency Response, Compensation and Liability Act (CERCLA) or Emergency Right-to-Know Act of 1986 (EPCRA).

FULL-SCALE SYSTEM APPLICATIONS

The criteria described in the bench-scale testing sections were used to design several full-scale oxidation systems completed during 1998 and 1999. The results of selected full-scale systems for differing soil conditions and oxidizing chemicals are summarized below.

Potassium Permanganate Treatment - Saturated Sands. This site is located in northern Ohio. The sandy soils extend to approximately eight feet below ground surface (bgs), above a dense clay aquitard. The depth to ground water is approximately four feet bgs. Ground water concentrations were 81 ug/l for 1,2-DCE. Testing showed the sand contained moderate organic interferences with an oxidant demand less than 200 milligrams of potassium permanganate per kilogram of soil. The chemical demand associated with dissolved humic acids and minerals was also tested and found to be negligible. Treatment was performed by placing 110 pounds of dry potassium permanganate salt at the interface of the water table above the impacted zone. Application in this manner minimized concerns of spreading contamination during liquid injection. Ground water samples were collected from monitoring wells within the treatment zone for a period of three months following potassium permanganate addition. Complete destruction of 1,2-DCE was achieved as a result of the treatment. Remedial actions required less than one week of on-site activities at a cost less than $10,000. The success of this project demonstrates that *in-situ* oxidation can be cost effective for ground water treatment when used under ideal site conditions which include high permeability sands with low organic and mineral interferences.

Potassium Permanganate Treatment - Unsaturated Sandy Clay. A release of chlorinated solvents into the subsurface occurred in a former drum storage area at a manufacturing facility in Michigan. The contaminants were contained within the upper ten feet of soil. Concentrations of chlorinated solvents were approximately 60 mg/kg PCE and 40 mg/kg TCE. Elevated organic carbon and iron concentrations within the soil resulted in an oxidant demand of approximately 15,000 to 20,000 mg/kg for potassium permanganate. Oxidant application was performed by backhoe-mixing the dry permanganate salt into the clay soils. Pre-testing of the soil showed the moisture content was sufficient to activate the permanganate without the addition of water. Treatment of 100 tons of contaminated soil was completed in one day. Sampling was conducted three weeks later. The results of soil analyses showed that average solvent concentrations had decreased to 0.04 mg/kg for PCE and 0.02 mg/kg for TCE. Greater than 99.9 percent removal was achieved. The resulting treatment cost was approximately $70 per ton of soil. This was significantly less than the costs associated with off-site treatment and disposal.

Fenton's Treatment - Saturated Sands. This application was performed at a manufacturing facility in Michigan. The subsurface was comprised of fine sandy soils with a low organic carbon content and high iron concentrations. Soil concentrations exceeded 500 mg/kg for TCA and 150 mg/kg of TCE. Potassium permanganate was eliminated from consideration due to the elevated iron concentrations and the presence of TCA. The organic carbon content, coupled with low calcium carbonate buffer capacity, indicated Fenton's oxidation could be a cost effective treatment option. Injection probes were strategically installed based on contaminant profile and geological conditions. The iron concentration in the soil was sufficient to promote Fenton's oxidation under controlled pH conditons. Hydrogen peroxide and sulfuric acid were mixed together with water to create a dilute solution of Caro's acid. Systematic injection was conducted in pulsed intervals, beginning at the perimeter of the impacted area and working inward. Approximately 10,000 gallons of oxidant solution were injected during a one week period, treating an estimated 3,000 tons of soil. Soil samples were collected approximately two weeks after the treatment was completed. The analyses showed reduction to average concentrations of 0.75 mg/kg for TCA and 1.4 mg/kg for TCE, an overall treatment efficiency greater than 99 percent removal. Total treatment costs were approximately $30 per ton.

Fenton's Treatment - Weathered Bedrock. A release of TCA at this automotive parts manufacturing facility in Michigan resulted in shallow, weathered-bedrock contamination. Contamination was restricted to the upper four feet of the shale and sandstone bedrock, located between eight and twelve feet bgs. Total VOC concentrations ranged from 500 to 1,000 mg/kg. The weathered shale contained high concentrations of iron, very little organic material, and low calcium carbonate concentrations. The subsurface chemistry was well-suited for Fenton's oxidation. To address the low-permeability conditions, an *in-situ* slurry mixing technique was developed. The overburden soil above the impacted area was removed, and the upper four feet of the bedrock was broken apart as diluted Fenton's solution was added. The resulting slurry was continuously agitated during oxidant addition. VOC screening and pH testing were conducted during remediation to determine remedial progress. Concentrations decreased by 80 percent per hour during the first three hours of mixing activities. The reaction became contact-limited after three hours, and VOC concentrations stabilized between one and two parts per million. The slurry was solidified with lime and compacted before replacing the overburden soil. The area was sampled one week after completing remediation activities. Results showed a decrease the average TCA concentrations to 2.6 mg/kg. Although complete destruction of the solvent could not be achieved due to chemical contact limitations, an overall treatment efficiency of greater than 99.9 percent was achieved. Costs were approximately $100 per ton of bedrock material which was significantly less than estimates for off-site treatment and disposal.

CONCLUSIONS

In-situ chemical oxidation has become a well-recognized technology for remediating contaminated soils and groundwater. However, technological limitations must be considered prior to proceeding with full-scale applications. The proper selection of an oxidant is critical since reaction effectiveness varies with contaminant type. An understanding of reaction times and the life span for reactants is also required to ensure that adequate contact time can be achieved. This involves not only testing at the bench scale level, but also field monitoring during oxidant application to evaluate reaction progress. Thorough site investigations must be completed to properly identify potential reaction interferences, and optimal conditions must be maintained when the field application is implemented. The consultant should work closely with the appropriate regulatory agencies to assure that all requirements are addressed throughout the remedial program. Finally, limitations to the reaction should be recognized. As illustrated in the site examples, organic and mineral interferences, coupled with site-specific geological conditions, play a major role in determining the cost effectiveness for the technology.

REFERENCES

Haines A.H. 1985. *Methods for the Oxidation of Organic Compounds.* Academic Press, Orlando, FL.

Lee D. G. 1980. *The Oxidation of Organic Compounds by Permanganate Ion and Hexavalent Chromium.* Open Court Pub. Co., La Salle, IL.

Patai S. 1964. *The Chemistry of Alkenes.* Interscience Publishers, New York, NY.

Stewart R. 1964. *Oxidation Mechanisms.* W. A. Benjamin, New York, NY.

A MULTISITE FIELD PERFORMANCE EVALUATION OF *IN-SITU* CHEMICAL OXIDATION USING PERMANGANATE

Wilson S. Clayton, Ph.D., (IT Corporation, Englewood, Colorado)
Bruce K. Marvin, (IT Corporation, Concord, California)
Timothy Pac (IT Corporation, Norwood, MA)
Ernest Mott-Smith, P.E, (IT Corporation, Tampa, Florida)

ABSTRACT: Field data are presented from multiple in-situ chemical oxidation projects where potassium permanganate ($KMnO_4$) or sodium permanganate ($NaMnO_4$) were used for the treatment of chlorinated ethenes and other recalcitrant organic contaminants. The large body of field data is used to illustrate and evaluate the primary performance characteristics of the technology, including (1) subsurface permanganate transport, (2) permanganate-contaminant reaction kinetics, (3) matrix interactions and other secondary geochemical effects, (4) overall permanganate consumption, and (5) contaminants treated and overall reductions achieved. Analytical modeling and field data show that oxidant demand and permanganate reaction kinetics both exert critical control over permanganate transport. Relevant matrix interactions during in-situ oxidation with permanganate can include permanganate consumption by oxidation of native materials (such as sulfides, natural organic matter, and reduced metals) and oxidation of native materials that are more mobile in an oxidized valence state. These metals tend to quickly attenuate, by sorption and natural chemical reduction. Geochemical data show that chemical oxidation using permanganate is a highly dynamic geochemical process. Dissolved Mn is ephemeral, and primarily correlates to the presence of MnO_4^-, which is consumed by the oxidation reactions.

INTRODUCTION

Background. In-situ chemical oxidation is an emerging technology for treatment of organic contaminants. It is attractive, because it destroys contaminants in-situ, generating relatively innocuous products. In-situ chemical oxidation does not require groundwater pumping, and therefore does not generate any waste stream requiring treatment or disposal. In-situ oxidation projects can commonly be completed in a period of several months, which can reduce overall remediation costs. Furthermore, in-situ oxidation can be extremely effective for hard to treat contaminants. Treatment results can be obtained that may have been previously considered technically impractical. There are three primary issues involved in managing in-situ oxidation: (1) determining that a selected oxidant will treat the target contaminants, (2) determining the oxidant mass required for treatment, and to overcome competing oxidant demand, such as by natural organic matter, and (3) providing adequate subsurface transport of the oxidant, so that all target contaminated areas are treated

There are three primary oxidants being used for in-situ oxidation, including (in decreasing order of oxidant strength): hydroxyl radical (*OH), ozone (O_3), hydrogen peroxide (H_2O_2), and permanganate ion (MnO_4^-). The primary considerations in selecting one oxidant versus another relate to trade-offs between oxidant strength and oxidant stability. Stronger oxidants are more aggressive and can treat more recalcitrant contaminants, but they are less stable, more short-lived, and more difficult to deliver in the subsurface.

This paper describes the field performance characteristics of in-situ chemical oxidation using permanganate. In light of the above background, the reader should recognize that permanganate has less oxidant strength than *OH, O_3, or H_2O_2, but is more persistent in the subsurface. Among the contaminants readily oxidized by permanganate are chlorinated ethenes. The reaction of $KMnO_4$ with tricholoroethene (TCE) follows: $2KMnO_4 + C_2HCl_3 \rightarrow 2CO_2 + 2MnO_2 + 2K^+ + 3Cl^- + H^+$.

The TCE oxidation products represented above are the reaction end points, and in reality the contaminant oxidation proceeds by a number of intermediate steps. Yan et al. (1998) showed that the intermediate compounds for TCE oxidation using permanganate were primarily cyclic esters (i.e. $MnO_4C_2HCl_3$) followed by $HMnO_3$ and carboxylic acids ($H_aC_bO_cOH_d$).

Scope and Objectives. Since in-situ oxidation is relatively new, many of the fundamental field performance characteristics are still being learned. The approach adopted in this paper was to identify the most relevant performance criteria for in-situ oxidation using permanganate, review the relevant basic theory of each, and use data from multiple field sites to determine a range of observed performance. The primary performance criteria identified include (1) subsurface permanganate transport, (2) permanganate-contaminant reaction kinetics, (3) matrix interactions and other secondary geochemical effects, (4) overall permanganate consumption, and (5) overall contaminant reductions achieved.

SUBSURFACE PERMANGANATE TRANSPORT

Effective subsurface permanganate transport is required to deliver the oxidant to contaminants in the subsurface. In-situ oxidation is dependent on oxidant transport, not contaminant transport. Clayton (1988) indicated that second order oxidant reactions with contaminants provide a limit on oxidant transport. Since permanganate reaction rates are faster in the presence of oxidizeable materials, we can expect more limited permanganate transport in areas of a site with higher contaminant concentrations. As treatment proceeds, and the mass of oxidizeable material is reduced, permanganate transport will be greater.

Permanganate injection tests that include a conservative tracer provide a means to evaluate the impact of reactive transport on the overall permanganate distribution. Figure 1 shows the results of a combined permanganate-tracer injection test performed at the IDC Cape Canaveral, Florida demonstration project (Mott-Smith et al., 2000). In this test, 1.5 % to 2% KMnO4 was injected for 2.5 days with a 2 ppm sodium fluoride tracer. The fluoride tracer is conservative, and its distribution represents the transport of particles injected. The fluoride transport in this test was up to 20 feet and greatly exceeded the permanganate transport. Permanganate transport was

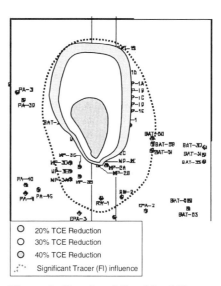

O   20% TCE Reduction
O   30% TCE Reduction
O   40% TCE Reduction
.·''·   Significant Tracer (FI) influence

Figure 1. Results of Combined Tracer-Permanganate Injection Test.

especially limited toward the south (bottom of figure), where initial TCE concentrations in soils were as high as 5,000 mg/kg, compared to 100 mg/kg TCE north of the injection point.

The tracer test results are consistent with the rapid consumption of permanganate in the presence of high TCE concentrations (i.e. DNAPL). In comparison, at a site in Maine, where TCE concentrations in water were approximately 1,500 ppb, and DNAPL was not present, permanganate transport was observed over distances in excess of 100 feet.

PERMANGANATE-CONTAMINANT REACTION KINETICS

The kinetics of reactions between permanganate and contaminants are obviously an important factor in the overall treatment success obtained. Huang, et al. (1999) showed that oxidation of TCE by KMnO4 was second order, with a fast second order rate constant of .083 +/- 0.05 M^{-1} s^{-1}. This rate constant applies to aqueous-phase reactions. Permanganate is essentially non-sorptive, while organic contaminants can exist in aqueous, sorbed, or NAPL phases. Permanganate-contaminant reactions occur in the aqueous phase, and NAPL and sorbed phases must be treated either by interfacial contact with the aqueous phase, or by mass transfer to the aqueous phase. The objective of this section is to consider the dynamics of permanganate-contaminant reactions, in terms of the presence and disappearance of reactants and products as well in terms of chemical equilibrium between contaminant phases.

An interesting observation from numerous field sites is that significant dissolved contaminant concentrations are generally not observed in the presence of aqueous permanganate. Figure 2 illustrates the breakthrough and

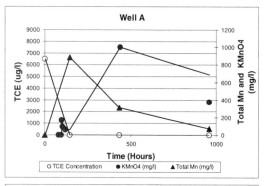

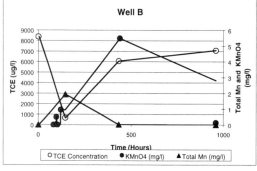

Figure 2. Dissolved TCE, KMnO4, and total Mn at two wells During a Field Pilot Test.

disappearance of permanganate in two monitoring wells at a site in Australia. Data from Wells A and B both show that Dissolved TCE levels are suppressed by the presence of permanganate. At Well A, TCE levels do not rebound, while at Well B TCE levels do rebound. This indicates that at Well A, a sufficient mass of permanganate was delivered to treat dissolved, and sorbed TCE, whereas at Well B, sorbed TCE was not completely oxidized. The untreated sorbed TCE mass at Well B provided the source of TCE mass transfer into the aqueous phase, as permanganate was consumed. Rebound of dissolved contaminant levels is common to field pilot tests, where the permanganate mass added may not achieve complete treatment. In many cases, full scale treatment is implemented in phases, with later phases focused on areas of greatest rebound. In this manner, the ultimate permanganate dose can be delivered most efficiently to the specific zones requiring treatment. This approach may be more cost effective than detailed characterization of the spatial variability of permanganate demand prior to treatment.

The total dissolved Mn shown in Figure 2 may include (1) dissolved manganese available from MnO_4^- ion, (2) manganese-containing intermediates such as cyclic esters (i.e. $MnO_4C_2HCl_3$) and $HMnO_3$, and (3) dissolved Mn^{2+} derived from MnO_2. Figure 2 indicates that total dissolved manganese goes toward zero as permanganate is consumed. This indicates that the measured dissolved Mn represents primarily (1) and (2), and that little or no dissolved Mn is derived from MnO_2. This is expected, since MnO_2 is a solid-phase colloidal precipitate, with very low aqueous solubility.

MATRIX INTERACTIONS AND GEOCHEMICAL EFFECTS

Relevant matrix interactions and geochemical effects involved in in-situ oxidation using permanganate include matrix oxidant demand, oxidation of native metals in soils, and changes in overall groundwater geochemistry related to redox state and ion exchange related to addition of Na^+ or K^+ with the permanganate salt

The overall permanganate consumption required to achieve contaminant treatment is the sum of permanganate demand from contaminant oxidation and matrix oxidant demand. Matrix oxidant demand refers to the oxidant consumption that can be attributed to background soil and groundwater conditions. Matrix demand can be derived from oxidation of natural organic matter, reduced metals, carbonates, sulfides, etc. Matrix demand can be highly variable, depending on background geochemical conditions, and is also highly dependent on permanganate solution concentration (R. Siegrist, pers. comm.), since permanganate reaction rates are second order.

Overall permanganate demand rates (i.e. contaminant demand plus matrix demand) across multiple sites have ranged from a few grams of permanganate per kg of soil (clean sands with dissolved contaminants) to as much as several hundred grams permanganate per kg of soil (organic clays containing non-aqueous phase liquid (NAPL) and 6% organic carbon content). In some cases, the overall permanganate demand may be driven more by matrix demand than by contaminant demand.

Oxidation of native metals in soils represents not only a source of matrix demand, but also potentially a mechanism by which groundwater quality can be degraded. The primary metals of concern include chromium, uranium, vanadium, selenium, and molybdenum. They may exist naturally in a chemically reduced, insoluble state at a particular site, and therefore not be detected in groundwater. However, because they are more soluble under oxidizing conditions, these metals can be mobilized by in-situ oxidation. Sites where this could be a potential problem can include sites where naturally occurring metals concentrations in soils are elevated, or sites where historical metals contamination was attenuated by natural chemical reduction processes. In such cases, it is desirable that natural geochemical conditions will attenuate the metals. Natural attenuation mechanisms can include sorption and chemical reduction back to an immobile valence. Metal sorption may occur at a range of different sites, including iron hydroxides, organic carbon, and MnO_2 produced by permanganate oxidation reactions.

The most commonly observed mobilization of metals is oxidation of Cr 3+ to Cr 6+. Preliminary data indicates that there is a strong correlation between total pre-treatment chromium concentrations in soil to the tendency to mobilize dissolved chrome during permanganate treatment. Table 1 shows data from three sites where elevated background chromium concentrations resulted in liberation of dissolved Cr6+. In all three cases, the dissolved chromium attenuated over time. Sites with background chromium in soils of less than about 5 mg/kg have not resulted in liberation of dissolved Cr6+ upon oxidation treatment.

Figure 3 shows the time trend of dissolved chromium for Site B (Table 1). Site hydrogeologic data and monitoring of surrounding wells indicated that the chromium attenuation is a result of a combination of sorption and chemical reduction and not attributable to dilution. The dissolved chromium half-life in this case was 6 days

Table 1. Lab and Field Data on Chromium Liberation and Attenuation.

Site	Lab/ Field	Pretreatment Total Cr in soil (mg/kg)	Maximum Dissolved Cr6+ Liberated (mg/l)	Cr6+ Attenuation Observed?
A	Lab	368	105 mg/l	Yes, 40 mg/l per pore volume of soil contact
B	Field	65	3 mg/l	Yes, to 0.007 mg/l in the field after 45 days
C	Field	28-94	1.5 mg/l	Yes, to 0.15 mg/l after 23 days

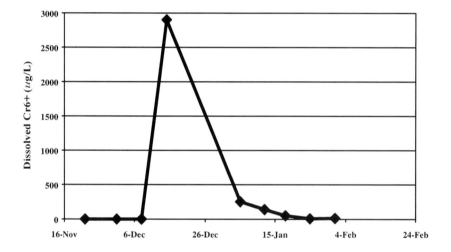

Figure 3.   Time-Series plot of dissolved Cr 6+ attenuation after permanganate treatment.

Permanganate oxidation can induce changes in overall groundwater geochemistry by influencing the redox state, and by inducing ion exchange. Oxidation reduction potential (ORP) values during in-situ permanganate treatment are typically > 200 mv within the zone of active treatment.

The addition of high concentrations of K^+ or Na^+ with the permanganate salt, can impact ion exchange for both cations and anions.   For example, following $KMnO_4$ addition, cations such as Mg^{2+} and Na^+ are commonly observed to increase as a result of cation exchange with K^+.   Anion concentrations can increase following $KMnO_4$ or $NaMnO_4$ addition, because as the $MnO4^-$ reacts, a net positive charge imbalance results, that can be offset by anion desorption (Struse, 1999).

CONTAMINANTS TREATED AND REDUCTIONS ACHIEVED

A wide range of contaminant types and concentrations have been subjected to in-situ oxidation using permanganate. Table 2 summarizes the combined lab and field results from multiple sites, many involving contaminant mixtures. These results show that while permanganate can treat a wide range of recalcitrant organics, not all organics can be treated effectively. Since permanganate is a weaker oxidant than ozone, hydrogen peroxide, or hydroxyl radical, these other oxidants may more be applicable to some of the more difficult to treat compounds listed here. Conversely, for the contaminants that are readily treated by permanganate, its stability and persistence in the subsurface facilitate advective and diffusive oxidant transport, potentially improving the overall level of treatment obtained.

Table 2. Summary of Contaminant Treatment Results in Lab and Field

Contaminant	Maximum Percent Treatment Achieved for Various Starting Contaminant Levels		
	Low Level Dissolved	High Level Dissolved/Sorbed	NAPL
chlorinated ethenes (PCE, TCE, DCE)	>99%	>99%	>99%
chlorinated ethanes (TCA, DCA)	>99%	40%	n.a.
vinyl chloride	>99%	n.a.	n.a.
chloroform	0%	0%	n.a.
methylene chloride	0%	0%	n.a.
carbon tetrachloride	0%	n.a.	n.a.
CFC 113	0%	0%	n.a.
toluene, xylene	>99%	n.a.	n.a.
benzene, ethylbenzene	30%	20%	n.a.
C6-C9 TPH	70%	50%	n.a.
C10-C14 TPH	0%	0%	n.a.
C15-C28 TPH	0%	0%	n.a.
Pentachlorophenol	>99%	>99%	n.a.
benzo(a)pyrene, and other PAHs	>96%	>96%	n.a.
MTBE	>99% *	>99% *	

notes: 1) n.a. – data not available
2) all data expressed as > represent treatment results below detection limits
3) * - MTBE transforms to tert-butyl alcohol (TBA), which does not degrade in the presence of permanganate.

SUMMARY AND CONCLUSIONS

In-situ chemical oxidation is an emerging remediation technology that shows great promise, but is not a "silver bullet" for treatment of all sites. Successful implementation of in-situ oxidation requires favorable oxidant-contaminant kinetics, delivery of an adequate mass of oxidant, and effective subsurface oxidant delivery and transport. While this paper has not dealt with hydrogeologic factors, ultimately geologic conditions may largely dictate success or failure at a given site.

Permanganate is highly effective for treatment of some compounds, such as chlorinated ethenes, marginally effective for others such as chlorinated ethanes, and somewhat ineffective on some contaminants such as benzene. Permanganate has been found effective for treatment of dissolved, sorbed, and NAPL contamination. Since treatment of NAPL requires a larger oxidant dose, all relevant performance criteria and design issues become more critical.

Many of the general observations and performance criteria that apply during in-situ chemical oxidation using permanganate also apply to other in-situ oxidants, including ozone and Fenton's reagent. We have attempted in this paper to avoid direct comparisons between oxidant performance. The reader is encouraged to critically examine the potential strengths and weaknesses of all in-situ oxidants.

REFERENCES

Clayton, W.S., 1998, Ozone and Contaminant Transport During In-Situ Ozonation, *in* Physical, Chemical, and Thermal Technologies, Remediation of Chlorinated and Recalcitrant Compounds, p. 389-395, eds. G.B. Wickramanayake, and R.E. Hinchee, Battelle Press, Columbus.

Huang, K, Hoag, G., Chheda, P., Woody, B., and Dobbs, G., 1999, *Kinetic Study of Oxidation of Trichloroethylene by Potassium Permanganate*, Env. Eng. Sc., (16)3. p. 265-274.

Mott-Smith, E., Leonard, W.C., Lewis, R., Clayton, W.S., Ramirez, J., and Brown, R., 2000, *In Situ Oxidation Of DNAPL Using Permanganate: IDC Cape Canaveral Demonstration,* 2nd International Conference on Remediation of Chlorinated and Recalcitrant Compounds.

Siegrist, Robert, Personal Communication, January, 2000.

Struse, A., 1999, Diffusive Transport of Potassium Permanganate in Low Permeability Media, M.S. Thesis, Department of Environmental Systems Engineering.

Yan and Schwartz, 1998, Trichloroethylene Oxidation by Potassium Permanganate, *in* Physical, Chemical, and Thermal Technologies, Remediation of Chlorinated and Recalcitrant Compounds, eds. G.B. Wickramanayake, and R.E. Hinchee, Battelle Press, Columbus.

A FIELD DEMONSTRATION OF TRICHLOROETHYLENE
OXIDATION USING POTASSIUM PERMANGANATE

Daniel J. McKay, Jeffrey A. Stark, Byron L. Young, John W. Govoni, Christopher M. Berini, Timothy J. Cronan, and Alan D. Hewitt
(U.S. Army Corps of Engineers, Hanover, New Hampshire, USA)

Abstract: A multi-year demonstration of in-situ oxidation using a solution of potassium permanganate is ongoing at the U.S. Army Cold Regions Research and Engineering Laboratory (Hanover, New Hampshire). Consistent with the Laboratory's mission to provide solutions for cold-related problems, chemical oxidation was selected to evaluate a low-cost alternative to bioremediation in cold climates. Mean annual temperature at the Hanover site is 7 °C. Treatment began in August 1999 and is expected to continue through at least part of 2001. Oxidant is delivered to unsaturated soils about 30 m above the water table in regions that contained residual-phase trichloroethylene (TCE). Treatment progress is monitored through periodic collections and analyses of the solid, liquid, and gas matrices within the treated volume. Here, an assessment of oxidant distribution in the heterogeneous formation is presented in the context of initial site conditions, treatment methodology, and preliminary results. Observations of permanganate distribution in post-treatment soil samples illustrate the importance of site characterization for improving the operational efficiency of an oxidative treatment system.

INTRODUCTION

Variably saturated, heterogeneous, low-permeability soils are challenging media for removal of sorbed organic contaminants. In settings of contrasting permeability, contaminants may accumulate above lower permeability strata, facilitating migration into areas otherwise inaccessible due to diffusion kinetics. Because of its characteristic stability in media that contain low levels of natural organic matter and reduced mineral species, potassium permanganate ($KMnO_4$) has the potential to access and to treat contaminated soils in transport-limited conditions.

Previous oxidation field studies have underscored the importance of a comprehensive, site-specific understanding of soil stratigraphy, soil properties, and contaminant distribution. One of the earliest field trials to evaluate soil treatment with permanganate was reported by Schnarr et al. (1998). There, $KMnO_4$ solution was injected over a period of several months into an aquifer containing residual sources of perchloroethylene (PCE). Noting horizontal bedding features on scale of millimeters to centimeters, it was concluded that the effectiveness of treatment with $KMnO_4$ depended strongly on non-aqueous-phase liquid (NAPL) distribution and its effects on dissolution. West et al. (1997) reported similar observations following permanganate injection within a heterogeneous aquifer containing TCE. They reported significant reduction of aqueous TCE for zones permeated by $KMnO_4$, but oxidant distribution appeared to be the controlling factor for overall remediation.

NAPL Oxidation in the Vadose Zone. Because $KMnO_4$ is delivered in solution, NAPL must be transferred to the solution for oxidation to occur. This process of dissolution depends strongly on the interfacial area between the non-aqueous and aqueous phases. The pore-scale spatial characteristics of a NAPL in two- and three-fluid porous systems are a consequence of many factors, including wetting behavior, spreading coefficient, fluid saturation ratios, and the discrepancy (between two- and three-fluid systems) of forces resisting an infiltrating NAPL. Scanning electron microscopy has revealed fundamental differences in the distribution of NAPL within two- and three-fluid systems. In a three-fluid soil system, Hayden and Voice (1993) observed continuous NAPL distribution, including pendular rings and films. Wilson et al. (1988) observed isolated NAPL globules in a two-fluid system characteristic of capillary entrapment, and interconnected NAPL films between water and air phases and within the air phase in three-fluid systems.

The net result of these contrasting spatial configurations can be a larger ratio of NAPL surface area to volume in unsaturated (versus saturated) conditions as demonstrated in qualitative experiments with PCE (Schwille, 1988). Consequently, unsaturated pores may provide larger interfacial areas between phases, enabling aqueous oxidation of residual NAPLs to be particularly effective in the vadose zone. Additionally, one may anticipate improved efficiency with oxidant distribution in the vadose zone owing to lesser $KMnO_4$ dilution with existing groundwater and the presence of capillary forces to facilitate oxidant migration in zones of reduced hydraulic conductivity.

TCE Oxidation by Permanganate. As potassium ion (K^+) is non-reactive in aqueous solution, reduction of the permanganate ion (MnO_4^-) is considered here. In the presence of reducing agents, the oxidation state of manganese is determined by the solution pH (Stewart, 1965). In basic or neutral aqueous solutions, the manganese precipitates as manganese dioxide (MnO_2), which is also a strong oxygen transfer agent:

$$MnO_4^- + 2H_2O + 3e^- \quad MnO_2\,(s) + 4OH^-. \tag{1}$$

Acidic solutions can reduce MnO_4^- to the manganese ion (Mn^{2+}):

$$MnO_4^- + 8H^+ + 5e^- \quad Mn^{2+} + 4H_2O. \tag{2}$$

Based on laboratory observation, mass balance and redox requirements, Schnarr (1992) proposed the following reaction equation for $KMnO_4$ and TCE:

$$C_2Cl_3H + 2\,KMnO_4 \quad 2\,MnO_2(s) + 2\,CO_2 + 2KCl + HCl \tag{3}$$

in which six electrons are transferred from carbon to manganese atoms per molecule of TCE.

Huang et al. (1999) confirmed the stoichiometry of Equation 3 based on measurements of consumption for $KMnO_4$ and TCE and determined a second order rate constant of 0.89 $M^{-1}s^{-1}$ at 20°C. Yan and Schwartz (1999) calculated a second order rate constant for TCE of k = 0.67 $M^{-1}s^{-1}$. The authors determined that pH in the range of 4–8 did not affect the rate of TCE disappearance but noted that decomposition of a cyclic complex was highly pH dependant, with competition for reaction among MnO_4^- and OH^- ions.

Site Description. In-situ treatment of vadose-zone soils is applied at two separate source areas as part of an overall approach to address TCE contamination in the groundwater at the Hanover, NH, location. These sites, separated by a distance of about 50 m, are identified as Areas of Concern 2 and 9 (AOC 2 and AOC 9). The source at AOC 2 was a leaking underground storage tank that was removed in 1972. TCE contamination at AOC 9 is attributed primarily to accidental surface spills during the 1970s and 1980s.

Each site encompasses an area of approximately 300 m^2. The majority of TCE is found within the first 10 m below the ground surface. Water table depth is approximately 40 m. Pre-treatment estimates of TCE mass, based on nearly 700 combined samples, indicated approximately 2500 and 2000 kg at AOC 2 and 9, respectively. However, additional samples collected during treatment at AOC 9 suggest that these initial estimates may have been low. The lacustrine sediments at both sites are stratigraphically similar, and contain comparable contaminant distribution profiles, owing to a pattern of inter-bedded clay lenses within a predominantly silt formation. Large numbers of particle size analyses were performed to map field-scale heterogeneity. The results indicated that both sites, taken as a whole, have nearly identical compositions consisting of 81% silts, 12% clays, and 7% sands. Because of these similarities and the limited space that is available here, only data from AOC 9 are presented hereafter.

Soils within the initial treatment zone at AOC 9, between 3.7 and 9.8 m below the ground surface, consist of an average of 92.8% fines (< 0.075 mm) based upon analyses (ASTM D 1140-92) of 415 samples (σ = 9.4). The mean calculated porosity based upon analyses (ASTM D 4404-84) of 76 samples is 43.3% (σ = 8.2%).

Background soil chemistry is summarized in Table 1.

TABLE 1. Background soil chemistry at AOC 9.

Analysis	Method	Mean	σ	Units	Sample Qty
potassium	3050/6010	1718.6	592.0	mg kg^{-1}	59
manganese	3050/6010	509.7	177.6	mg kg^{-1}	59
chloride (detected)	9251	94.1	145.9	mg kg^{-1}	37
chloride (non-detected)	9251	< 51		mg kg^{-1}	22
pH	9040/9045	9.0	0.7	pH	59
cation exchange capacity	9081	9.8	3.9	meq $100g^{-1}$	59

Clay lenses within the silt formation, typically 4–10 cm thick, appear at regular intervals of several tens of centimeters. The highest concentrations of residual TCE, on the order of tens of thousands of milligrams per kilogram, are found directly above many of these lenses (Table 2). Concentrations within and directly below the clay formations are often two and three orders of magnitude lower, respectively. These data appear to reflect years of diffusion of TCE across otherwise impermeable regions. As previously stated, $KMnO_4$ may also diffuse into these areas, owing to its stability, the corresponding kinetic processes of TCE dissolution and oxidation, and resultant increase in concentration gradients for both aqueous-phase TCE and permanganate (Schnarr et al. 1998).

TABLE 2. TCE in sub-samples of soil cores obtained from AOC 9.

Core I.D.	Sample Depth (cm)	Soil Classification	TCE (mg kg^{-1})
5-1	528	silt	48,900
5-1	533	clay	574
5-1	543	silt	57
5-2	586	silt	34,800
5-2	591	clay	459
5-2	601	silt	23
6-2	583	silt	61,000
6-2	588	clay	517
6-2	598	silt	16
6-3	645	silt	38,000
6-3	650	clay	311
6-3	655	silt	92

MATERIALS AND METHODS

Construction and Operation. The in-situ soil treatment system consists of an oxidant mixing plant, two satellite buildings to manage distribution to individual wells, and 32 injection wells (as of this submission). A material-flow schematic is shown in Figure 1. Currently, batch mixtures consist of 75 kg $KMnO_4$ and 5000 L of water (1.5% $KMnO_4$). Operating capacity at the present concentration is 400 kg of $KMnO_4$ per day, though lower delivery rates are normally applied. Solution flow rate to individual wells is typically 4–6 L min^{-1} (1–1.5 g s^{-1} of $KMnO_4$) at pressures ranging from 0.5–1 bar. Both source areas are treated simultaneously with each site having one injection point in use at any given time. A system of air-actuated valves is programmed to sequentially step through the injection arrays at pre-set intervals.

Permanganate solution is delivered to the subsurface through an array of 5.1-cm-diameter well screens placed inside 7.6-cm-diameter boreholes (Figure 2). Each well screen is 1.2 m long and isolated above by an inflatable borehole packer that enables delivery of the oxidant under pressure. All near-surface piping is buried beneath insulation board and contains internal heat tracer for freeze protection.

The permanganate mixing plant houses two 11,300-L solution tanks, two gravimetric screw feeders with hoppers, an overhead crane for material handling, a ventilation and dust control system, a wash area for cleaning used drums and equipment, emergency and personal protection equipment, controls, alarms, and data acquisition hardware. The hopper/feeders and solution tanks were con-

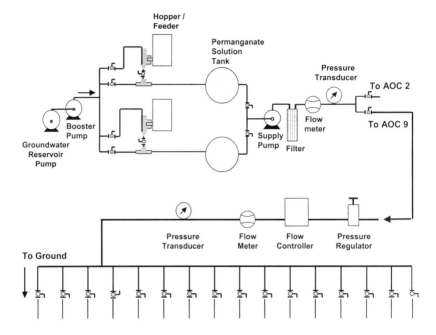

FIGURE 1. Material flow schematic.

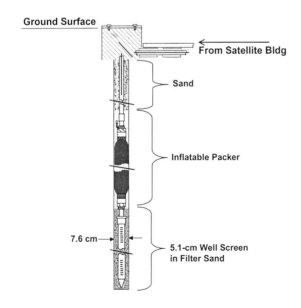

FIGURE 2. Typical subsurface construction.

structed in duplicate to allow simultaneous batch preparation and consumption. Material flow processes including rates, pressures and temperatures are monitored within the oxidant mixing plant and each satellite distribution building. All aboveground material handling hardware, at both the mixing plant and the satellite buildings, is completed within lined containment walls to prevent surface spills to the environment.

Site Monitoring. The primary measures of treatment progress are visual inspection of soil cores for evidence of $KMnO_4$ permeation and analyses of soil samples to determine TCE concentrations. Lysimeters, soil gas implants, and neutron moisture probe access pipes were installed to monitor secondary effects including pore water chemistry, soil gas composition, and moisture distribution.

Soil samples are collected using a 2.5-cm direct-push sampler. Continuous profiles are collected at 0.6-m intervals, and then subdivided to 0.3-m lengths. Each section is split along the length of the sample to produce quarter sections of the original core. Half of each 0.3 m section is preserved in HPLC-grade methanol for laboratory analysis. A small quantity of sodium thiosulfate is added to the methanol to reduce any permanganate present in the sample. The remainder of soil is used to determine moisture content and to perform additional analyses.

RESULTS AND DISCUSSION

In the alkaline soil environment of AOC 9 (see Table 1), $MnO_2(s)$ will precipitate during the reduction of MnO_4^- per Equation 1. For soil density of 1.4 g mL^{-1}, soil porosity of 0.43, and $MnO_2(s)$ density of 1.2 g mL^{-1}, the hypothetical maximum pore volume reduction by $MnO_2(s)$ is 0.36% per 1000 mg kg^{-1} of TCE. In regions having TCE concentrations of several tens of thousands of milligrams per kilogram (see Table 2), localized reductions in permeability may be significant. Schnarr (1992) measured a decrease in pore volume of 5.6% following injection of $KMnO_4$ into a PCE-contaminated soil, compared to 1% volume change attributed to the contaminant. As NAPL occupies a specific range of pore sizes determined by initial moisture content (Schaefer et al., 1999), the pore-scale distribution of the $MnO_2(s)$ may be more important than quantity if accumulations in pore throats weaken pore connectivity. Mercury intrusion tests on soils containing extensive MnO_2 staining are thus far inconclusive as to changes in pore size distribution.

Soil cores containing unreduced $KMnO_4$ and MnO_2 staining on TCE-contaminated soil provide visual evidence of $KMnO_4$ migration pathways. Permeated soils that contained high initial concentrations of TCE are indicated by dark brown discolorations caused by accumulation of MnO_2. The intensity of discoloration is assumed to be comparative to the amount of MnO_2 and hence is a visual indicator of original TCE distribution. Figure 3 shows a section (6.5- to 6.7-m depth) from a 2.5-cm soil core collected approximately 1 m from an injection well (4.1- to 5.3-m depth) following the addition of 200 kg $KMnO_4$. Dark brown staining directly above a clay lense indicates TCE distribution characteristic of that found in Table 2. The sample described in this example contained a composite average of 1900 mg kg^{-1} TCE—less than the 5700 mg kg^{-1} of a neighboring sample collected prior to treatment. $KMnO_4$ did not appear to have permeated the clay layer. As previously suggested, the stability of $KMnO_4$ may enable passive, long-term treatment of the clay soil because of the parallel diffusion processes of TCE and $KMnO_4$. Inspection of numerous other samples (not shown) found $KMnO_4$ permeation through veins of relatively course-grained material between layers of finer material. Isolated layers of light brown staining caused by MnO_2, bordered above and below by thin layers of dark brown staining, suggest $KMnO_4$ permeation through the heterogeneous formation may have followed discrete pathways preceded by TCE.

Figure 3. 2.5-cm-diameter soil core collected near AOC 9 injection well. Depth shown is 6.5–6.7 m (left to right). Section to the left of dark band showed light purple to brown discoloration, indicating $KMnO_4$ permeation and light oxidation. Dark band indicated extensive oxidation in area where TCE was concentrated above clay lense.

Preliminary results indicate that regions of relatively low initial TCE concentrations can be effectively treated if flooded by $KMnO_4$ solution. Areas within AOC 9 that contained TCE concentrations on the order of ten to hundreds of milligrams per kilogram were treated to levels near or below the detection limit of about 100 $\mu g\ kg^{-1}$ following single events of intensive $KMnO_4$ additions to nearby injection wells. However, mass balance estimates and initial results indicate that areas containing TCE concentrations on the order of several thousand milligrams per kilogram will require extensive and prolonged applications of $KMnO_4$ treatments. For optimum efficiency, the rate of these applications should not greatly exceed the rate of TCE dissolution. Consequently, we are modifying our approach to oxidant delivery in these areas to include low-flow, direct-push injection points that will enable localized control of oxidant distribution.

CONCLUSIONS

Preliminary analytical data and inspection of post-treatment samples have highlighted the importance of site characterization for proper design of efficient in-situ oxidative treatment systems. Post-treatment soil cores provided visual evidence that $KMnO_4$ permeation through the vadose zone might often have traced isolated paths of TCE migration. In areas with relatively low TCE concentrations, broad, nonspecific application of $KMnO_4$ appears to be effective. Regions of concentrated NAPL will require contaminant delineation and localized delivery.

REFERENCES

Hayden, N.J., and T.C. Voice. 1993. "Microscopic Observation of a NAPL in a Three-Fluid-Phase Soil System." *Journal of Contaminant Hydrology.* 12: 217–226.

Huang, K-C., G.E. Hoag, P. Chheda, B.A. Woody, and G.M. Dobbs. 1999. "Kinetic Study of Oxidation of Trichloroethylene by Potassium Permanganate." *Environmental Engineering Science.* 16: 265–274.

Schaefer, C.E., R.R. Arands, D.S. Kosson. 1999. "Measurement of Pore Connectivity to Describe Diffusion Through a Nonaqueous Phase in Unsaturated Soils." *Journal of Contaminant Hydrology.* 40: 221–238.

Schnarr, M.J. 1992. "An In-Situ Oxidative Technique to Remove Residual DNAPL from Soils." M.S. Thesis, University of Waterloo, Waterloo, Ontario, Canada.

Schnarr, M., C. Truax, G. Farquhar, E. Hood, T. Gonullu, and B. Stickney. 1998. "Laboratory and Controlled Field Experiments Using Potassium Permanganate to Remediate Trichloroethylene and Perchloroethylene DNAPLs in Porous Media." *Journal of Contaminant Hydrology.* 29: 205–224.

Schwille, F. 1988. *Dense Chlorinated Solvents in Porous and Fractured Media.* Lewis Publishers, Chelsea, Michigan.

Stewart, R. 1965. In: K.Wiberg (Ed.). *Oxidation in Organic Chemistry: Oxidation by Permanganate.* Academic Press, New York, NY.

West, O.R., S. Cline, W. Holden, F. Gardner, B. Schlosser, J. Thate, D. Pickering and T. Houk. 1997. *A Full-Scale Demonstration of In-Situ Chemical Oxidation Through Recirculation at the X-701B Site: Field Operations and TCE Degradation.* ORNL/TM-13556. Prepared by Oak Ridge National Laboratory for the U.S. Department of Energy.

Wilson, J.L., S.H. Conrad, E. Hagan, W.R. Mason, and W. Peplinski. 1988. "The Pore Level Spatial Distribution and Saturation of Organic Liquids in Porous Media." In: *Proceedings of the NWWA Conference on Petroleum Hydrocarbons and Organic Chemicals in Ground Water—Prevention, Detection, and Restoration.* Dublin, OH.

Yan, Y.E. and F. Schwartz. 1999. "Oxidative Degradation and Kinetics of Chlorinated Ethylenes by Potassium Permanganate." *Journal of Contaminant Hydrology.* 37: 343–365.

PERMANGANATE INJECTION FOR SOURCE ZONE TREATMENT OF TCE DNAPL

Michael Moes, P.E. (Erler & Kalinowski, Inc., Englewood, Colorado, USA)
Carey Peabody, Ph.D. (Erler & Kalinowski, Inc., San Mateo, California, USA)
Robert Siegrist, Ph.D., P.E. (Colorado School of Mines, Golden, Colorado, USA)
Michael Urynowicz, P.E. (Colorado School of Mines, Golden, Colorado, USA)

ABSTRACT: Potassium permanganate ($KMnO_4$) solution was injected into a low permeability formation at a former industrial site in California in March 1999 to test the effectiveness of *in situ* chemical oxidation (ICO) in treating trichloroethene (TCE) in high salinity groundwater. TCE was detected at concentrations as high as 260 mg/L. For a field-scale pilot study, approximately 550 kg (1,200 pounds) of $KMnO_4$ were injected into the saturated zone where the TCE may occur as dense non-aqueous phase liquid (DNAPL) residual in the predominantly clay/silt sediments. The TCE concentration in groundwater was initially reduced to below detection limits at locations where oxidant was effectively distributed; however, oxidant distribution in the fine-grained sediments was not sufficiently uniform to achieve overall treatment performance goals. Naturally occurring inorganic compounds such as chromium and selenium were apparently oxidized by $KMnO_4$ to higher and more mobile oxidation states (e.g., Cr^{+3} to Cr^{+6}), although such oxidations are considered transient and reversible. Trihalomethanes (THMs) were formed in groundwater, presumably from the interaction of $KMnO_4$, natural organic material, and chloride and bromide ions in the saline groundwater, but dissipated naturally after several weeks. This project demonstrated (a) private sector and regulatory agency acceptance of ICO treatment using $KMnO_4$, (b) *in situ* treatment of TCE and other chlorinated ethenes in groundwater where oxidant was effectively distributed, (c) the need to further develop effective oxidant distribution techniques for fine-grained media, and (d) the occurrence of oxidized inorganic compounds and THMs following ICO treatment under the test conditions at the Site.

INTRODUCTION

Chlorinated volatile organic compounds (VOCs), including trichloroethene (TCE), cis-1,2-dichloroethene (c12DCE), vinyl chloride (VC), carbon tetrachloride (CTC), and 1,1,2-trichloro-1,2,2-trifluoroethane (CFC113) were present in groundwater at a former industrial site near San Francisco Bay, California. TCE was detected at concentrations as high as 260 mg/L, indicating TCE may occur in the formation as DNAPL residual that could persist at the Site for decades if not remediated. Although the groundwater is not a potable water source, the VOCs in groundwater could potentially present long term human health risks if a vapor migration and inhalation exposure pathway was complete.

There are no commercially proven *in situ* technologies for remediating chlorinated VOC DNAPLs in the type of saline, clay-rich sediments that occur at the Site. As such, the final remedy selected for the Site was (a) removal of shallow soil containing VOCs, and (b) long-term engineering and institutional controls to reduce the potential for exposures to VOCs in groundwater. In addition, it was decided to perform ICO treatment to reduce the VOC mass and concentration in groundwater at the Site, subject to demonstrating ICO treatment feasibility during a pilot study. To be considered feasible, it was necessary to demonstrate that the sustained TCE concentration in groundwater could be reduced to a Site-specific risk-based action level of 4 mg/L.

ICO Treatment "Side Effects" and Obtaining Regulatory Approval. $KMnO_4$ is a strong oxidant, so the potential "side effects" of injecting it into groundwater were evaluated. In addition to oxidizing the target VOCs, $KMnO_4$ can also oxidize naturally occurring organic carbon and minerals in soil. Organic carbon oxidation can be beneficial by depleting over time excess $KMnO_4$ injected during treatment. As such, recovery of excess $KMnO_4$ was not required at the Site.

Oxidation of naturally occurring minerals can potentially be of concern. Specifically, the oxidation of trivalent chromium (Cr^{+3}) to the more soluble and more toxic hexavalent chromium (Cr^{+6}), and the oxidation of selenium from Se^{+4} to the more soluble Se^{+6}, are undesirable reactions that may occur. However, in sediments with natural reducing capacity, the oxidation of chromium and selenium is reversible (a) as reducing conditions re-establish over time in treated areas, and (b) as groundwater containing the oxidized compounds migrates into untreated sediments with reducing capacity. Therefore, particularly at sites where there is no direct exposure to groundwater, the transient occurrence of oxidized metals may be an acceptable "side effect" if the technology can mitigate the long-term human health risks associated with the VOCs in the groundwater.

As with any emerging technology, obtaining regulatory approval was a concern for this project. However, the San Francisco Bay Regional Water Quality Control Board (RWQCB) was presented with both (a) potential benefits of the ICO treatment (i.e., VOC mass reduction and risk reduction), and (b) the potential side effects described above. The RWQCB approved ICO as a portion of the final remedy, subject to completing a successful field demonstration.

Site Conditions. Groundwater occurs at the Site beginning at approximately 2 to 2.5 m below ground surface (bgs). The saturated sediments are predominantly clay and silt. ICO testing was performed in an area centered on well MW-1 (see Figure 1). TCE was detected at 38 mg/L in groundwater from MW-1, and at 260 mg/L in a grab groundwater sample collected from 6 m bgs in boring SB-1, located 1 m from well MW-1. Well MW-1 is screened from 5.5 to 8.5 m bgs.

The groundwater at the Site is saline, with total dissolved solids (TDS) levels ranging from 2,100 to 154,000 mg/L and grading from lower to

TABLE 1. Pre-Treatment Groundwater at MW-1

Parameter	Value
TCE	38 mg/L
cDCE	<0.5 mg/L
VC	<0.5 mg/L
TDS	110,000 mg/L
HCO_3^-	350 mg/L
Manganese	19 mg/L
Potassium	210 mg/L
Eh	380 mV
pH	6.0

higher TDS with depth. Selected pre-treatment groundwater parameters at well MW-1 are listed in Table 1. The redox potential relative to the standard hydrogen electrode (Eh) was 380 millivolts (mV), and bicarbonate alkalinity (HCO_3^-) was 350 mg/L. Although present elsewhere at the site, cDCE and VC were not detected in the pre-treatment groundwater sample from MW-1, possibly due to elevated detection limits related to the presence of TCE, and possibly due to high TDS locally limiting microbial reductive dechlorination processes.

Selection of ICO Using KMnO₄. ICO using $KMnO_4$ was selected for testing considering that, from a risk-based perspective, the primary chemicals of concern at the Site were TCE and vinyl chloride. Those compounds, as well as c12DCE, are readily oxidized by $KMnO_4$ in water (Siegrist et al., 1999). As expected, bench-scale testing (see below) indicated $KMnO_4$ would not be an effective treatment for CTC and CFC113, compounds that are also detected at the Site. However, the CTC is limited in lateral extent and at levels that may be acceptable for a risk-based closure, and CTC113 does not significantly contribute to estimated human health risks at the Site. Therefore, the target VOCs for ICO treatment at the Site were TCE, c12DCE, and VC, i.e., acknowledging that CTC and CFC113 would not be substantially affected. In fact, the recalcitrance of CTC aided in evaluating the treatment effectiveness of the target compounds.

The stoichiometric mineralization of TCE (C_2HCl_3) with the permanganate ion (MnO_4^-) can be written as follows:

$$C_2HCl_3 + 2\,MnO_4^- \rightarrow 2\,CO_2 + 2\,MnO_2 + 3\,Cl^- + H^+$$

The reaction products are carbon dioxide, manganese dioxide (insoluble), chloride ion, and hydrogen ion. Because MnO_4^- can also oxidize dissolved divalent manganese in groundwater, the dissolved manganese concentration in groundwater may decrease following treatment.

BENCH-SCALE TESTING

Bench-scale testing was performed at the Colorado School of Mines in Golden, Colorado, in 1998, using soil samples collected from 2.5 and 6 m bgs and groundwater samples collected from three wells at the Site. Objectives of the testing were (a) to estimate the natural oxidant demand (NOD) of the native soil at the Site, (b) to observe the TCE degradation kinetics in soil and groundwater samples from the Site, and (c) to confirm treatability of TCE, c12DCE and VC in high TDS groundwater.

To test the NOD, soil samples from the Site were air-dried to remove VOCs and re-moisturized with deionized water. $KMnO_4$ solution was then added to a series of subsamples at net doses of 1 and 10 milligrams of $KMnO_4$ per gram of dry soil (mg/g). Residual $KMnO_4$ was measured in sacrificial samples over a period of three weeks. The total organic carbon (TOC) concentration in the soil ranged from non-detect to 3,600 mg/kg, and the measured NOD was higher in samples with higher TOC. At the $KMnO_4$ dose of 1 mg/g, all oxidant was depleted within 2 days to 2 weeks, varying with the TOC level of the sample. At the $KMnO_4$ dose of 10 mg/g, all oxidant was depleted in samples with higher

TOC within 7 days, and approximately one-half the oxidant was depleted in lower-TOC samples after 3 weeks of testing. The NOD tests demonstrated that at an oxidant loading between 1 and 5 mg/g, the oxidant would persist in the sediments for several days (i.e., allowing TCE oxidation) and would eventually be depleted. Based in part on these results, a target oxidant loading of 2.2 mg/g was selected for the field-scale pilot study.

To test Site-specific TCE degradation kinetics, $KMnO_4$ solution was added to a series of well-mixed samples of (a) groundwater alone, and (b) slurried mixtures of groundwater and soil. Initial TCE concentrations ranged form 3 to 34 mg/L, and the TCE concentration in each sample was measured over time for two hours. Control studies indicated no appreciable loss of TCE in the absence of $KMnO_4$. In all tests with $KMnO_4$, the TCE concentration was reduced over 99 percent within a time period of 7 minutes or less when using a $KMnO_4$ loading of 5,000 mg/l. At a lower $KMnO_4$ loading of 500 mg/l, TCE removal ranged from 78 to 97 percent during the first 30 minutes. The results confirmed that rapid TCE oxidation (i.e., relative to excess oxidant consumption) was feasible by adding $KMnO_4$ to well-mixed samples of soil and groundwater, including the high TDS samples collected from the Site.

To observe the treatment effectiveness on a broader suite of VOCs, 500 mg/L $KMnO_4$ was added to a groundwater sample collected from well MW-2 (located outside the area shown on Figure 1). Analytical results from before and after the addition of $KMnO_4$ are listed in Table 2. Results indicated excellent destruction of chlorinated ethenes, including the target compounds TCE, c12DCE, and VC. As expected, oxidation of the chloromethanes (CTC and chloroform) and CFC113 was limited or non-existent, as the observed concentration decreases of less than 20 percent in the sample may or may not be indicative of oxidation.

TABLE 2. Laboratory Results for Permanganate Treatment of VOCs in Groundwater from Well MW-2 (500 mg/L $KMnO_4$)

VOC	Concentration (mg/L)		
	Pre-$KMnO_4$	Post-$KMnO_4$	%Decrease
Carbon tetrachloride	0.030	0.027	10%
Chloroform	0.054	0.046	15%
1,1-Dichloroethene	0.008	<0.005	>37%
cis-1,2-Dichloroethene	8.7	<0.005	>99.9%
trans-1,2-Dichloroethene	0.018	<0.005	>72%
Trichloroethene	43	0.009	99.98%
Vinyl chloride	0.039	<0.010	>74%
CFC-113	0.053	0.043	19%

FIELD-SCALE PILOT STUDY

Although bench-scale testing clearly indicated the ability of $KMnO_4$ to rapidly oxidize TCE in well-mixed samples, it was recognized that the primary technical challenge at the site was to effectively deliver the oxidant to the contaminated sediments *in situ*. Therefore, a field-scale pilot study was performed at a location of the Site where TCE concentrations were the highest. The predominantly clay sediments at the Site meant that transmitting a KMnO₄ solution uniformly through the soil any significant distance was infeasible. Therefore, it was decided to inject KMnO₄ solution in as concentrated a manner as feasible, i.e., to attempt to place the oxidant solution directly throughout the entire targeted treatment zone. Solution was injected using 2.5 cm diameter lances that were pushed into the ground on a staggered 1.2-m lateral spacing, injecting KMnO₄ across a depth range of 1.5 to 8.5 m bgs. The grid of injection locations is shown on Figure 1. The lances were pushed into the ground in 0.6 to 0.9 m vertical steps. Solution was injected as the lances were pushed into the ground, and, at the bottom of each step, a packer located 0.9 m above the lance tip was inflated and KMnO₄ solution was injected into the sediments.

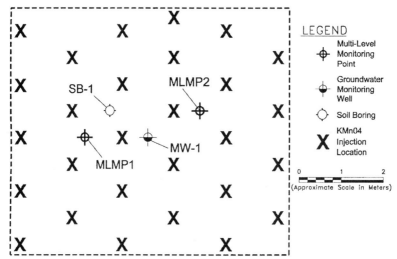

FIGURE 1. ICO Pilot Study Injection Grid and Monitoring Locations

The effectiveness of the injection process was spatially non-uniform. At some injection locations and depths, essentially all of the injected solution would be observed flowing back up the outside of the lance to the ground surface, i.e., rejected by extremely low permeability sediments in the treatment zone. At other locations no fluid rejection was observed, i.e., all of the solution was retained in the treatment zone. Similar spatial non-uniformity was observed in the post-treatment monitoring results, as described below.

A total of 550 kg of KMnO₄ were injected during the pilot study at a concentration of approximately 50,000 to 60,000 mg/L. Injections occurred on 9 days between 18 and 31 March 1999. The target oxidant loading for the pilot study was 2.2 mg/g. Due to the localized injection variability and oxidant solution rejection by low permeability sediments (see above), the average oxidant loading attained was only 0.9 mg/g.

Three days after KMnO₄ injections were completed, 5-cm diameter multi-level monitoring points MLMP1 and MLMP2 (see Figure 1) were installed using direct-push drilling with a 9 cm drive casing. Each MLMP had 90-cm screened sampling intervals centered at 3, 6, and 9 m bgs, accessed through a 1.5-cm wide dedicated channel inside the casing. During installation, visual observations of the purple $KMnO_4$ solution in the soil cores indicated widely varying distribution of $KMnO_4$ had occurred during the injection process.

Post-treatment groundwater monitoring was performed for 12 weeks at MW-1, MLMP1 and MLMP2. Groundwater samples were analyzed for MnO_4^- (as $KMnO_4$), Eh, pH, VOCs, dissolved metals, and other parameters.

RESULTS AND DISCUSSION

VOCs. Selected pre- and post-treatment groundwater monitoring data are listed in Table 3 to illustrate some of the findings from the pilot study. The primary finding was that the $KMnO_4$ injection process was not sufficiently effective over the entire treatment region. The treatment objective was 4 mg/L TCE in groundwater in well MW-1. The TCE concentration decreased to 1.4 mg/L at MW-1 one day after the injections were completed, but after 12 weeks the TCE concentration was 35 mg/L, i.e., near its pre-treatment level of 38 mg/L.

At MLMP2-20 (i.e., the 6-m (20-foot) deep sampling interval at MLMP2), TCE was not detected in groundwater until 8 weeks after the completion of injections. Corresponding to the lack of TCE, the initial post-treatment MnO_4^- concentration and Eh measured at MLMP2-20 were the highest measured during the test, indicating $KMnO_4$ solution was effectively injected in the localized vicinity of MLMP2-20. CTC behaved somewhat as a recalcitrant tracer, and was routinely detected in samples from MLMP2-20 at concentrations typical of groundwater in that area of the Site. Overall the data suggest effective treatment of TCE in the vicinity of MLMP2-20.

At MLMP1-20 (i.e., the 6-m deep sampling interval at MLMP1), the results are less obvious. The $KMnO_4$ and Eh data reached maximum levels 2 to 3 weeks after injections were completed, suggesting some post-injection diffusion or advection of MnO_4^-. Correspondingly, TCE concentrations decreased from 180 mg/L one week post-injection to 52 mg/L 2 weeks post-injection, and stayed in the 52 to 89 mg/L range during the period from 2 weeks to 12 weeks post-injection. A pre-treatment grab groundwater sample from 6 m bgs at location SB-1 (located 1 m from MLMP1) contained 260 mg/L TCE, suggesting the pre-treatment TCE level at MLMP1-20 may have been in that range. As such, the reduction to 51.9 mg/L at the end of the monitoring period may represent a localized TCE concentration reduction as high as 80%.

Brominated THMs, including bromoform (BF), bromodichloromethane (BDCM), and dibromochloromethane (DBCM), were not previously detected at the Site but were detected in some post-treatment groundwater samples. Measured BF concentrations (Table 3) were highest during the first four weeks post-treatment at location MLMP2-20, coincident with the sampling location where $KMnO_4$ concentrations were also highest. However, the BF dissipated naturally and was not observed 8- and 12-weeks post-treatment. Although not listed in Table 3, BDCM was detected in one sample (0.031 mg/L), and DBCM was detected in 7 samples (0.068 to 0.68 mg/L) during post-treatment monitoring.

TABLE 3. Selected Pre- and Post-Injection Groundwater Monitoring Data

Location	Date[a]	KMnO$_4$ (mg/L)	Eh (mV)	pH	VOC[b] Concentration (mg/L)		
					TCE	CTC	BF
MW-1	12/10/98	0	383	6.1	**38**	**1.70**	<0.5
	4/1/99	178	924	5.9	**1.4**	**1.28**	**0.780**
	4/7/99	12	890	5.9	**25.3**	**1.48**	**0.542**
	4/14/99	24	--	6.1	**24.2**	**1.40**	<0.40
	4/22/99	0.8	559	5.9	**25.7**	**1.25**	<0.33
	4/29/99	--	556	6.1	**25.8**	**1.03**	<1.0
	5/27/99	0	--	6.0	**27.8**	**0.87**	<0.50
	6/23/99	0	394	5.9	**35.8**	**1.87**	<0.50
MLMP1-20	4/7/99	21	681	5.3	**180**	<0.5	<0.50
	4/14/99	135	--	6.1	**52**	<2.5	<2.5
	4/22/99	51	890	6.1	**89**	<2.5	<2.5
	4/29/99	35	865	6.2	**63.9**	<2.5	<2.5
	5/27/99	6	--	6.1	**71.3**	<1.25	<1.25
	6/23/99	5	421	6.0	**51.9**	<1.25	<1.25
MLMP2-20	4/7/99	630	909	7.0	<0.025	**3.20**	**0.43**
	4/14/99	962	--	6.8	<0.125	**3.30**	**3.0**
	4/22/99	577	838	6.6	<0.125	**3.30**	**6.5**
	4/29/99	274	851	6.6	<0.100	**2.76**	**0.17**
	5/27/99	10	--	6.5	**2.20**	**2.99**	<0.125
	6/23/99	3	563	6.4	**2.69**	**3.52**	<0.063

(a) KMnO$_4$ was injected during 9 days between 18 and 31 March 1999. (b) VC and c12DCE were not detected at the monitoring locations listed in the table.

Dissolved Metals. Groundwater samples collected from well MW-1 were filtered and analyzed for dissolved metals pre-treatment and post-treatment. Metals detected are listed in Table 4. Dissolved metals detected post-treatment that were not detected pre-treatment include chromium, mercury, nickel, and selenium.

TABLE 4. Dissolved Metals (mg/L) Detected[a] in Groundwater at MW-1

Date	Cd	Cr(tot)	Cr^{+6}	Hg	Ni	Se	Ag	Zn
12/10/98	0.13	<0.02	<0.01	<0.0002	<0.02	<0.08	0.051	0.058
4/22/99	0.11	0.48	0.53	0.0032	0.098	0.57	0.097	0.056
5/27/99	0.048	0.38	0.35	0.0026	0.11	0.53	0.11	0.047
6/23/99	0.11	0.48	0.41	0.0027	0.14	0.63	0.11	0.024

(a) Metals analyzed for but not detected included Sb, As, Be, Pb, and Th.

Chromium and selenium are known to be more soluble in their higher oxidation states (i.e., Cr^{+6} vs. Cr^{+3}, and Se^{+6} vs. Se^{+4}) so the post-treatment occurrence of dissolved hexavalent chromium and dissolved selenium was anticipated in planning the test. Also, chromium (and other metals) occur as impurities in technical grade KMnO$_4$ and may contribute to dissolved metals concentrations in groundwater. It is anticipated that chromium and selenium will revert to their lower oxidation states of lower solubility (a) as the Eh of the groundwater decreases from the elevated post-treatment levels (see Table 3), and (b) as groundwater migrates from the treatment area into untreated sediments under more reducing conditions.

Other observations in groundwater at MW-1 include (a) dissolved manganese concentrations decreased from 19 mg/L pre-treatment to 8 mg/L 12 weeks post-treatment, and (b) dissolved potassium concentrations increased from 210 mg/L pre-treatment to 330 mg/L 12 weeks post-treatment.

CONCLUSIONS

The Site where this ICO test was performed is on the very difficult end of the spectrum of sites at which any *in situ* remediation might be attempted. The massive clay sediments presented a formidable challenge in delivering the oxidant, and in the end the oxidant delivery was not as effective as would be needed to remediate the Site. Treatment was effective in localized areas where oxidant was effectively distributed, but overall the treatment did not meet remedial objectives. Other oxidant delivery methods, possibly including *in situ* soil mixing or closely-spaced solid-phase $KMnO_4$ placement (i.e., relying on diffusive transport from the solid phase), may achieve success at these types of sites. Clearly, oxidant delivery is a primary determinant of treatment success.

The ICO treatment at this Site resulted in increased concentrations of some dissolved metals in groundwater, including hexavalent chromium and selenium, as well as the transient formation of trihalomethanes. However, ICO treatment was approved at this Site based in part on the recognition that such "side effects," should they occur, would not affect human health risks at a Site where groundwater is not used for water supply, and that the occurrence of oxidized metals may be short-lived relative to the anticipated decades-long lifespan of the DNAPLs at the Site. Further, the destruction of sufficient TCE and VC by ICO treatment would significantly reduce potential future human health risks that might otherwise exist if a VOC vapor migration and inhalation exposure pathway was complete at the Site. Therefore, although results from field testing demonstrated increased concentrations of hexavalent chromium and other metals in groundwater, ICO treatment technology can still be an important tool for risk-based corrective action at Sites where oxidant can be effectively delivered.

Overall, the ICO testing at this Site demonstrated (a) private sector and regulatory acceptance of the technology, even when considering potential "side effects", (b) the potential for successful treatment in situations where the oxidant can be effectively distributed, (c) the need for better techniques at delivering oxidant solutions to fine-grained media, and (d) the occurrence of oxidized inorganic compounds and THMs following treatment.

REFERENCES

Siegrist, R.L., M.A. Urynowicz, and O.R. West. 1999. "An Overview of In Situ Chemical Oxidation Technology Features and Applications." *Proceedings of the Conf. on Abiotic In-Situ Technologies for Groundwater Remediation.* August 31-September 2, 1999, Dallas Texas. U.S. EPA Center for Environmental Research Information. Cincinnati, Ohio.

IN SITU OXIDATION OF DNAPL USING PERMANGANATE: IDC CAPE CANAVERAL DEMONSTRATION

Ernest Mott-Smith, P.E, (IT Corporation, Tampa, Florida)
Wendy C. Leonard, P.G. (IT Corporation, Miami, Florida)
Richard Lewis, C.P.G, (IT Corporation, Norwood, Massachusetts)
Wilson S. Clayton, Ph.D., (IT Corporation, Englewood, Colorado)
Jorge Ramirez, E.I.T (IT Corporation, Tampa, Florida)
Richard Brown, Ph.D., (Environmental Resource Management, Ewing, New Jersey)

ABSTRACT: Sites contaminated with chlorinated solvents are a major problem for the United States Department of Energy (DOE) and Department of Defense (DOD). As a result, DOE and DOD, along with the USEPA and NASA formed the Interagency DNAPL Consortium (IDC) to jointly fund a large scale field demonstration of chlorinated solvent remediation technologies, at NASA's Launch Complex 34, located at Cape Canaveral, Florida. One technology demonstrated at Cape Canaveral is *in situ* chemical oxidation using potassium permanganate ($KMnO_4$). This paper presents the procedures and results of the permanganate demonstration project, which is taking place from August 1999 through March 2000.

The primary objectives of the permanganate demonstration project were to document the field-scale performance of the technology under realistic field conditions, and develop reliable cost and performance data for remediation of dissolved, sorbed, and residual DNAPL trichloroethylene (TCE) contamination. TCE was present before treatment at up to 19,000 mg/kg in soil, and 1,500 mg/l in groundwater (approximately 6,100 kg. of TCE).

The results to date indicate that the delivery approach is on target to reach the goal of 90% reduction of TCE. A cost and performance model is being developed to provide a tool for future permanganate project screening and design.

INTRODUCTION

The Interagency DNAPL Consortium (IDC) is sponsoring the demonstration of three innovative technologies for remediation of trichloroethylene (TCE) dense non-aqueous phase liquid (DNAPL) at Launch Complex 34 at Cape Canaveral Air Station. The three technologies being demonstrated are *in situ* chemical oxidation using potassium permanganate ($KMnO_4$), Six-Phase Heating, and Steam Injection. The three technologies are being implemented at three adjacent 15 by 23 m test cells. IT Corporation was awarded the contract to design, install and operate the field demonstration for *in situ* chemical oxidation using potassium permanganate.

The primary objectives of the technology demonstrations are to document the field-scale performance of each technology under realistic field conditions, and develop reliable cost and performance data for use at other sites. The target cleanup goal for the demonstration was 90% reduction in TCE levels.

Chemical oxidation offers the following advantages over other treatment technologies: it is a destructive technology; it has a short time frame of treatment (hours to months); it has minimal energy and equipment requirements; it does not require vapor phase treatment; and it generates minimal waste. Many oxidants are currently being examined for treatment of DNAPLs. These include: ozone, H_2O_2, Fenton's reagent, $NaMnO_4$, $KMnO_4$ and oxygen. Potassium permanganate was selected due to its relative stability and persistence, ease in handling, relatively low cost (~$3.00/kg), and the ability to visually see results of the application.

$KMnO_4$ has been widely used for over 40 years by the drinking water, wastewater and chemical manufacturing industries. The reaction chemistry is well known. It works over a wide pH range and does not require an additional catalyst as with Fenton's reagent. It reacts relatively quickly without the generation of heat, and completely oxidizes a wide range of common recalcitrant organic contaminants including: TCE, PCE, DCE isomers, vinyl chloride and phenols. With chlorinated ethenes such as TCE it reacts with the double bond and ultimately breaks down the compound into carbon dioxide, MnO_2, potassium ions, chloride ions and H^+. The reaction pathways and transient intermediate products identified by Yan and Schwartz (1998) include carboxylic acids, primarily formic and oxalic acid.

Groundwater subjected to permanganate treatment can be visually observed to change color from purple to brown to clear upon complete oxidation (the soluble permanganate ion is purple, and MnO_2 forms a brown colloid that ultimately settles out of solution). This makes visual observation an easy way to qualitatively monitor treatment.

The overall effectiveness of *in situ* permanganate treatment is a function of the reaction kinetics, the transport and contact between $KMnO_4$ and the contaminant, and competitive reactions with other oxidizable species (e.g. iron, natural organics). The effective use of this technology lies in creating an engineered approach for maximizing the contact between $KMnO_4$ and the contaminant being oxidized. Therefore, a good understanding of the site hydrogeology and contaminant mass distribution is essential for an efficient treatment design.

SITE CONDITIONS

The LC-34 site was used as a launch site for Apollo rockets from 1960 to 1968. TCE was historically used for rocket engine flushing, metal cleaning and equipment degreasing. Figure 1 shows a cross section through the oxidation cell. Site lithology varies from sandy clay to shell hash. Three primary lithologic units have been defined: (1) the Upper Sand Unit (USU), which consists of fine sands; (2) the Middle Fine Grained Unit (MFGU), which consists of silty fine sand with lenses of sandy clay; and (3) the Lower Sand Unit (LSU), which consists of interbeded layers of shell hash, fine sand, silty fine sand, and sandy clay. The boundaries between these three zones occur at 7.6 m and 10.7 m below grade in the oxidation cell, and the LSU is underlain by clay at 13.7 m. The average hydraulic conductivity values of the units ranges from 1.8 to 4.6 ft/day (0.55 to 1.4 m/day) (Battelle 9/99); however the hydraulic conductivity of individual lenses within the units varies significantly due to shell hash lenses and silty clay stringers.

The mass distribution of TCE was defined by baseline sampling (Battelle 9/99). The estimated mass of TCE in the USU, MFGU, and LSU was 846 kg, 1,048 kg, and 4,228 kg, respectively (6,122 kg total). TCE contamination starts at 4.6 m below grade in the USU, and the overall vertical TCE profile is highly influenced by the layered lithology. The areal

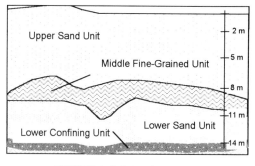

FIGURE 1. LC-34 Lithology

distribution is also variable across the cell, with the highest TCE mass located adjacent to suspected historical TCE sources.

DESIGN BASIS

A bench scale test was run to derive the usage rate of $KMnO_4$ for the contaminant mass and background (matrix) demand of the soil. The results of this test indicated that a 5x use factor be applied to the 2.41 stoichiometric ratio of $KMnO_4$ to TCE. The resultant use ratio used for the design basis was 12 kg $KMnO_4$ /kg TCE. This ratio is both contaminant and site specific.

SYSTEM DESIGN

At this site, a direct push pressure injection (i.e. lance permeation) was selected for solution injection. This system was designed to deliver permanganate at precise dosages over 0.6 m vertical intervals. This design strategy was chosen based on the fact that formation permeabilities and contaminant mass distribution vary both laterally and vertically. The permanganate dose and application rate and duration were adjusted at each point to reflect the corresponding level of contaminant mass at that location and vertical interval. Initially, eleven separate injection rods and injection tips were driven to the desired depths by a direct

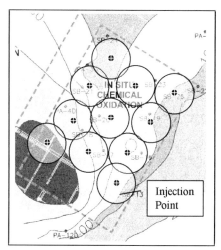

FIGURE 2. Injection Layout

push rig. Figure 2 shows the injection network. The injection tip consisted of a 0.6 m long perforated drive stem, with ¼- inch (6.35 mm) diameter holes and a 0.010-inch (0.25 mm) slot continuous wire wound stainless steel screen (Figure 3). The permanganate solution was mixed using an automated portable bulk feed system developed by Carus Chemical Company. This system allowed the

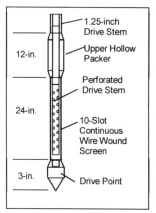

FIGURE 3. Injection Tip

permanganate concentration and flow rate to be automatically produced and controlled using a PLC (programmable logic controller) unit and minimal operator oversight. The permanganate was delivered to the site via bulk shipments, of approximately 20,411 kg of free-flowing grade and transferred into a bulk feed silo (Figure 4). From there the KMnO$_4$ was conveyed to the hopper via a rotary screw where it gravity fed into the helix feeder screw that metered the material into the mix tank. Maximum throughput of the system was 159 kg/hr of permanganate. The solution was flash mixed to the programmed concentration from 0.1% to 3% by weight with make up water supplied by an on-site fire hydrant. A high-pressure dual chemical feed pump and injection head manifold setup was used to convey KMnO$_4$ from the mix tank to the injection points. A sand filter was installed between the mix tank and injection manifolds to remove the 1% quartz sand component of the free flowing permanganate before it entered the injection screen. Each manifold line contained a totalizer flow meter, pressure gauge, ball valve, check valve, and PVC hosing. The flow rates per point ranged from 0.1 to 6.1 gpm (0.006 – 0.4 L/s) at wellhead pressures of 15-55 psig (103 – 379 kPa). A process flow diagram is shown in Figure 5.

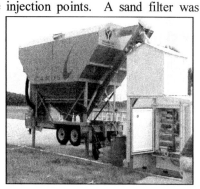

FIGURE 4. CARUS KMnO$_4$ Feed System

MONITORING METHODS

Twelve multi-level clustered monitoring well locations within the cell were used to monitor the injected fluid and treatment results. A total of 46 discrete sample intervals existed within the cell with an additional 15 discrete sample intervals from 5 perimeter wells outside the cell. Soil samples were collected at various times during treatment to provide samples for visual and analytical evaluation. Analytical techniques employed included an on-site mobile lab to analyze fluoride, MnO$_4^-$ and VOCs from groundwater and soil samples.

Field parameters of temperature, dissolved oxygen, oxidation/reduction potential, specific conductivity, and pH were recorded using a Hydrolab Minisonde 4 Multiprobe with a Hydrolab Surveyor 4 Data Display (Figure 6). The

FIGURE 5. Multi-probe Meter

multiprobe was deployed downhole in the wells with a diameter of 5 cm or larger. In smaller diameter wells, a flow-through cell was adapted to the multiprobe. Color was also monitored visually and quantified using a Hach model DR/2010 portable datalogging spectrophotometer. The readings from the spectrophotometer were plotted against a calibration curve developed using potassium permanganate standards, to obtain the $KMnO_4$ concentration.

Treatment performance monitoring was performed by Battelle, and included VOC analysis from all site monitoring wells, as well other parameters, such as iron, manganese, chloride, total dissolved solids and trace metals. IT had supplemental soil and groundwater VOC samples collected and analyzed using the on-site geoprobe rig used for the injection and the on-site mobile lab. As the treatment demonstration is not complete, the post-demonstration cell wide performance monitoring has not yet been performed.

Operational monitoring consisted of tracking the total system and individual injection point pressures, injection concentrations, injection flow rates and duration of permanganate injection in order to compare actual vs. target injection rates per each 0.6 m injection interval. The spectrophotometer was used to quantify the permanganate concentration and to verify the concentration from the feed system. An interactive Excel spreadsheet was used to update this information on a daily basis and to determine the future injection schedule.

TREATMENT OPERATIONS

The initial treatment operations consisted of a tracer test performed at injection point IP-3 in order to determine the effectiveness of the lance permeation approach with respect to *in situ* permanganate transport and reaction rates. A sodium fluoride tracer (2% solution), and permanganate solution (624 kg of a 1.5 –2% solution) were injected into a (3 m) interval of the MFGU for a three day period. The tracer test results (Clayton et. al, 2000) determined that the lance permeation, without groundwater pumping, was adequate to transport the injected fluid over a practical effective radius. The tracer test also demonstrated the significant consumptive use of $KMnO_4$ by the DNAPL phase. Other design basis parameters, such as the injection flow rate and pressure, were confirmed by this test.

The *in situ* permanganate treatment was conducted in three phases, two of which were competed as of this writing. The treatment was divided into phases so that interim results obtained could be used to focus subsequent injections into areas that required further treatment. The first full cell injection was initiated on September 8 and ended October 29. This phase extended across the entire treatment zone (4.6 - 13.7 m below grade) at eleven locations (refer to Figure 2). The second phase ran from November 17 – 24, and treatment was focused on specific areas that required further treatment. The third phase of treatment is scheduled to be conducted during March 2000. A total of 40 days of injection has occurred through the second injection phase, with approximately 35,340 kg of permanganate injected. The amount injected is approximately 50% of the target amount needed to reach the 90% TCE reduction goal. 1,911 kg have been injected in the USU (108 % of target). 10,448 kg have been injected in the MFGU

(83% of target), and 13,904 kg have been injected in the LSU (27% of target). The LSU did not receive injection during the second phase, and will be a primary target for the final treatment phase.

RESULTS

The effectiveness monitoring results were evaluated according to lithologic units (USU, MFGU and LSU). The TCE concentrations in the groundwater collected in the USU showed reductions of > 99% in 7 of the 11 wells sampled. The remaining wells showed greater than 62% TCE reduction. Purple color was observed in all the cell wells. Soil samples in the USU showed greater than 90% TCE reductions (Figure 7). Table 1 shows the analytical soil sampling results for each zone for Week 8 (Week 12 for LSU). Table 2 summarizes the groundwater analytical data table for each zone. Figure 8 presents the contour map of %TCE reduction in the USU, MFGU, and LSU.

The TCE % reductions observed in the MFGU groundwater samples indicated that 7 of 15 wells showed greater than a 90% TCE decrease. All of

these water samples exhibited a purple color. The other reductions ranged from 4.3% to 88.6%. The soil from ·the MFGU showed TCE reductions of greater than 80% in 7 of 11 samples.

The %TCE reductions observed in the LSU groundwater samples indicated reductions ranging from -14% to 95% with 4 of 15 wells showing greater than 70% reductions. Because only 27% of the design target for permanganate has been added to this

FIGURE 6. Purple-Stained Soil Cores

unit, and the soil data is not as reliable as the other units, additional injection and sampling will be needed to document the effectiveness. Field parameter monitoring was useful in identifying the permanganate front. An increase in specific conductivity was the first indicator of the permanganate front, followed by an increase in ORP, followed by a sustained brown, then purple color.

SUMMARY AND CONCLUSIONS

Final soil sampling and mass estimate comparisons across the entire cell in all units will be performed at the end of this next injection to document the final system effectiveness. While the final results are not available, some conclusions can clearly be drawn from the data:

- Significant TCE mass destruction was achieved.
- TCE concentrations in groundwater are a good indicator of treatment effectiveness, but are very dynamic during treatment. These results can reflect transient TCE decreases due to mass destruction and localized displacement, as well as transient TCE rebound due to chemical re-equilibration and localized displacement.

- TCE soil data is the most reliable to document mass destruction, but may not always reflect an overall improvement in short-term groundwater quality.
- No hazardous chemical by-products were generated by the reaction. Evaluation of the impact on secondary drinking water standard constituents such Mn, Cl, and salts is ongoing.
- Significant formation plugging was not evident. The injection flowrates remained decreased less than 10% through the first two phases.
- Lance permeation injection provides the advantage of targeting permanganate dosages to specific intervals and minimizes capital and operating costs, but may provide more limited mixing of injected solutions.
- A cost model is being developed which will incorporate the site demonstration cost and effectiveness to allow for prediction to other sites.
- Field parameter monitoring of specific conductivity, ORP and color were useful in identifying the permanganate front.

TABLE 1. Soil Analytical Results

Sample ID	Depth (ft. bls)	Baseline TCE ppb (ug/Kg)	TCE ppb (ug/Kg)	Percent Reduction TCE	Color Present
B1	15-19	400	<20	95.00%	Purple
B1 ⊙	17-19	540	<20	96.30%	Purple
B1	20-21	940	419	55.43%	None
B2	20-21	19,150	5,870	69.35%	Brown
B3	20-21	30,550	24	99.92%	Purple
B4	20-21	48,000	42	99.91%	Purple
B5	20-21	10,430	138	98.68%	Brown
B6	19-20	9,000	<20	99.78%	Purple
B6	22-23	15,000	<20	99.87%	Brown
B1 ⊙	24-27	110,000	22,000	80.00%	Brown
B3	25-26	104,500	43,400	58.47%	Brown
B6	28-29	170,000	126	99.93%	No
B1	29-30	180,000	242,500	-34.72%	No
B2	30-31	120,000	15,300,000	-12650%	No
B3	30-31	160,000	1,040	99.35%	No
B4	30-31	315,000	57,100	81.87%	Brown
B5	30-31	11,000,000	1,810,000	83.55%	Brown
B6	32-33	200,000	<20	99.99%	Purple
B3	34.5-35	120,000	80	99.93%	Brown
B5	34-35	350,000	236,000	32.57%	Brown
B1 ⊙	35-39	119,000	14,000	88.24%	Brown
B5 ⊙	38-39	10,000,000	15,400	99.85%	Brown
B5 ⊙	40-41	6,800,000	10,000	99.85%	Brown/Gray
B5 ⊙	41-43	1,800,000	25,200	98.60%	Brown
B1	43-45	86,000	78,600	8.60%	Brown
B5 ⊙	44-45	2,300,000	16,600	99.28%	Brown/Gray

Note:
Calculated TCE reductions are based on soil borings collected at close locations from Pre-demo soil boring locations.
⊙ = Semi-quantitative results (exceeded sample hold times, samples frozen)

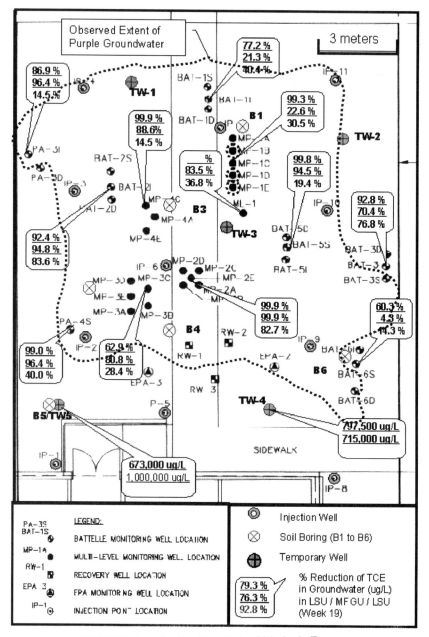

FIGURE 7. Oxidation Results by Lithologic Zone

TABLE 2. Groundwater Analytical Results

Sample ID	Baseline	Week 3 TCE (ug/L)	Week 8 TCE (ug/L)	Week 19 TCE (ug/L)	TCE % reduction	Color
Interior Wells - Upper Saturated Unit						
BAT-1S	1,142,500	940,000	1,010,000	260,000	77.2%	Purple/Br
BAT-2S	1,112,500	14,000	457	84,600	92.4%	Purple/Br
BAT-3S	1,105,000	229,000	262,000	79,400	92.8%	Purple/Br
BAT-5S	297,500	47,750	1,240,000	555	99.8%	Purple/Cl
BAT-6S	1,087,500	122	1,990	432,000	60.3%	Purple/Br
ML-1-19	61,750	203,000	Dry	ns		Purple
MP-1A	777,500	6,490	630,000	5,420	99.3%	Purple
MP-2A	427,500	5,050	<2	180	99.96%	Purple
MP-3A	515,000	36	<2	191,000	62.9%	Purple/Cl
MP-4A	745,000	0	<2	176	99.98%	Purple/Cl
PA-3S	652,500	950,000	580,000	85,800	86.9%	Purple/Cl
PA-4S	690,000	0	<2	7,070	99.0%	Purple/Br
Perimeter Wells - Upper Saturated Unit						
PA-5S	197	84,500	9,600	750,000	-380611%	Purple/Cl
PA-6S	290	993	10,800	68,400	-23486%	
PA-8S	5,730	15,300	115,000	79,300	-1284%	
PA-9S	790,000	1,095,000	1,200,000	1,060,000	-34.2%	
PA-12S	482,500	870,000	1,240,000	ns		
Interior Wells - Middle Fine Grained Unit						
BAT-1I	1,055,000		360,000	830,000	21.3%	Purple/Br
BAT-2I	970,000		68,800	50,000	94.8%	Purple/Br
BAT-3I	990,000		1,060,000	293,000	70.4%	Purple/Br
BAT-5I	867,500		985,000	47,800	94.5%	Purple/Cl
BAT-5I-D	897,500			955,000	-6.4%	Purple/Br
BAT-6I	997,500		42,500	9,850	99.0%	
ML-1-25	982,500	1,010,000	545,000	240,000	75.6%	
ML-1-31	750,000		595,000	775,000	-3.3%	
MP-1B	877,500		965,000	540,000	38.5%	
MP-1C	812,500		590,000			Purple/Cl
MP-2B	760,000		290	17	99.998%	Purple/Cl
MP-2C	695,000		265	49,700	92.8%	Purple/Cl
MP-3B	800,000		60,000	247,000	69.1%	Purple/Br
MP-3C	767,500		9	92,200	88.0%	Purple/Br
MP-4C	810,000		2,980	39,300	95.1%	Purple/Br
PA-3I	1,097,500	1,150,000	600,000	42,500	96.1%	Purple/Br
PA-4I	1,192,500		274			
Perimeter Wells - Middle Fine Grained Unit						
PA-5I	17,250	71,000	114,000	670,000	-3784.1%	Purple/Cl
PA-6I	1,007,500	1,050,000	1,280,000	955,000	5.2%	Clear
PA-8I	987,500	1,040,000	1,000,000	805,000	18.5%	Clear
PA-9I	967,500	1,035,000	900,000	790,000	18.3%	Clear
PA-12I	1,035,000	1,210,000	1,320,000			Clear
Interior Wells - Lower Sand Unit		**Week 11**				
BAT-1D	1,132,500		610,000	675,000	40.4%	Purple/Br

Br = color changes from purple to brown when purged, Cl = clears when purged

TABLE 2. Groundwater Analytical Results (Continued)

Sample ID	Baseline	Week 3 TCE (ug/L)	Week 8 TCE (ug/L)	Week 19 TCE (ug/L)	TCE % reduction	Color
Interior Wells - Lower Sand Unit			Week 11			
BAT-2D	1,160,000		835,000	190,000	83.6%	Purple/Br
BAT-3D	962,500		94,200	223,000	76.8%	
BAT-5D	1,135,000		730,000	915,000	19.4%	Purple/Cl
BAT-6D	752,500		164,000	860,000	-14.3%	Purple/Br
ML-1-37	595,000		610,000	273,000	54.1%	Purple/Br
ML-1-43	435,000		775,000	350,000	19.5%	Purple/Br
MP-1D	607,500		603,000	484,000	20.3%	
MP-1E	627,500		965,000	372,000	40.7%	
MP-2D	635,000		1,300	190,000	70.1%	Purple/Cl
MP-2E	622,500		2,640	29,700	95.2%	Purple/Cl
MP-3D	527,500		127,000	432,000	18.1%	Purple/Cl
MP-3E	557,500		420,000	341,000	38.8%	Purple/Cl
MP-4E	830,000		338,000	710,000	14.5%	Purple/Cl
PA-3D	1,082,500	1,130,000	585,000	650,000	40.0%	Purple/Cl
PA-4D	1,157,500		1,050,000			
Perimeter Wells - Lower Sand Unit						
PA-5D	183,250	170,500	258,000	570,000	-211.1%	Clear
PA-6D	987,500	405,500	665,000	860,000	12.9%	Clear
PA-8D	477,500	625,000	900,000	960,000	-101.0%	
PA-9D	287,500	295,000	400,000	580,000	-101.7%	Clear
PA-12D	565,000	685,000	945,000			Clear

Br = color changes from purple to brown when purged, Cl = clears when purged

- Displacement of TCE contaminated groundwater up to 28 feet from the oxidation plot was observed. This displacement is primarily a result of the requirement to treat up to the cell boundary. However, displacement of contaminated groundwater can be contained within the treatment zone for a full-scale application if necessary.
- The total project cost is estimated to be $978,000 upon completion (including all costs except performance monitoring by Battelle and IDC oversight costs). The unit cost for *in situ* soil treatment is $237/yd^3 ($307/m^3). The project cost per kg of TCE is expected to be approximately $177/kg of TCE (assuming 90% reduction is obtained). These costs reflect the research and demonstration nature for the project. Final unit cost factors will be prepared at the project completion to represent the technology cost both with and without the additional costs related to the demonstration project.

REFERENCES

Battelle. September 1999. Pre-Demonstration Assessment of the Treatment Plots at Launch Complex 34, Cape Canaveral, Florida, Part 1: Soil Analysis Results and Field Measurements.

Clayton, W. 1999. *A Multisite Field Performance Evaluation of In Situ Oxidation Using Permanganate.* 2nd International Conference on Remediation of Chlorinated and Recalcitrant Compounds

PASSIVE DESTRUCTION OF PCE DNAPL BY POTASSIUM PERMANGANATE IN A SANDY AQUIFER

Matthew D. Nelson, Beth L. Parker and John A. Cherry (Department of Earth Sciences, University of Waterloo, Canada), Tom Al (Department of Geology, University of New Brunswick, Canada)

ABSTRACT: A drive-point tool for injecting reactive fluids into permeable sandy aquifers was developed and tested for in-situ remediation of contamination from a previous PCE DNAPL experiment in a shallow unconfined sand aquifer. The goal of this approach is to inject aqueous reactive chemicals such as potassium permanganate ($KMnO_4$) under moderate pressures (400-600 kPa) over a short time period, targeted at multiple depths in a borehole to create stacked coalescing treatment zones. Additional coverage is accomplished using multiple boreholes. The injections may be repeated periodically at intervals of 2 to 3 months until the desired degree of in situ treatment is achieved. Two separate injection episodes were conducted in which 14.7 kg of $KMnO_4$ were delivered in 860 L of solution. The monitoring showed that the distribution of $KMnO_4$ reached over 1.37 meters away from the injection points and that most of the consumption of $KMnO_4$ occurs within 2 months. Results suggest that the direct-push tool is time-efficient and effective for distributing reactive solutions. Diffusion and density driven advection, both sinking and spreading along lower permeability layers, further enhance the ability to deliver $KMnO_4$ into zones where DNAPL persists.

INTRODUCTION

Persistent plumes of dissolved chlorinated solvent contamination in sandy aquifers are common due to subsurface accumulations of dense non-aqueous phase liquids (DNAPLs). For permanent aquifer restoration the DNAPL must be removed or destroyed in-situ. One approach for in-situ destruction of chlorinated ethenes is chemical oxidation using potassium permanganate ($KMnO_4$), which was first studied by Schnarr and Farquhar (1992). Schnarr et al. (1998) showed the effectiveness of $KMnO_4$ for destruction of tetrachloroethylene (PCE) and trichloroethylene (TCE) in batch and column experiments using DNAPL mixed in sand, and in field experiments at Canadian Forces Base Borden, Ontario. Other investigators reported encouraging results based on field trials (Hood et al., 1997; McKay et al., 1998; Siegreist et al., 1999).

For sandy aquifers and some types of aquitards, an alternative to the use of recirculating forced-advection for the delivery of dissolved $KMnO_4$ to DNAPL zones is the inject-and-leave approach (Parker and Cherry, 1997). In this method, a high concentration $KMnO_4$ solution is injected rapidly into the DNAPL zone followed by a long period during which the $KMnO_4$ redistributes with no repeated injection until the oxidation by $KMnO_4$ ceases. For this study a direct-push drive-point tool for $KMnO_4$ injection at discrete depths was developed to deliver the $KMnO_4$ to useful distances radially from the injection point in the Borden aquifer.

The 9 x 9 m sheet pile cell used in this study at Canadian Forces Base Borden, was utilized previously for a geophysical experiment to evaluate the effectiveness of several different geophysical techniques for observing the movement and distribution of dense immiscible-phase organic solvents in the subsurface (Brewster et al., 1995). In 1991, 770 L of PCE, dyed red using the hydrophobic dye Sudan IV, were released in the 9 x 9 m cell. The DNAPL formed layers distributed from 1 m depth to the bottom of the aquifer at 3.5 m. In subsequent months, 425 L were removed by pumping 5 cm diameter wells at the bottom of the aquifer (Morrison, 1998). Soil vapor extraction (Flynn, 1994) and air-sparging (Tomlinson, 1999) removed part of the PCE in the aquifer, however an estimated 185-275 L of DNAPL remain inside the cell, primarily in the bottom meter of the aquifer and at the aquifer-aquitard interface (Morrison, 1998; Nelson, 1999).

The layered nature of DNAPL distribution makes many remediation schemes ineffective or inefficient because treatment chemicals transported by advection can circum-navigate lower permeability zones where much of the mass is present. Clean-up times will be long because mass transfer rates of contaminant and oxidants will control the time frame for destruction. The Drive Point Delivery System (DPDS) was designed to deliver a reactive chemical solution close to these discrete DNAPL layers within the aquifer. The goal is to inject the chemical reactant at targeted depths to distribute the $KMnO_4$ solution in discs that radiate away from the injection hole into the DNAPL zones. After the injection, diffusion and density-driven advection of the $KMnO_4$ solution provides more contact between the reactive chemical and the DNAPL mass.

To achieve maximum solvent mass destruction from each injection episode, the injected solution must be at or near the $KMnO_4$ solubility limit of 64 g/L at 20°C (Perry et al., 1984). A $KMnO_4$ solution at or near the solubility limit has a density relative to distilled water of approximately 1.03. Schincariol and Schwartz (1990) used laboratory sand tank experiments with NaCl solutions with relative densities in the range of 1.001 to 1.02 to study density driven advection. In a tank with layered permeable media they found that changes in hydraulic conductivity from K_1=2.5 x 10^{-5} to K_2=1.4 x 10^{-5} m/s, caused a 10 000 mg/L solution of NaCl to flow through the higher permeability layer then mound on the lower permeability layer. By analogy to these experiments, it is expected that high concentration $KMnO_4$ solutions, 25 to 50% $KMnO_4$ saturation with densities in the range of 1.01 to 1.02, will sink and have a strong tendency to spread laterally along the top of sediment layers with slightly lower permeability. In the 9 x 9 cell the hydraulic conductivity of the aquifer ranges between 4 x 10^{-3} to 4 x 10^{-7} m/s.

The second transport mechanism relied upon is aqueous diffusion of the reactants. Rapid solvent destruction rates ensure high concentration gradients to drive diffusion at the $KMnO_4$-contaminant interface. In the field experiment presented here, the changes in $KMnO_4$ distribution caused by the combined influences of density and diffusion were monitored using a dense network of multi-level groundwater samplers and subsamples from continuous cores. The aquifer is fully saturated however no groundwater flow occurs inside the sheet pile cell except as a result of the injection of the dense aqueous $KMnO_4$ solution.

MATERIALS AND METHODS

Two $KMnO_4$ injection episodes were conducted in the unconfined Borden sand aquifer contained within a 9 x 9 m sheet pile cell keyed into an underlying clay aquitard, one in November 1997 and the second in July 1998 (Fig. 1). The injections were accomplished by driving a 7 cm long stainless steel, wire-wrapped well screen fitted to AW drill rod and connected to a 27 L pressure vessel (the Drive-Point Delivery System, herein referred to as the DPDS). The injections were made in a small experimental volume of the cell approximately 1.5 x 3 m in plan view situated in the bottom meter of the aquifer, 2.35 to 3.35 m depth. Two injection locations were used in each episode. In the first injection episode injections occurred at two depths at each location and in the second injection episode injections occurred at three depths at each location.

During the first injection episode, 5.7 kg of $KMnO_4$ were injected in 427 L of solution. The average injection rate was 3.8 L per minute at 470 kPa. The injection pressures varied between 380 and 515 kPa and rates between 3.5 and 4.5 L per minute. In the first hole, INJ-1, 99 L were released at 2.70 m depth, and 103 L at 3.30 m depth. In the second hole, INJ-2, 84 L were released at 2.70 m depth, and 98 L at 3.30 m depth. The rest of the 43 L of fluid was injected during the pushing of the injection tip to desired depths. The injection took approximately 8 hours including equipment setup and disassembly.

During the second injection episode, which took 7 hours, 9 kg of $KMnO_4$ were delivered in 429 L at an average injection rate of 3.7 L per minute, with pressures between 430 and 590 kPa. As in the first experiment two injection holes were used within 30 cm from the first injection locations. In the first hole, INJ-3, 104 L were released at 2.63 m depth, 90 L at 3.15 m depth, and 21 L at 3.35 m depth. In the second hole, INJ-4, 105 L were released at 2.50 m depth, 82 L at 3.15 m depth, and 39 L at 3.30 m depth.

Groundwater samples were collected from a detailed 3D sampling grid of multi-level samplers constructed out of 3.18 mm (1/8") tubing and analyzed for $KMnO_4$, chlorinated ethenes (PCE, TCE, and DCE), pH, Eh, and conductivity. Continuous cores were taken at 23 locations in 1.5 m long, 5 cm diameter aluminum tubes using a piston coring device (Starr and Ingleton, 1992). The cores were cut longitudinally and sampled at specific depths based on visual inspection for PCE and MnO_2 occurrence.

RESULTS AND DISCUSSION

Each injection episode created 4 $KMnO_4$ ellipsoids, two in the upper aquifer and two at the aquifer-aquitard interface. The ellipsoids in the upper aquifer had a diameter to thickness ratio of approximately 3. The ellipsoids created at the aquifer-aquitard interface had a larger diameter resulting in a diameter to thickness ratio of 4-5. The horizontally elongated nature of the KMnO$_4$ emplacements is most likely due to anisotropy in the aquifer hydraulic conductivity (K$_{horizaontal}$ > K$_{vertical}$). Falling head tests on repacked core segments, 5 cm long, show only minor K variability (1x10^{-6} to 1x10^{-4} m/s) inside the cell (Fig. 1), however horizontal beds are visible at the mm to cm scale illustrating the scale of hydraulic conductivity variability.

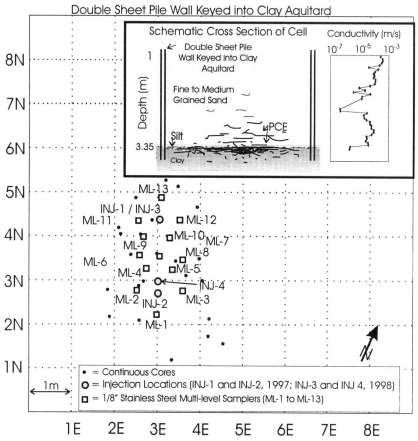

FIGURE 1. Plan and cross-section views of the 9 x 9 m cell and layout of the KMnO$_4$ experimental area. Each multi-level sampler is constructed of 1/8" diameter stainless steel tubes terminating at 11 depths vertically spaced 10 cm apart. INJ 1 and 2 were used in KMnO$_4$ injection episode 1 and INJ 3 and 4 in episode 2. A typical profile of permeability determined at a 5-10 cm sample interval by falling head permeameter shows the layering of the Borden aquifer.

Groundwater sampling showed post-injection movement of KMnO$_4$ in the aquifer. Spreading of KMnO$_4$ was evident in 5 multi-level samplers and was indicated by the arrival of KMnO$_4$ in some multi-level samplers at late time. In bundles 6, 8, 9, 10, and 12, KMnO$_4$ was absent at all sample depths until late time when it appeared at the 3.05 and 3.15 meter depths, the depth of the sand/silt interface. The KMnO$_4$ on the aquifer-aquitard interface spread laterally away from the injection hole locations with time (Fig. 2). The maximum coverage was at 23 days after the injections, coinciding with the arrival of KMnO$_4$ from the upper injection levels (Fig. 3). The spreading of KMnO$_4$ along the aquifer-aquitard interface enabled the treatment of soil up to a 1.37 m radius and with

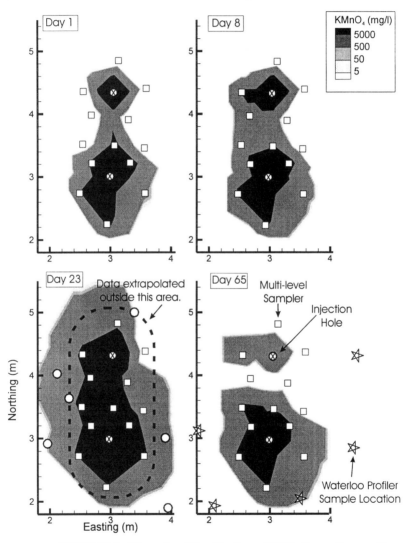

FIGURE 2. Plan view Distribution of KMnO$_4$ on the aquifer-aquitard interface (3.05 - 3.35 m depth) after the second injection episode. The contours are extrapolated beyond the multi-level samplers using data from soil cores (Day 101/102) and Waterloo Profiler (Day 92/93) results.

more complete coverage than in the overlying aquifer. Because prior to injection no groundwater flow occurred in the cell, the principal mechanism causing the post-injection migration of $KMnO_4$ on this time scale was density-driven advection.

As the $KMnO_4$ reacted with PCE, the Mn precipitated as Mn-oxide, and Cl^- was released from the destruction of PCE. The decline in MnO_4^- mass and Cl^- production of over time were measured in aqueous samples to track the reaction progress. By Day 85 MnO_4^- had decreased and chloride had increased from background levels of <10 mg/L to over 5000 mg/L which confirmed PCE destruction.

Cores taken after the injection episodes showed discrete layers of dark manganese precipitation at the mm to cm scale. These layers were most prominent, and appeared more consistently, above the silt layers at the aquifer-aquitard interface where layers of MnO_2 coatings ranged from 0.5 to 3 cm thick. These layers of extensive manganese precipitation reflect zones where PCE NAPL was initially present, indicating that the post-injection movement of $KMnO_4$ did contact DNAPL layers located above the finer-grained silt layers.

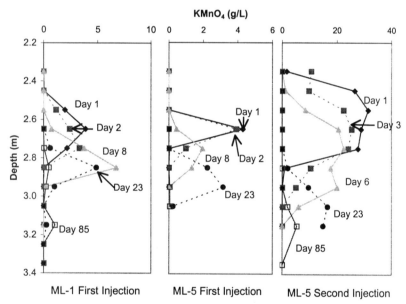

FIGURE 3. Evidence for $KMnO_4$ solution migration due to density effects based on downward progression of peak concentrations observed in multi-level samplers over time.

In the upper injection levels, in the absence of distinct silt or clay beds, the $KMnO_4$ migrated downward. In these zones the $KMnO_4$ ellipsoids descended through the aquifer gaining an additional 30 to 60 cm in diameter over a 60 cm vertical drop. Multi-level sampler profiles indicate the increase in depth of the peak concentration of $KMnO_4$ over time at a rate between 0.3 and 2.5 cm/d, with an average of 1.5 cm/d (Fig. 3). This density-driven advection maximizes the contact between KMnO₄ and the aquifer.

Schindcariol and Schwartz (1990) noted that the mounding of dense miscible fluids such as salt solutions on lower permeability units is similar to the behaviour of DNAPL. This post-injection re-distribution of KMnO₄, both downward migration and lateral spreading along lower permeability layers, allows the KMnO₄ to achieve contact with the DNAPL that resides at the bottoms of aquifers (Fig. 4).

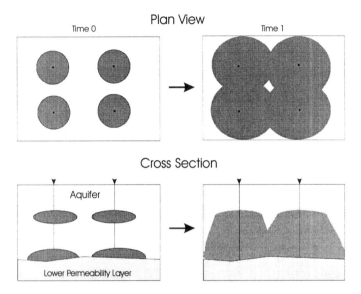

FIGURE 4. Conceptual representation of observations where solutions of KMnO₄ flow down and along lower permeability beds due to density-driven advection targeting the areas where DNAPL is most likely to exist.

CONCLUSIONS

During each injection episode 430 L of KMnO₄ solution were delivered to targeted areas over a period of 8 hours. The injections resulted in elliptical or disc-like emplacements of KMnO₄ that reached 1.4 m laterally from the injection points. Density flow caused the downward migration of KMnO₄ through the sand and lateral spreading along the silt layers at the aquitard-aquifer interface. This lateral spreading of KMnO₄ increased the volume of aquifer treated, especially along the silty beds. The redistribution of KMnO₄ resulted in treatment of a cylindrical volume of sand that extended 1 m radially from each injection hole.

The redistribution of the dense miscible KMnO₄ by density-driven advection during passive conditions is advantageous because it allows the KMnO₄ to contact the DNAPL situated along lower permeability layers at the bottom of the aquifer. Therefore, this injection method accomplished the delivery of KMnO₄ to targeted zones using minimal equipment and avoids continuous energy input, system operation and maintenance. Consequently, the inject-and-leave approach can result in reduced remediation costs.

Acknowledgements

Many individuals have supported this work in the lab and field including: E. Acquarone, D. Bassett, M. Gorecka, H. Groenevelt, B. Ingleton, J. Ingleton, P. Johnson, T. Jung, K. Laukonen, D. Thomson, S. Turcotte, and C. Turner. Special thanks are extended to Paul Johnson, Bob Ingleton, and Hester Groenevelt, as this project would not have been possible without their knowledge and support.

Funding was provided by ICI and NSERC-IOR to B. Parker and J. Cherry, and a graduate student scholarship to M. Nelson from Geomatrix.

References

Brewster, M.L., Annan, A.P., Greenhouse, J.P., Kueper, B.H., Olhoeft, G.R., Redman, J.D., and Sander, K.A. 1995. "Observed Migration of a Controlled DNAPL Release by Geophysical Methods." *Ground Water.* v.33, no. 6: 977-987.

Flynn, D.J. 1994. *Soil Vacuum Extraction to Remove Residual and Pools of Perchloroethylene from a Heterogenous Sand Aquifer: A Field Trial.* M.A.Sc. Thesis, University of Waterloo, Waterloo, ON, Canada.

Hood, E.D., Thomsom, N.R., and Farquhar, G.J. 1997. "In Situ Oxidation: An Innovative Treatment Strategy to Remediate Trichloroethylene and Perchloroethylene DNAPLs in Porous Media." 6th Symposium on Groundwater and Soil Remediation, Montreal, Quebec. March, 1997.

McKay, D., Hewitt, A., Reitsma, S., LaChance, J., and Baker, R. 1998. "In-situ Oxidation of Trichloroethylene using Potassium Permanganate: Part 2. Pilot Study". Remediation of Chlorinated and Recalcitrant Compounds. Monterey, California. May, 1998.

Morrison, W.E. 1998. *Hydrogeological Controls on Flow and Fate of PCE DNAPL in a Fractured and Layered Clayey Aquitard: A Borden Experiment.* M.Sc. Thesis, University of Waterloo, Dept. of Earth Sciences, Waterloo, Ontario, Canada.

Nelson, M.D. 1999. *The Geochemical Reactions and Density Effects Resulting from the Injection of KMnO₄ for PCE DNAPL Oxidation in a Sandy Aquifer.* M.Sc. Thesis, University of Waterloo, Dept. of Earth Sciences, Waterloo, Ontario, Canada.

Parker, B.L., and Cherry, J.A. 1997. *Treatment of Contaminated Water in Clays and the Like.* US Patent no. 5 641 020.

Perry, R.H., Green, D.G., and Maloney, T.O. 1994. *Perry's Chemical Engineer's Handbook.* McGraw Hill, Toronto, ON., Canada.

Schincariol, R.A., and Schwartz, F.W. 1990. "An Experimental Investigation of Variable Density Flow and Mixing in Homogeneous and Heterogeneous Media." *Water Resources Research*, vol. 26, no. 10: 2317-2329.

Schnarr, M., and Farquhar, G. 1992. "An In-Situ Oxidation Technique to Destroy Residual DNAPL from Soil." Subsurface Restoration Conference, The Third Conference on Ground Water Quality. Dallas, Texas. June, 1992.

Schnarr, M., Traux, C., Farquhar, G., Hood, E., Gonullu, T., and Stickney, B. 1998. "Laboratory and controlled field experiments using potassium permanganate to remediate trichloroethylene and perchloroethylene DNAPLs in porous media." *Journal of Contaminant Hydrology*, vol. 29: 205-224.

Siegrest, R.L., Lowe, K.S., Murdoch, L.C., Case, T.L., and Pickering, D.A. 1999. "In Situ Oxidation by Fracture Emplaced Reactive Solids." *Journal of Environmental Engineering.* vol. 125, no. 5: 429-440.

Starr, R.C., and Ingleton, R.A. 1992. A new method for collecting core samples without a drilling rig. *Ground Water Monitoring Research*, vol. 12: 91-95.

Tomlinson. D.W. 1999. *Performance Assessment of In Situ Air Sparging within a Layered Sand Aquifer.* M.A.Sc Thesis, University of Waterloo, Dept. of Civil Engineering, Waterloo, Ontario, Canada.

PILOT-SCALE STUDY OF IN-SITU CHEMICAL OXIDATION OF TRICHLOROETHENE WITH SODIUM PERMANGANATE

Kun-Chang Huang, Pradeep Chheda and George E. Hoag (University of Connecticut, Storrs, Connecticut)

Bernard A. Woody and Gregory M. Dobbs (United Technologies Research Center, East Hartford, Connecticut)

ABSTRACT: This study indicates that the dissolved-phase TCE was rapidly oxidized by permanganate while the destruction of source free phase TCE appeared to be controlled by the rate of mass transfer of both permanganate and TCE toward their interface. The oxidation of dissolved-phase TCE by permanganate had caused a significant decrease in pH [e.g., from 6.5 (background) to 3.5 (at ~1 pore volume)] and an increase in chloride [e.g., from 83 mg/L (background) to 380 mg/L (at ~1 pore volume] in the extracted water in the beginning of the test. The downward migration of permanganate fronts near the aquitard provides evidences for the destruction of source free phase TCE. Two sets of soil cores collected near the aquitard at different times reveal that MnO_2 layers were formed above the source TCE zones and increased as the test proceeded.

INTRODUCTION

In the subsurface, trichloroethylene (TCE) dissolves in aqueous phase and partitions into gaseous and soil phases, in addition to existing as a free phase (i.e., exhibiting an ability to flow as a pure liquid under sufficient potentiometric head) or as a residual phase (i.e., existing as a pure phase but not able to flow as a pure liquid under normally encountered potentiometric heads). Because of its nature, pure TCE tends to be distributed in the saturated zone of the subsurface in a manner of fluids with multiphase flow phenomena, fingered and discontinuous or pooled accumulations at low permeability layers, such as a TCE-contaminated aquifer found in Connecticut.

Remediation of TCE-contaminated aquifers is a great challenge to environmental scientists and engineers. The literature contains few successful examples where actual free phase TCE in the saturated zone has been remediated to the extent that closure criteria have been met, or even, in many cases, where gross mass removal was achieved to any significant extent. In situ chemical oxidation with permanganate is an innovative technology for the remediation of TCE- and tetrachloroethylene (PCE)-contaminated aquifers. This remediation alternative has recently been studied through both laboratory-scale and pilot-scale studies. In aqueous phase, permanganate can completely mineralize or dechlorinate TCE and PCE depending on the reaction conditions (Huang and Hoag, 2000). The kinetics and mechanism of permanganate oxidation of TCE and PCE have been thoroughly investigated by Yan and Schwartz (1998) and by Huang and Hoag (2000), respectively. Moreover, several early pilot studies indicate that $KMnO_4$ is an effective oxidant for the destruction of TCE and PCE in the subsurface (Cline et al., 1997; West et al., 1997; Schnarr et al., 1998). These early studies reveal the feasibility of this technology for the remediation of TCE- and/or PCE-contaminated sites.

Objectives. In order to improve the performance of application of in situ permanganate oxidation, this study was conducted with primary objectives of: (1) investigating the processes and variables that predominately determine the effectiveness of destruction of source free phase TCE with permanganate in the subsurface, (2) examining the variations in selected parameters (e.g., pH, ORP, conductivity, and temperature) and reaction products (e.g., Cl⁻ and MnO_2) during the remediation event and (3) obtaining engineering data and experience for the permanganate remediation technology.

METHODS

The pilot study underway at a TCE-contaminated site in Connecticut involved injecting a sodium permanganate ($NaMnO_4$) solution of ~10 g/L at flow rate of ~0.6 L/min or ~0.3 L/min into a sheet-pile isolated cell to destroy TCE both in dissolved phase and free phase in the cell. The oxidant was continuously injected into the cell until the residual concentration of $NaMnO_4$ in the extracted water essentially remained unchanged when the injection of $NaMnO_4$ was stopped (to date, this test has included three periods of injection of $NaMnO_4$).

The cell contains two injection wells (IW-1 and IW-2), an extraction well (EW-1), eleven multi-level monitoring wells (MW-1 to MW-11) and three water level monitoring wells (WL-1 to WL-3). A layout of the cell showing locations of various wells is presented in Figure 1. The $NaMnO_4$ solution was injected into the cell, right above the aquitard through two injection wells installed on one side of the cell. Groundwater was continuously extracted at the same elevation and flow rate as the injection through the extraction well installed on the other side of the cell. The extracted groundwater was monitored for selected operation parameters using an on-line monitoring system before it was pumped into a treatment system for volatile organic compounds (VOCs) and discharged. The eleven multi-level monitoring wells were made to monitor the reaction conditions occurring in the cell during the test.

Site Description. The site studied is contaminated with TCE. Free phase TCE was found at various locations in discrete forms, mostly near the aquitard-aquifer interface within the test area, which is isolated by a Waterloo Barrier™ steel sheet pile (~ 1.2 m x 3.8 m x 9.3 m deep). The water table inside and outside the cell occurs at ~ 4 m below the ground surface. There is approximately 9.3 m of uniform sand (referred to as aquifer) in the cell, which overlies a thick deposit of clay (referred to as aquitard). Approximately 4 cm of pooled free phase TCE product was found immediately above the aquitard.

System Description. The test system included an injection system and an extraction system. The injection system comprised of a mass flow controller to regulate and monitor the tap water flow, a metering pump to add 40% $NaMnO_4$ at a controlled rate to achieve a desired $NaMnO_4$ concentration in the injectant, an in-line static mixer and a timer with a solenoid valve to direct injectant flow alternatively between two injection wells. The injectant was injected into the ground through the 0.9 m long x 2.5 cm diameter spiral screen (stainless steel) with 0.5 mm slot size located at the bottom of IW-1 and IW-2.

The extraction system comprised of a lift pump to extract groundwater through EW-1, a stainless steel chamber to facilitate on-line monitoring of pH, redox and temperature of the extracted water, an on-line flow meter, a glass column to trap any free phase TCE that might be present in the extracted water, an air stripper with a blower to remove any TCE from groundwater that may be present, especially during the initial phase of the test, two granular activated carbon canisters placed in series to trap TCE in the air coming from the air stripper, a pH neutralization system to control the pH of the extracted water in the air stripper, a discharge pump to transfer extracted groundwater from the air stripper to the sanitary sewer and a flow totalizer to measure the total volume of water discharged.

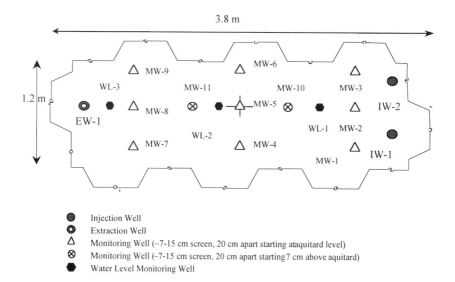

Figure 1. Layout of various wells within the sheet pile test cell.

RESULTS AND DISCUSSION

This pilot study was started by injecting a $NaMnO_4$ solution of ~10 g/L into the cell at a flow rate of 0.6 L/min (at the first injection period of ~2 days). The pH dropped from 6.5 to 3.5 and the chloride increased from ~ 82 mg/L to 380 mg/L in the extracted groundwater within ~2 days after the test began, most likely due to the reaction between dissolved TCE and permanganate. As the test proceeded, the chloride reached a peak concentration and then decreased along with the increase in pH, implying that the magnitude of TCE-permanganate reactions decreased. The reactions probably became limiting at the area near the interface of free TCE phase and aqueous permanganate phase. The pH values in all of the wells turned into basic conditions prior to the second injection period. The increase in alkalinity in the cell is most likely a result of many side reactions (e.g., equations 1 and 2), which may increase both the alkalinity and oxidant demand. These side reactions (e.g., $NaMnO_4$ with MnO_2 or ferrous ions originated from the dissolution of steel sheet pile under acidic conditions) appear to result in significant oxidant consumption in the test.

$$4MnO_4^- + 2H_2O \xrightarrow{\text{light}/MnO_2} 4MnO_{2(s)} + 3O_{2(g)} + 4OH^- \qquad (1)$$

$$2H_2O + 3Fe^{2+} + MnO_4^- \rightarrow 3Fe^{3+} + MnO_{2(s)} + 4OH^- \qquad (2)$$

During the second injection (~ 12 days) and the third injection (~13 dyas), most of the wells behaved as expected, reflecting the two separate injection events with two distinct peaks in $NaMnO_4$ and conductivity. A typical graph showing the variation of $NaMnO_4$, pH, chloride, conductivity, ORP and temperature in MW-5 (located at the middle of the cell) is presented in Figure 2. Although there was a definite lag time between the three rows of monitoring wells (i.e., MW-1 to MW-3, MW-4 to MW-6, and MW-7 to MW-9) which was caused by the difference in distance the flow had to travel from the injection points to the respective monitoring well rows, the parameters monitored varied in a similar pattern in these wells. As shown in Figure 2a, $NaMnO_4$ level decreased faster and to a lower concentration in the upper level section of the well when compared to the middle section, indicating that the oxidant demand was higher at the upper portion of the well. Permanganate consumption appears to be controlled by side reactions rather than by dissolution of free phase TCE.

Similar to the pH change in MW-5 as shown in Figure 2b, the pH variation in most of the wells increased during the second and third injection period and did not return to the background pH. The pH levels were higher during the third injection event. The chloride concentration increased after the termination of the injection events (Figure 2c), indicating that the remaining permanganate continued to oxidize the residual TCE in the cell.

The values of ORP and conductivity in the cell increased with the increase in $NaMnO_4$ concentration (Figure 2d and Figure 2e). The increase in ORP indicates that the once reducing environment had been replaced by an oxidizing environment. The ORP reading remained steady in most of the wells, although some peaks were observed and seemed to correspond with one of the injection events. Temperature remained consistent, with the only variation caused by the ambient air temperature during sampling events (Figure 2f).

During the test, TCE was only detected in those wells that did not show any permanganate concentrations and at the lower levels of the wells MW1-9 (the lower level rows). Most of the lower levels of these wells contained free product, and were not sampled regularly.

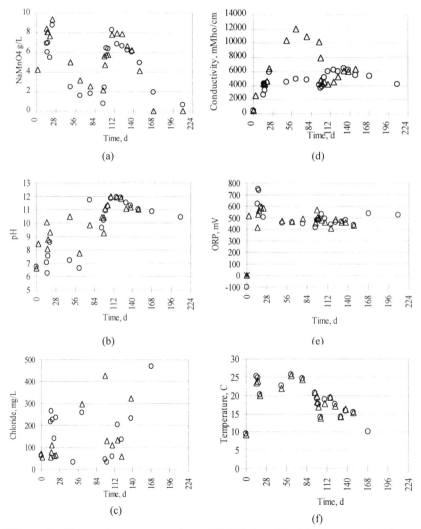

Figure 2. Change in concentration of NaMnO$_4$, pH, chloride, conductivity, ORP and temperature over test time in the extracted groundwater from MW-5. Empty circles: data from upper level of MW-5; Empty triangles: data from middle level

Two sets of soil cores (DC 1 to DC-3 and DC 4 to DC-6) collected during the second and third injection period, respectively, provide clear evidences for the destruction of source free phase TCE by the oxidant. The results of analysis for $NaMnO_4$ and pH & chloride for the sub-samples collected from these soil cores using the drainable core technique are shown in Figure 3 and Figure 4, respectively. It is evident from Figure 3a that the permanganate front gradually moved downward and through where source pure phase TCE existed as the test proceeded, indicating that the source pure TCE near MW-2 where the DC-2 and DC-4 were collected was destroyed by permanganate. Moreover, based on the relatively low chloride concentration and high pH as shown in Figure 4a, it may be reasonably presumed that most of the free phase TCE near the aquitard of MW-2 area has been destroyed.

Comparisons of DC-1 with DC-5 (both collected near MW-5) and DC-3 with DC-6 (both collected near MW-8) for $NaMnO_4$ concentration are shown in Figure 3b and Figure 3c, respectively. The only slight downward movement of permanganate fronts near the aquitard at both MW-5 and MW-8 and the very high chloride concentration and low pH conditions (Figure 4b and Figure 4c) indicate that pure TCE remained in these areas and the reactions continued. Furthermore, it was found by visual observation of the profile of the split soil cores that dark brown MnO_2 layers were formed near the bottom portions (near the aquitard) of the soil cores. The MnO_2 layers increased (e.g., 4 cm in DC-2 and 7 cm in DC-4) toward the clay layer as the test proceeded. Furthermore, it is evident from Figure 4 that the reactions between TCE and permanganate had rendered the area around free phase TCE zones very acidic when most other regions in the cell were in alkaline conditions. This may be considered as an advantage for the permanganate remediation technology, for both the permanganate and MnO_2 have greater oxidation strength to oxidize organic compounds under acidic conditions.

CONCLUSIONS

This pilot study shows that $NaMnO_4$ rapidly oxidized the dissolved-phase TCE in the test cell. Downward migration of permanganate fronts near the aquitard was observed, indicating the destruction of source pure TCE with permanganate. The efficiency of destruction of source free phase TCE appears to be controlled by the mass transfer of both permanganate and TCE toward the aquitard-aquifer interfaces. MnO_2 layers, a result of TCE-permanganate reactions, at the interface were observed and increased toward the free TCE zones as the test proceeded. The efficiency of mass transfer of both TCE and permanganate may decrease as more MnO_2 particles precipitate at the mass transfer paths.

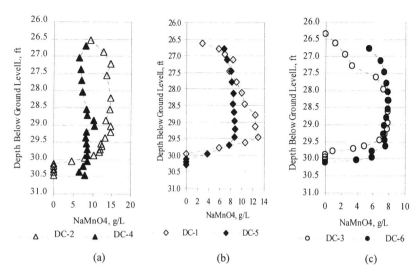

Figure 3. Change in NaMnO₄ concentration with depth below ground level for two sets of soil cores collected right above the aquitard during the second and the third phase of injection. a) DC-2 and DC-4 both collected near MW-2; b) DC-1 and DC-5 both collected near MW-5; c) DC-3 and DC-6 both near MW-8

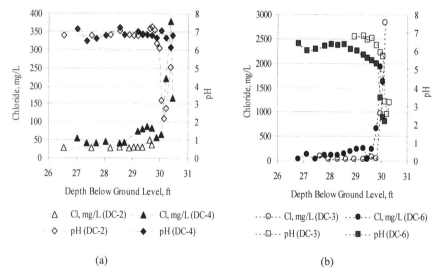

Figure 4. Change in pH and chloride with depth below ground level for two sets of soil cores collected right above the aquitard during the second and the third phase of injection. a) DC-2 and DC-4 collected near MW-2; b) DC-3 and DC-6 collected near MW-8

REFERENCES

Cline, S. R., O. R. West, R. L. Siegrist, and W. L. Holden. 1997. "Performance of In Situ Chemical Oxidation Field Demonstration at DOS Sites." presented at the In Situ Remediation of the Geoenvironment Conference, Minneapolis, MI.

Huang, K. C., G. E. Hoag, G. E. 2000. "A Study of Oxidation of Chlorinated Ethenes with Permanganate in Aqueous and Porous Media." Ph.D. dissertation. The University of Connecticut, Storrs, CT.

Schnarr, M., C. Truax, G. Farquhar, E. Hood, T. Gonulla, and B. Stickney. 1998. "Laboratory and Controlled Field Experiments Using Potassium Permanganate to Remediate Trichloroethylene and Tetrachloroethylene DNAPLs in Porous Media." *J. Contam. Hydrol.* 29:205-224.

West, O. R., S. R. Cline, W. L. Holden, F. G. Gardner, B. M. Schlosser, J. E. Thate, D. A. Pickering, and T. C. Houk. 1997. "A Full-Scale Demonstration of In Situ Chemical Oxidation through Recirculation At X-701b Site." *Oak Ridge National Laboratory*, Oak Ridge, TN.

Yen, Y. E., and F. W. Schwartz. 1998. "Abiotic Remediation of Ground Water Contaminated by Chlorinated Solvents." Ph.D. dissertation. The Ohio State University, Columbus, OH.

FULL-SCALE SOIL REMEDIATION OF CHORINATED SOLVENTS BY IN SITU CHEMICAL OXIDATION

Richard Levin, Edward Kellar (Environmental Science & Engineering, Inc., Gainesville, Florida)
James Wilson (Geo-Cleanse International, Inc., Kenilworth, New Jersey)
Leslie Ware [Anniston Army Depot (SIOAN-RK), Anniston, Alabama]
Joseph Findley, John Baehr (U.S. Army Corps of Engineers, Mobile, Alabama)

ABSTRACT: In July 1997, the first large scale use of the patented Geo-Cleanse process for removal of DNAPL chlorinated solvents and hydrocarbons was begun over an area of approximately 2 acres at a former clay backfilled lagoon site on Anniston Army Depot (ANAD), Anniston, Alabama.

This rapid *in situ* technology is being used to reduce volatile organic compounds to approved risk based cleanup levels on over 43,125 cubic yards (yd^3) of soils containing up to 31-percent trichloroethene, methylene chloride, daughter products, and BTEX. Soil injection extends from 6 feet (ft) to at least 25 ft below grade. Groundwater pockets of high strength DNAPL in residual sludge layers and associated clay smear zone are being eliminated selectively from 29 to 71 ft (top of bedrock) to prevent overlying soil recontamination.

This *in situ* chemical oxidation relies on the well documented Fenton's Chemistry for creation of hydroxyl radicals from hydrogen peroxide to non-specifically "mineralize" solvents, oil and grease, and naturally occurring organics to water, carbon dioxide, oxygen, and inorganic salts through an aggressive exothermic reaction. Treatment proceeds from shallow to deep zones through an array of patented steel injectors screened at specifically designed depths to treat organics present in both saturated and unsaturated zones. Injection quantities are based on pollutant mass present.

The full-scale 155-day injection of up to 132,925 gallons (gal) of peroxide into 255 injectors is complete. Soil concentrations of up to 21,760,000 micrograms per kilogram (mg/kg) have been reduced to below detection. Operating data to date indicate no adverse migration of organics to surrounding soils or groundwater. Project test data, remedial design and operations summaries, *in situ* monitoring, and pre- and post-treatment results are presented.

BACKGROUND

The U.S. Environmental Protection Agency (EPA) has designated 44 sites at Anniston Army Depot (ANAD), Anniston, Alabama, as Resource Conservation and Recovery Act (RCRA) Solid Waste Management Units (SWMUs). The former Facility 414 Old Lagoons have been designated as SWMU12.

SWMU12, consisted of three unlined industrial waste lagoons that cover an area of approximately 470 by 300 ft. Industrial wastes placed into the lagoons included: abrasive dust waste containing cadmium and lead, petroleum hydrocarbons, solvents, including trichloroethene (TCE) degreasers and methylene chloride paint stripper sludges, alkaline cleaners, grease trap wastes, and Industrial Wastewater Treatment Plant (IWTP) sludge.

During closure of the lagoons in 1978, the majority of sludge were removed from the lagoons and local clays were used as fill; however, the closure was incomplete and a significant amount of solvent and waste oil remained in the site soils and aquifer matrix, particularly in the middle lagoon. Subsequent RI/FS activities led to a decision to perform the emergency removal of 2,000 yd^3 of contaminated soils from three isolated areas.

Additional delineation was conducted increasing the area of treatment needed. The majority of VOCs were found at depths of 8 ft and greater. High groundwater levels have been encountered at levels of 14 feet below land surface (ft-bls). TCE accounts for approximately 85 percent of the 71,000-pounds VOC total. The total surface area of impacted soil is approximately 45,000 square feet (ft^2) (220 ft x 120 ft and 120 ft x 120 ft). The total affected volume (exceeding SSLs for VOCs) was increased from the original 2,000 yd^3 to 7,200 yd^3 above the mean high water table (14 ft-bls) and 6,500 yd^3 below groundwater. The highest concentrations of TCE, 20,100 mg/kg, occurred in samples taken from the 10-ft depth interval. However, concentrations greater than 9,000 mg/kg TCE at depths down to 24 ft.

PURPOSE AND OBJECTIVES

The emergency removal (ER) objective at SWMU12 was to use *in situ* chemical oxidation on soils to treat (remove) waste chemical constituents, volatile organic compounds (VOC), that may be contributing significantly to exceedences of health-based concentration limits in onpost soils and area groundwater. Specific objectives of the removal included:

- Destruction of organics (VOCs) in soils contributing to groundwater contamination,
- Containment of chemical constituents within the zone of treatment until destruction was complete,
- Prevention of mobilization of inorganics, and
- Assurance that treatment was complete by monitoring groundwater and soils chemical concentrations.

Based on a detailed site assessment performed according to current EPA Soil Screening Levels (SSL) Screening Guidance using all of the newly developed relevant site data, the site specific approved soil SSLs, as soil cleanup criteria were developed. These were calculated to be protective of groundwater at the Depot boundary.

- Trichloroethene—41 mg/kg,   • Methylene Chloride—63 mg/kg,
- 1,2-dichloroethane—51 mg/kg,   • Tetrachloroethene — 5.2 mg/kg.
- Vinyl Chloride—23 mg/kg,

Following extensive site characterization, risk assessment, and a pilot demonstration program, full-scale soil remediation injection is proceeding under an Emergency Removal Plan.

METHODOLOGY

Process Description. *In situ* chemical oxidation of organic contaminants is achieved by injection of hydrogen peroxide and a catalyst formulation into the affected media under carefully controlled conditions. The Geo-Cleanse® Process used, is a patented technology in which hydrogen peroxide and trace quantities of metallic salts are injected into the impacted media. This *in situ* oxidation system is capable of rapid complete, non-selective oxidation of organic compounds in soil and groundwater. The basic reaction in the Geo-Cleanse® Process is simplified below:

Hydrogen Peroxide	+	Organic Contaminant	---->	Carbon Dioxide	+	Water

The Geo-Cleanse® Process delivers a calculated charge of hydrogen peroxide and catalyst to the contaminated region via a patented injection methodology and equipment. This process maximizes the dispersion and diffusion of the reagent through the soil and/or the affected aquifer. The patented injectors are specially designed to withstand the elevated temperatures and pressures resulting from Fenton's reaction, while achieving maximum dispersion of the reagents through the subsurface.

The injection of hydrogen peroxide and the catalytic system results in an exothermic subsurface reaction that generates heat, pressure, oxygen, and carbon dioxide. During the reaction sequence, the organic compounds are successively converted to shorter chain mono- and di carboxylic (fatty) acids. These compounds are non-hazardous, naturally occurring substances, and are further degraded into carbon dioxide, chlorides, and water by subsequent reactions.

The actual oxidation is driven by formation of a free hydroxyl radical via Fenton's reaction chemistry. This methodology for the treatment of organic compounds in wastewater has been widely studied, utilized, and proven effective by the wastewater industry.

The preferred Fenton's Reaction is:

H_2O_2	+	Fe^{2+}	---->	OH.	+	OH^-	+	Fe^{3+}
Hydrogen peroxide		ferrous iron		hydroxyl radical		hydroxylferric ion		iron

The hydroxyl free radical (OH.) is an extremely powerful oxidizer organic compounds. Residual hydrogen peroxide, due to its unstable characteristics, rapidly decomposes to water and oxygen in the subsurface environment. Soluble iron amendments added to the subsurface during the Geo-Cleanse® Process in trace quantities are precipitated out during conversion to ferric iron.

Within SWMU12, the organic-contaminated waste was treated *in situ* to achieve health-based remedial action goals protective of onsite groundwater as a drinking water source. Treatment consisted of *in situ* organic destruction accomplished by a controlled sequence of catalyzed peroxide injections into an engineered array of approximately 255 injectors. Chemical oxidation of the contaminated soils proceeded over a 5-month period (146 injection days) and entailed the closely monitoring of the establishment of Fenton's chemistry conditions during injection of an estimated 132,925 gal of 50-percent hydrogen peroxide.

Injection System Design. Three types of injectors were used targeting three distinct depth intervals. Single shallow injectors (screened from 8 to 14 ft) were installed in areas where contamination is above 15 ft. The top of the injection interval for the shallow injectors was established to provide a minimum hydraulic confinement to induce acceptable horizontal migration of the reagents and minimizing the potential of reagents migrating to the surface. Cluster injectors consisting of a shallow injector paired in the same borehole with a deep soil injector (screened from 20 to 26 ft) were constructed in areas where contamination was found at both deep and shallow depths. Single intermediate injectors were placed in areas where contamination existed between 14 and 20 ft. A total of 255 injectors were installed.

Shallow and cluster injectors were installed approximately 20 ft apart. Intermediate injectors were placed between the shallow injectors approximately 20 ft from each other and 10 ft from the nearest shallow or cluster injector.

In addition, 25 deep groundwater injectors screened in the saturated zone were installed as monitor wells during chemical oxidation. The groundwater monitor wells (deep injectors) were located throughout the contaminated area approximately 60 ft apart. These injector wells were used to detect effects on the groundwater from the *in situ* treatment. A line of groundwater monitor wells were located on a line paralleling the eastern margin of the SWMU to detect any downgradient effect of the treatment immediately outside the SWMU boundary. Finally, one monitor well was located within the SWMU, but upgradient of the contaminated area. All monitor wells were used regularly and frequently to define the potentiometric gradient at the site to monitor injection conditions.

Each of the deep groundwater injector wells were constructed in the uppermost water-bearing strata. In most locations, this strata occurs in the weathered zone at the top of the bedrock or within a few feet of the top of the rock. The wells penetrated a minimum of 5 ft into the bedrock or approximately 50 ft.

PROGRESS AND RESULTS

Injection Control and *In Situ* Monitoring. Catalyst solution (ferrous sulfate) and sulfuric acid (for pH control) were injected to prepare the soils for peroxide injection. Surrounding wells were monitored for the following parameters to determine readi-ness for peroxide:
- pH (maintain near 5.5),
- Fe^{3+} (catalyst ion availability); and
- Conductivity

Reaction of peroxide with ferrous sulfate and organics yields oxygen, carbon dioxide, chlorides, and water as products. After commencement of peroxide injec-tion, the following parameters were monitored in the vent well off-gas to determine the progress of the reaction:
- O_2 (reaction product),
- H_2O_2 (excess reactant), and
- CO_2 (reaction product).

Injection proceeded at a rate (0.25 gal per minute) and pressure (1 - 5 psi) that maximizes the radius of influence of the reaction without causing preferential flow or undue venting. The injection rate and pressure was recorded continuously to ensure accurate records of peroxide volume at each injector. Hourly reading of carbon dioxide and breathing zone volatiles using an OVM were monitored from the surrounding injectors to monitor the progress of the reaction. If during monitoring it is determined that there is a preferential flow direction of peroxide reaction, surrounding injectors were capped to prevent the release of pressure, causing the peroxide to disperse in another direction.

As the rate of reaction with available organics peaks, carbon dioxide, and oxygen levels in the off-gas also peaked. During monitoring, carbon dioxide concentrations peaked at 10 to 14 percent and oxygen levels peaked at approximately 40 percent. As the available organics was consumed, the concentration of carbon dioxide and oxygen in the off-gas began to decline and the concentration of excess peroxide increased.

Each sub-array of injectors receiving peroxide at one time were treated with the minimum target chemical quantities and/or the *in situ* monitoring indicates adequate endpoints for the operational parameters (oxygen, carbon dioxide, and hydrogen peroxide) in the involved and neighboring vents and injectors. Following treatment of a sub-array and pre-treatment catalyst adjustment of the subsurface environment, peroxide connections was moved to the succeeding sub-array position.

Monitoring Groundwater. In addition to carbon dioxide monitoring, the monitor wells located in SWMU12 were monitored daily for peroxide, pH, iron, conductivity, temperature, and chlorides. The monitoring data was collected and tabulated to determine effects on the groundwater from soil treatment.

Vent Flow Balance. A vent flow balance (VFB) system was installed to aid in maintaining an effective radial dispersion of catalyst and peroxide. During catalyst and peroxide injection, air lines were connected to up to twelve injector/monitors adjacent to the injectors being used. By monitoring the pressures and carbon dioxide measurements of the adjacent injectors, it could be determined if the injected fluids are dispersing into the soils radially. Ideally, the injection should disperse from the injector radially up to approximately 15 to 20 ft from the injector. If monitoring indicates that there was a preferential direction of flow, the VFB system was used to effectively disperse the peroxide in a radial direction by creating a slightly negative pressure differences near the injector.

Post-Treatment Sampling and Confirmation. To determine the effectiveness of the oxidation process, post-treatment sampling commenced while the full-scale treatment program was active. The injection program started in the far west portion of the middle lagoon and proceed to the east. Once the injection program reached the "O" line on the injection grid, post-treatment sampling began.

The locations of the post-treatment samples were chosen to ensure that the edges of the contaminated area remain clean and that the injection treated soils were below the SSLs. By beginning the post-treatment sampling during treatment, site condition changes could be effected based on sampling results. Soil samples were collected using a Geoprobe®. Samples will be collected at intervals corresponding to the intervals with the highest concentrations in the nearest delineation boring and/or injector locations. Samples were sent to ESE's laboratory in Gainesville and analyzed for VOCs using Method 8010/8020. When the sample analyses show that the soil concentrations are below the SSLs, the area was deemed clean (Table 1). The injection has reduced the total VOC concentration by up to <90 percent from original concentrations. Confirmatory samples were taken at selected location and sent to the Gainesville laboratory for complete VOC and metals analysis. If the screening samples showed that contamination concentrations remain above SSLs, the location was re-treated with peroxide for a polishing treatment. After polishing treatment, another set of screening samples were taken until the analyses showed concentrations of VOCs below SSLs.

Table 1. Soil Analysis (Pre and Post Sampling) - Trichloroethene (TCE)

Sample Location	Depth (FT BLS)	Pre-Injection Concentration	Post-Injection Concentration
K46	10	35	1
K46	12	1	1
O48	8	63	21
O48	10	905	24
O48	12	1430	24
S32	12	1400	4
S40	10	4230	9
S40	12	1130	3
S48	8	136	11
S48	10	3270	18
U20	8	10100	4
U20	12	7630	52
U24	8	11400	37
U24	12	5900	3
U28	10	510	34
U28	14	780	187
U32	8	4420	20
U32	10	20100	95
U32	12	7110	411
U48	10	1500	43
U52	8	9030	22
U52	10	1290	11
W20	8	3400	12
W20	10	3500	120
W28	8	29000	1014
W28	10	1500	489
W28	12	7300	340
W32	8	220	26
W32	12	450	65
W44	10	140	3
W44	12	15	1
W48	8	672	<1
W48	10	50	13
Y24	6	1000	19
Y28	6	6300	340

Target   Concentration   41   (mg/kg)

CONCLUSIONS

In situ chemical oxidation using hydrogen peroxide was proven to be effective in reducing the contaminant concentrations in clays and groundwater at the Anniston Army Depot to below the SSLs. The use of *in situ* chemical oxidation at ANAD was the first large scale use of hydrogen peroxide to treat chlorinated solvents. The objectives of the project were met by destruction of the organics (VOCs) in soils which were contributing to groundwater contamination, being able to contain the chemical constituents within the zone of treatment until destruction was complete, prevented the mobilization of inorganics, and documented that treatment was complete by monitoring groundwater and soils chemical concentrations. The cost data is being evaluated for the project, however, it appears that the cost of the *in situ* chemical oxidation treatment is one-fourth the cost of excavation and disposal of the contaminated soils.

To date there has been no evidence of adverse pollutant mobility produced by the process. Conversely, adequate dispersion in the tight clays, while difficult in several locations, has proceeded a reasonably productive rates and low pressures. The target volumes of peroxide have been adequate to treat the organics except where localized dispersion is affected. Approximately 10 to 20 percent of the injectors locations needed polishing through either existing injectors or the construction of additional injectors. The process has proven to capability to reduce soil concentrations to below detection limits.

REMEDIATION OF CHLORINATED COMPOUNDS BY CHEMICAL OXIDATION

Darlene E. Coons, M. Talaat Balba, Cindy Lin, Susan Scrocchi, Alan Weston
(Conestoga-Rovers & Associates, Niagara Falls, New York)

ABSTRACT: Five laboratory treatability studies were conducted to assess the potential application of chemical oxidation for the remediation of chlorinated compounds present in three different media (neat DNAPL, contaminated groundwater, and contaminated soil). The oxidizing agents tested were potassium permanganate and Fenton's Reagent. The media contained mixtures of chemicals including chlorinated aliphatics (PCE, TCE, TCA, DCA) and chlorinated aromatics, simple aromatics (BTEX), and PCBs. The results obtained showed that both potassium permanganate and Fenton's Reagent were effective in reducing the contaminant concentrations in all media. In Study #1, greater than 95 percent reduction in PCE, TCE, 1,1,2,2-TCA, carbon tetrachloride, chloroform, and 1,2-DCA concentrations in DNAPL was observed after 3 weeks of potassium permanganate treatment. Additionally, over 75 percent removal of PCBs within three weeks using either potassium permanganate or Fenton's Reagent was observed. In Study #2, both oxidants degraded total VOCs in DNAPL to 1% or less in three weeks, a reduction in VOC concentration of over 98%. Study #3 showed that potassium permanganate was more effective in reducing VOC concentrations in soil whereas Fenton's Reagent was more effective in reducing both VOC and SVOC concentrations in groundwater. They were equally effective in reducing SVOCs in soil. Study #4 indicated that while both oxidants were capable of reducing aromatic concentrations in groundwater to levels below the detection limits, treatment with Fenton's Reagent achieved lower detection limits. Study #5 showed that potassium permanganate degraded PCE, TCE, and vinyl chloride monomer in groundwater to levels below their detection limits. Based on these promising results, Conestoga-Rovers & Associates is planning pilot tests of chemical oxidation treatment at several sites.

INTRODUCTION

The contamination of soil and groundwater with chlorinated compounds is a common problem that poses daunting challenges for environmental remediation. Conventional treatment technologies are effective but can be costly and may require lengthy treatment periods. Quicker, less expensive alternatives are desired. Chemical oxidation can be cost-competitive with other treatment technologies and may be a more attractive alternative because it can be performed in situ and the process destroys the contaminants rather than transferring them to another media.

Releases of chemicals at industrial sites also typically result in the formation of localized multi-component dense nonaqueous phase liquid (DNAPL) pockets in the subsurface soil and groundwater. Intensive research efforts have been recently focussed on the development of appropriate methods for the delineation and remediation of DNAPL contamination. Chemical oxidation is one of a few emerging technologies that has shown great potential for in situ treatment of DNAPL.

In a chemical oxidation process, the oxidizing agent reduces the chlorinated compounds to nonhazardous or less toxic compounds, ultimately to carbon dioxide and water. Two oxidizing agents with potential applications in remediation are potassium permanganate ($KMnO_4$) and Fenton's Reagent, a mixture of hydrogen peroxide with an iron catalyst and sulfuric acid. They are readily available in bulk quantities and are relatively easy and safe to handle.

Objective. Five laboratory treatability studies were conducted to assess the potential application of chemical oxidation utilizing potassium permanganate or Fenton's reagent for the remediation of DNAPL, contaminated soil, and contaminated groundwater. The targeted contaminants were mixtures of chemicals including chlorinated aliphatics such as tetrachloroethene, trichloroethene, tetrachloroethane, and dichloroethane, chlorinated aromatics, simple aromatics (benzene, toluene, ethylbenzene, and xylene [BTEX]), and PCBs.

MATERIALS AND METHODS

Each of the five laboratory treatability studies was conducted on environmental samples (contaminated soil, groundwater, or DNAPL) obtained from different Resource Conservation and Recovery Act (RCRA) or Superfund sites. Fenton's Reagent used in these studies was prepared using 50% hydrogen peroxide (H_2O_2) in a mixture of 10% v/v sulfuric acid solution and 2% w/v ferrous sulfate solution.

For VOC/SVOC analyses performed in-house, the untreated DNAPL samples were shaken with methanol to extract the organics from the DNAPL to the aqueous phase. For treated samples, methanol was added to extract the organics into the aqueous/methanol phase. In both cases, the aqueous/methanol phase was then analyzed for volatile organic compounds utilizing a purge and trap connected to a gas chromatograph equipped with a flame ionization detector (FID). The analysis was performed under the guidance of SW-846 5030/8021.

For PCB analysis performed in-house, the entire sample was poured into a separatory funnel and extracted with approximately 30mL of prepared hexane/methylene chloride (85:15 v/v). The solution was separated by draining off the aqueous layer. The organic extract was placed into a flask/concentrator apparatus and was concentrated to 10mL, adding hexane if needed. The analysis was performed under the guidance of SW-846 Method 8082.

The detailed methods used for each study and the results obtained are summarized in the following section.

REMEDIATION STUDIES AND RESULTS

Study #1. This study was conducted on DNAPL contaminated primarily with tetrachloroethene (PCE), trichloroethene (TCE), 1,1,2,2-tetrachloroethane (TCA), carbon tetrachloride (CT), chloroform, 1,2-dichloroethane (DCA), and PCBs. The first phase of this treatability study was performed to determine the potential for $KMnO_4$ to treat the volatile compounds present in DNAPL. Initially, several reaction conditions were visually studied to determine the effect of KMnO₄ in oxidizing the DNAPL. Solid KMnO₄ with neat DNAPL, varying concentrations of aqueous solutions of KMnO₄ (4-16% w/v) with DNAPL, and varying reaction times were tested. Using KMnO₄ in aqueous form (4-6% w/v) was observed to be most effective, possibly due to the more complete solubility of the reagent. The effect of incubation time was then studied. Test vials were prepared with DNAPL in 4-6% solutions of KMnO₄. Sacrificial samples were analyzed by GC methods in-house at the CRA Treatability Laboratory in Niagara Falls, NY at 48 hours, 6 days, and 2 weeks. Based on these preliminary results, a sample of DNAPL was treated with a 6% solution of KMnO₄, incubated for 3 weeks, and analyzed for VOCs.

This treatability study showed that KMnO₄ was effective in oxidizing the volatile organics present in the DNAPL. The analytical results are shown graphically in Figures 1-1 and 1-2.

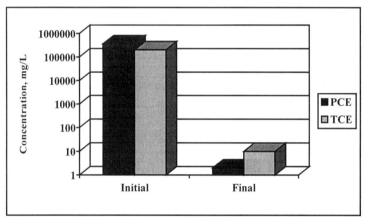

FIGURE 1-1. Reduction in Concentration of Chlorinated Alkenes in DNAPL after 3 weeks of Treatment with Potassium Permanganate.

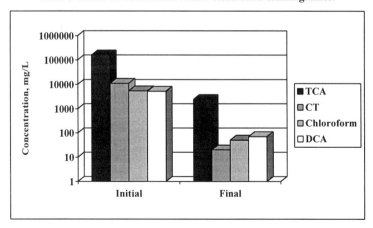

FIGURE 1-2. Reduction in Concentration of Chlorinated Alkanes in DNAPL after 3 weeks of Treatment using Potassium Permanganate.

Total organic carbon was also monitored and was reduced from 30,400 mg/L to approximately 260 mg/L. This indicates that the chemical oxidation process was effective in destroying most of the organic compounds, including the chlorinated solvents. Under the condition tested, the concentrations of the volatile compounds were reduced more than 95 percent after three weeks of treatment.

The second phase of this treatability study was performed to determine the potential for either $KMnO_4$ or Fenton's Reagent to treat the PCBs present in the DNAPL. A series of reactions were set up in glass serum bottles using neat DNAPL: a) a 6% solution of $KMnO_4$, and b) an 8% solution of Fenton's Reagent. Control samples were prepared using similar amounts of distilled water instead of oxidant solution. The treated samples were incubated for 2 weeks and then analyzed for residual PCBs.

Results of PCB degradation testing showed 83 percent reduction in PCB concentration using $KMnO_4$ and 75 percent with H_2O_2 after 2 weeks.

Study #2. This study was conducted on heavy DNAPL samples containing primarily PCE and chlorinated benzenes. Preliminary tests were carried out to determine the potential for $KMnO_4$ or Fenton's Reagent to treat both DNAPL samples. The tests consisted of four different aqueous solutions of $KMnO_4$ and Fenton's Reagent (2, 4, 6, and 8% w/v). The test samples were prepared by weighing the DNAPL into glass sample bottles and determining final solutions based on specific gravity of the DNAPL. Each sample bottle was tightly sealed with Teflon screw top lids and visually observed for indications of reactivity. Based on these observations, DNAPL samples treated with 8% solutions of $KMnO_4$ and Fenton's Reagent were prepared and incubated on a shaker for 3 weeks. Control samples were prepared using distilled water instead of an oxidizing agent. Final VOC analysis was performed using GC methods.

The analytical results indicate that after a 3-week incubation period, the total detected VOCs concentrations in all treated samples for both DNAPLs were at or below one percent. The results are summarized in Tables 2-1 and 2-2.

TABLE 2-1. Percent Concentrations of Target Compounds in Control versus Treated Samples - DNAPL #1

Parameter	Control	$KMnO_4$	% Removal	Fenton's Reagent	% Removal
Tetrachlorobenzene	18.3	0.06	99.7	0.03	99.8
1,2,4-Trichlorobenzene	17.4	0.15	99.1	0.09	99.5
Tetrachloroethene	8.6	non-detect	99.9	0.09	99
Chlorotoluene	5.1	0.17	96.7	0.11	97.8
Chloroform	2.5	non-detect	99.9	non-detect	99.9
Xylenes	2.1	non-detect	99.9	0.05	97.6
Ethylbenzene	1.7	0.07	95.9	0.06	96.4
Sum of all other detected VOCs present at <1%	5.1	0.2	96.1	0.11	97.8
Total detected VOCs	*60.6*	*0.65*	*98.9*	*0.52*	*99.1*

TABLE 2-2. Percent Concentrations of Target Compounds in Control versus Treated Samples - DNAPL #2

Parameter	Control	KMnO₄	% Removal	Fenton's Reagent	% Removal
Tetrachlorobenzene	29.8	0.09	99.7	0.03	99.9
Tetrachloroethene	17.2	non-detect	99.9	0.35	98
1,2,4-Trichlorobenzene	13.3	0.34	97.4	0.18	98.6
Chlorotoluene	3.2	0.01	99.7	0.03	99.1
Ethylbenzene	3.1	0.14	95.5	0.05	98.4
Chloroform	2.6	0.08	96.9	non-detect	99.9
Chlorobenzene	1.8	0.16	91.1	non-detect	99.9
Carbon tetrachloride	1.5	0.08	94.7	0.05	96.7
Sum of all other detected VOCs present at <1%	3.6	0.14	96.1	0.08	97.8
Total detected VOCs	*76*	*1.02*	*98.7*	*0.75*	*99*

Study #3. This study was performed on groundwater and soil contaminated with volatile organic compounds (VOCs) and semi-volatile organic compounds (SVOCs). One percent and 5% solutions of KMnO₄ and Fenton's Reagent were prepared. For the groundwater study, the groundwater was placed in glass bottles and predetermined amounts of the prepared oxidant solutions were added so that the final concentrations of the oxidizing agent in the treated groundwater samples would be 0.1% and 0.5%, respectively. No headspace was allowed and the bottles were sealed immediately to avoid loss of volatiles. For the soil study, 100g of soil was weighed into glass jars. 10mL of the prepared oxidant solutions (1%, 5%) was added to the jars and each jar was capped as quickly as possible to avoid loss of volatiles, though some headspace was allowed in each soil sample. Control samples were prepared for each test using distilled water instead of oxidizing agent. All treated samples were allowed to incubate for three weeks. The soil and groundwater samples treated with 5% solution and the corresponding control samples were analyzed for VOCs.

The treatability study showed that both KMnO₄ and Fenton's Reagent were effective in oxidizing the compounds detected in the soil and groundwater. However, KMnO₄ was more effective in reducing VOC concentrations in soil and Fenton's Reagent was more effective in reducing both VOC and SVOC concentrations in groundwater. Both oxidants were equally effective in reducing SVOCs present in soil. The results are shown in Table 3-1.

TABLE 3-1. Percent Reduction in VOC and SVOC Concentrations After a 3-Week Treatment Period

Parameter	KMnO₄	Fenton's Reagent
Soil (μg/Kg)		
VOCs	79-86	32-53
SVOCs	65-73	62-74
Groundwater (μg/L)		
VOCs	55	75
SVOCs	up to 29	74-97

Study #4. This study was conducted on groundwater contaminated primarily with xylenes, ethylbenzene, and ketones. Chemical oxidation tests using $KMnO_4$ and Fenton's Reagent were conducted. Two sets of each of the treatments were carried out. The first treatment tested consisted of 4% solution of $KMnO_4$ in groundwater. The second treatment tested consisted of adjusting groundwater pH to 5 by addition of 0.1% sulfuric acid solution and adding 4% solution of Fenton's Reagent (50% H_2O_2 with 2% w/v ferrous sulfate solution). Control samples were prepared for each test using distilled water. The treated samples along with controls were shaken for 5 minutes and then incubated statically at room temperature for 3 weeks and analyzed for VOCs.

    The treatability study results showed that while both $KMnO_4$ or Fenton's Reagent reduced the concentrations of the volatile compounds present in the contaminated groundwater, treatment with Fenton's Reagent achieved lower levels. The results are summarized in Table 4-1.

TABLE 4-1. Initial and Final Concentrations (µg/L) of Volatile Compounds in Groundwater

Parameter	Average of 2 Controls	$KMnO_4$	Fenton's Reagent
2-Butanone (MEK)	395	ND(190)*	17
Ethylbenzene	415	ND(20)*	ND(0.80)*
Xylenes	1825	ND(50)*	ND(2.0)*
Toluene	13	ND(23)*	ND(0.90)*
4-Methyl-2-pentanone (MIBK)	123	ND(240)*	ND(9.7)*

*ND(x) Non-detect at or above x.

Study #5. This study was performed on groundwater containing vinyl chloride, DCE, TCE, and PCE. Three batches of untreated groundwater samples were prepared for the study. Groundwater was placed into glass serum bottles with teflon-lined septa and aluminum crimp caps to prevent VOC losses by volatilization. A 1% solution of $KMnO_4$ was prepared and varying amounts were injected by syringe into the serum bottles. Each test was carried out in duplicate. The treated samples along with controls were placed on an orbital shaker for one hour to ensure homogenization. The samples were then incubated at room temperature. Samples were analyzed after 1 and 7 weeks.

    VCM, DCE, and TCE were oxidized to below their detection limits by all the $KMnO_4$ solutions. PCE was oxidized by all the solutions, but a small amount remained in the two samples treated with 0.5mL of the 1% solution. Further study of the minimum concentration of $KMnO_4$ needed for VCM removal was conducted with samples spiked with 2ppm VCM. The results showed that an 8ppm solution oxidized 99% of the VCM after 7.5 weeks.

LARGE SCALE FIELD APPLICATIONS

    The results of the laboratory treatability studies indicated that both $KMnO_4$ and Fenton's Reagent were effective in reducing concentrations of chlorinated compounds in DNAPL, contaminated soil and groundwater. Up to 95 percent of the chlorinated aliphatic compounds and over 75 percent of the PCBs were degraded as a result of chemical oxidation treatment. Based on these results, Conestoga-Rovers & Associates (CRA) is currently moving forward to field pilot testing chemical oxidation at several sites.

CRA has completed the conceptual design and work plan for remediation of PCBs and PCE in the vadose zone of soils at a site in Long Island, NY. The design involves flushing the subsurface soil (5 to 30 ft feet below ground surface) with solutions of potassium permanganate over a period of 6 weeks.

CRA is preparing to implement this technology for the treatment of DNAPL at a chemical manufacturing facility in Louisiana that consists primarily of a mixture of PCE and TCE. The intent is to treat and prevent further migration offsite of a DNAPL plume by injection of potassium permanganate through a well curtain.

Potassium permanganate treatment will also be implemented at a third chemical manufacturing plant in Western New York to treat residual contamination under a building. An infiltration system around the building involving the use of perforated pipe will be used. The intent is to reduce the source of contamination that is currently affecting bedrock groundwater downgradient of the source and thereby reduce the length of time needed to meet the clean-up criteria.

At another site in Western New York, potassium permanganate will be used to clean up an isolated area of contamination that has been detected outside a slurry wall by injection into a series of wells.

REFERENCES

Andrews, T., D. Zervas, and R. S. Greenberg. July 1997. "Oxidizing agent can finish cleanup where other systems taper off." Reprinted with permission from *Soil and Groundwater Cleanup Magazine.*

Bryant, J. D. and J. T. Wilson. 1998. "In-Situ Fenton's Reagent Chemical Oxidation of Hydrocarbon Contamination in Soil and Groundwater." Reprinted with permission from *Remediation.*

Bryant, J. D. and J. T. Wilson. 1998. "Rapid Delivery System Completes Oxidation Picture." *Soil and Groundwater Cleanup Magazine*, August/September 1998, pp. 6-11.

Environmental Security Technology Certification Program. November 1999. *Technology Status Review In Situ Oxidation.*

Jerome, K. M., B. B. Looney, and J. Wilson. 1998. "Field Demonstration of In Situ Fenton's Destruction of DNAPLs." In Wickramanayake, G. B. and R. E. Hinchee (Eds.), *Physical, Chemical, and Thermal Technologies: Remediation of Chlorinated and Recalcitrant Compounds,* pp. 353-358. Batelle Press, Columbus, Ohio.

LaChance, J. C., S. Reitsma, D. McKay, and R. Baker. 1998. "In Situ Oxidation of Trichloroethene using Potassium Permanganate. Part 1: Theory and Design." In Wickramanayake, G. B. and R. E. Hinchee (Eds.), *Physical, Chemical, and Thermal Technologies: Remediation of Chlorinated and Recalcitrant Compounds,* pp. 397-402. Batelle Press, Columbus, Ohio.

McKay, D., A. Hewitt, S. Reitsma, J. LaChance, and R. Baker. 1998. "In Situ Oxidation of Trichloroethylene using Potassium Permanganate: Part 2. Pilot Study." In Wickramanayake, G. B. and R. E. Hinchee (Eds.), *Physical, Chemical, and Thermal Technologies: Remediation of Chlorinated and Recalcitrant Compounds,* pp. 377-382. Batelle Press, Columbus, Ohio.

Schnarr, M., C. Truax, G. Farquhar, E. Hood, T. Gonullu, and B. Stickney. 1998. "Laboratory and controlled field experiments using potassium permanganate to remediate trichloroethylene and perchloroethylene DNAPLs in porous media." *Journal of Contaminant Hydrology, 29:* 205-224.

United States Environmental Protection Agency, Solid Waste and Emergency Response. 1998. *Field Applications of In Situ Remediation Technologies: Chemical Oxidation.* EPA 542-R-98-008.

Watts, R. J. 1992. "Hydrogen Peroxide for Physicochemically Degrading Petroleum-Contaminated Soils." *Remediation,* August 1992, pp. 413-425.

IN-SITU CHEMICAL OXIDATION
LIMITED BY SITE CONDITIONS - A CASE STUDY

Eric P. Roberts, P.E., P.G. (Excalibur Group. LLC. Germantown, Maryland)
Nicholas Bauer (Saltire Industrial, Reston, Virginia)

ABSTRACT: This paper presents a case study in which the authors selected, implemented and completed follow-up effectiveness monitoring of *in-situ* chemical oxidation using Fenton's reaction chemistry to address a 0.25-acre "hot spot" portion of a dissolved chlorinated solvent plume at a former electroplating and manufacturing plant between April 1998 and July 1999. The site is located on the northern flank of a drumlin and the unconsolidated material beneath the site consists of poorly sorted till deposits. In April 1998, a total of 3,500 gallons of hydrogen peroxide and 3,200 gallons of catalyst were pumped into the subsurface through 12 injectors at two depths. TCE concentrations near the center of the hot spot were reduced from a concentration of 1,970 $\mu g/L$ to 15.6 $\mu g/L$ one week after injection and to less than 1 $\mu g/L$ four weeks after injection. However, after 5 months, the TCE concentration rebounded to 2,490 $\mu g/L$. After 15 months, TCE in five monitoring wells rebounded to levels at or significantly above baseline and historical average concentrations. The measured groundwater TCE concentrations indicate that TCE mass destruction occurred predominantly along preferential flow paths and that the remedy failed to adequately treat the residual source. In addition, application of the technology at this site may have, at least temporarily, changed the configuration of the groundwater contaminant plume. This study illustrates the critical importance of long-term follow-up groundwater sampling in the evaluation of the performance of the remedial technology and highlights some important limitations of the technology.

INTRODUCTION

In-situ chemical oxidation (CO) technology applications based on Fenton's reaction chemistry are emerging as promising means of addressing source areas contaminated with a wide range of organic compounds (Bryan and Wilson, 1999; USEPA, 1998). The technology applications are generally applicable for relatively small contamination problems where expedited cleanup is desired or where long-term solutions are not economically or logistically viable. However, the technology has limitations and potential risks, some of which are illustrated in this field-scale case study. The authors employed the technology with the specific objective of reducing TCE contamination in groundwater in a 0.25-acre area of a site to below a regulatory threshold of 2,340 $\mu g/L$ within a relatively short period of time. A discussion of the site, the application of the technology to the site conditions and how and why the technology failed to achieve the project objectives is described below.

Site Description. The site was used between 1969 and 1987 for the manufacturing of small metal parts. Various electroplating processes were used as part of the operations including metal cleaning, stamping and painting. The facility also operated its own plating wastewater treatment system. Chlorinated solvents, which were used for cleaning and degreasing, later became the primary focus of environmental investigation and remediation, including the remedial action that is the subject of this paper.

Geology/hydrogeology. Situated on the northern flank of a drumlin, the stratigraphy of the unconsolidated material beneath the site consists primarily of glacial till deposits. The till is generally comprised of a compact, poorly sorted mixture of silt, clay, sand and gravel with occasional sand lenses. Organic material has been found to comprise less than 1% of site soil. Weathered bedrock metasediments, oligoclase-quartz granulite and mica-oligoclase quartz gneiss with some schist underlie the unconsolidated sediments between 30 and 50 feet below the ground surface. Unconfined groundwater is present beneath the site at approximately 20 to 25 feet below grade and flows toward the northeast. Figure 1 depicts the localized stratigraphy in the chemical oxidation remedial area.

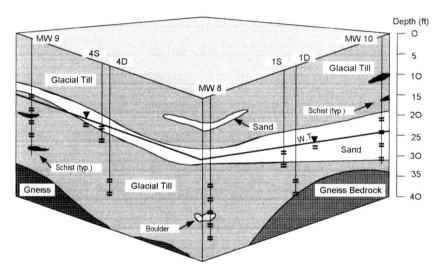

FIGURE 1. Cross-section showing subsurface stratigraphy and monitoring/injection well placement in CO remedial area.

Contaminant of Concern. Trichloroethylene (TCE) is dissolved in groundwater in a localized (<0.25-acre) circular to oblong plume, downgradient of the suspected contaminant source area. Dissolved TCE levels had historically been measured as high as 4,700 µg/L and as low as 17 µg/L in various wells located this hot spot area. More recent aquifer sampling had suggested that dissolved TCE levels were moderating but groundwater from certain wells continued to contain TCE above the regulatory threshold. Numerous field investigations failed to detect any significant levels of organic contamination in unsaturated soil above the TCE plume, consistent with a non-source area.

The localized nature of the groundwater contamination, the proximity of the site TCE concentrations in groundwater to the site threshold, and the need to implement an expedited and economical solution lead the authors to select a Fenton's reaction chemistry CO remedial approach. This CO approach relies on the *in-situ* production of the hydroxyl free radical (OH•), a powerful organic compound oxidizer, and contact between the oxidizer and the contaminant. Ideally, OH• is generated in accordance with Fenton's reaction:

$$H_2O_2 \quad + \quad Fe^{+2} \quad \longrightarrow \quad OH\bullet \quad + \quad OH^- \quad + \quad Fe^{+3}$$

The Fenton's CO injection program was implemented in April 1998. Methods employed and program results are presented below.

CHEMICAL OXIDATION METHODS

A total of 14 individual injector wells were installed to enable the delivery of the CO reactants into the affected aquifer. The injectors were concentrated in the target treatment area defined by the baseline TCE data and generally spaced under the assumption that each would treat TCE in groundwater in a 15- to 20-foot radius about each injector. Most of the individual injectors were installed as "shallow" and "deep" pairs in order to efficiently deliver the reactants over the entire saturated thickness. In two locations where the aquifer thickness was less than 5-7 feet (3D and 5D), only single-depth injectors were installed. Figure 1 illustrates the vertical spacing of 4S/4D and 1S/1D injectors, as typical well pairs, relative to the site geology.

The CO injection remedial program was conducted during daylight hours over a 6-day period in April 1998. During this time, Fenton's CO reactants were alternately injected under pressure (up to 50 psi) into one or more of the injector wells and aquifer responses to the injections were monitored. A total of approximately 3,500 gallons of 50% hydrogen peroxide (H_2O_2) and 3,200 gallons of a ferrous sulfate (iron sulfate heptahydrate) catalyst solution (10 mg/L to 100 mg/L) were injected into the aquifer. H_2O_2, catalyst and compressed air were independently metered and pumped to each active injector using a three-way manifold tee stick-up, equipped with isolation and check valves and a pressure gauge. The greatest fraction of the H_2O_2 used during the remedial program was injected into 1S (37%). Significant fractions were also injected into 5D (14%), 4D (13%), 4S (13%), 7D (10%) with lesser amounts of H_2O_2 delivered to the other injectors.

Parameters monitored included: delivered H_2O_2 concentrations and injection rates; delivered catalyst formulations and injection rates; injection pressures; groundwater elevations; water quality measurements for chlorides, iron, pH, and oxidation-reduction potential (E_H); and off-gas CO_2, O_2 and VOC concentrations. Observations of bubbling and frothing within wells and on the ground surface were also used to help make adjustments to the injection program.

Early in the CO injection program, a vacuum extraction system consisting of a 1-horsepower regenerative blower was installed and operated to help control unexpected fugitive VOC emissions at the ground surface which presented a health and safety concern. Flexible piping was connected to MW9 and MW17 and activated to collect and monitor vapors in the unsaturated soils. The vacuum extraction system discharge also provided an integrated picture of offgas composition, VOC mass volatilization rates and compositional changes over the course of the injection.

Post application monitoring of groundwater quality has been conducted since the completion of the 6-day CO injection program. Groundwater samples were collected and analyzed for VOCs by USEPA SW-846 Method 8260 at the following intervals following the injection: 1 week; 4 weeks; 5 months; and 15 months, with the most recent samples having been collected in July 1999. Groundwater sampling methods used have been consistent throughout the project.

RESULTS AND DISCUSSION

Indirect Performance Indicators. H_2O_2 levels and E_H were detected in area monitoring wells at increasingly high levels throughout the injection period with concentrations peaking on the 5th day. The highest concentration of H_2O_2 (>100 mg/L) was detected in MWs 2, 8, 13, 16 and 17 at various times during the later half of the injection program. E_H was increased from a low of about 150 mV to between 400 and 500 mV in some wells. Groundwater samples with highest E_H levels generally coincided with those collected from monitoring wells containing the highest concentrations of H_2O_2. Initial pH levels in the 6 to 7 range to more uniform pH levels in the 5 to 6 range by the second half of the injection program.

No definitive trends in measured chloride levels were observed over the course of the injection program due to the relatively low concentration of TCE in the target area groundwater and the limitations of the chloride test equipment (capable of measuring concentrations in the tens of mg/L). However, the absence of any detected increase in chlorides also suggests that pockets of more highly contaminated soil and groundwater (e.g., dense non-aqueous phase liquids) are either not present and/or were not successfully contacted or treated by the H_2O_2.

Groundwater was observed to mound in all 11 of the monitoring wells that were gauged with the magnitude of mounding generally tapering with distance from the remedial area. One notable exception to this general observation was the significant mounding (i.e., over 6 feet) observed in MW3, located approximately 60 feet from the nearest injector and approximately 90 feet from the closest active injector (5D) at the time of the mounding measurement. Increased mounding in this location suggests that a hydraulic conduit (i.e., enhanced permeability pathway) may have formed or pre-existed between the MW3 vicinity and the target treatment area.

O_2 levels in monitoring well head spaces and in the vapor extraction system discharge generally increased throughout the injection program. By the 5th day of injection, all O_2 levels were at or above ambient conditions. CO_2 concentrations generally decreased with time in the monitoring wells head spaces and in the vapor extraction system discharge, consistent with a decreasing supply of readily oxidized organic material. A notable exception to this pattern were the CO_2 concentrations measured in MW3 which were consistently the highest of all the monitoring wells and remained as high at the end of the injection (19.8%) as at the beginning (19.6%).

TCE concentrations in the vapor extraction system off-gas decreased from a high concentration of approximately 975µg/L (~180 ppm) to 300 µg/L (~60 ppm) over a three-day period spanning the middle portion of the CO injection program. Given that TCE contamination has not been found in site unsaturated soils, the measured TCE in off-gas suggests a saturated zone origin and that the TCE and other VOCs dissolved in groundwater and adsorbed to saturated soils were stripped and liberated by the injected air and reaction-released CO_2 and O_2.

Direct Performance Measurement. TCE concentrations in groundwater samples collected from target treatment area monitoring wells before and after completing the CO injection program are depicted in Figure 2. The figure illustrates that, with the exception of MW3, a significant drop in TCE concentrations from baseline or previous sampling results occurred in all of the monitoring wells in the days immediately following the CO injection. However, Figure 2 also shows that the decreases in TCE concentrations did not last.

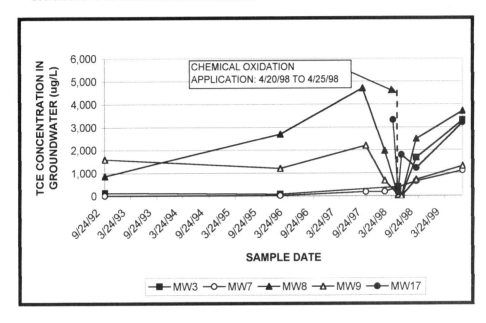

FIGURE 2: TCE concentrations in groundwater samples collected from five monitoring wells before and after completion of the CO injection program

One week following the CO injection program, TCE levels in MW8, MW9 and MW17 were measured to be 1 to 2 orders of magnitude less than baseline levels (1,970 µg/L, 674 µg/L and 3,320 µg/L, respectively). Four weeks later TCE concentrations in MW8 and MW9 were reduced by an additional order of magnitude (to the limits of laboratory detection), however, TCE in MW17 began to rebound by increasing to approximately 50% of its baseline concentration (i.e., to 1,798 µg/L).

Approximately 5 months following the CO injection program, TCE concentrations in MW3, MW7, MW8 and MW9 also rebounded dramatically. In fact, TCE concentrations measured in MW3 were 1 order of magnitude higher than had ever been historically measured in the well over the previous 9 years and TCE in MW7 was 3 times higher than had historically been measured. TCE in MW8 reached over 50% of a historical maximum concentration level (4,700 µg/L) and in MW9 reached over 30% of a historical maximum concentration level (2,200 µg/L).

The most recent groundwater sampling shows that 15 months after the CO injection program, TCE levels even further increased in all 5 wells. The TCE level in: MW3 was found to be almost 20 times higher than the historical pre-injection maximum concentration; MW7 was almost 6 times higher than its baseline level and about 14 times higher than the mean historical TCE concentration; MW8 was almost 2 times higher than its baseline level and about 1.6 times higher than the mean historical TCE level; MW9 was almost 2 times higher than its baseline level and approximately 90% of the mean historical TCE level; and MW17 was roughly the same as its baseline concentration.

The baseline and most recent TCE concentration data from Figure 2 are graphically overlain with interpolated isoconcentration contours on the site plan in Figure 3. This figure clearly shows not only that the CO application was unsuccessful in permanently reducing levels of TCE in groundwater, but also that the CO application may have, at least temporarily, changed the configuration of the TCE contamination plume.

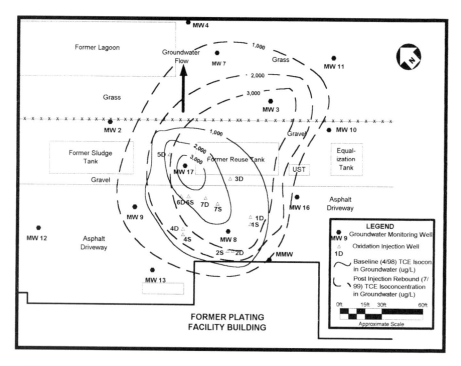

FIGURE 3: Site map showing location of TCE plume in groundwater before and after chemical oxidation treatment.

Explanations for the rebound of TCE in site monitoring wells that we entertained and briefly explored include the following: (1) the CO injection program successfully treated the hot spot but contaminated groundwater flowed back into the area from an upgradient source; (2) a continuing source of groundwater contamination in the hot spot area (e.g., NAPLs globules attached to soil particles) was not completely addressed by the CO injection and the source re-impacted the groundwater; (3) the effective radius of influence of the CO injection program around each injector was less than anticipated so intermediate areas were not addressed; and (4) the CO injection successfully treated TCE contamination along preferential flow paths but failed to address dissolved and adsorbed TCE in low permeability zones.

The rapid pace and pattern of rebounding TCE concentrations does not support the upgradient groundwater re-contamination explanation. As described above, the first monitoring well in the test area to rebound was MW17, located on the downgradient portion of the TCE plume and downgradient of monitoring wells which did not rebound until later. Furthermore, the order of magnitude rebound in MW17 occurred less than 5 weeks following the CO injection program, most likely an insufficient length of time for groundwater flow/chemical transport given the relatively low permeability of the site soils and the estimated average overburden groundwater velocity of about 13 feet per year. Finally, the rate of rebound for MWs 3, 7, 8, 9 and 17 were all roughly the same between sampling events conducted 5 months and 15 months after the CO injection in contrast to what would be expected if the wells were being re-contaminated from migrating groundwater.

Testing the Fenton's CO injection approach at the Department of Energy's Savannah River Site (SRS) similarly showed rebounding contaminate concentrations in groundwater following application of the technology. At SRS, the rebound was attributed to residual DNAPL globules in the soil (Jerome et al. 1997). However, this explanation for the technology's failure at the subject site is discounted because: (1) the site historical dissolved TCE concentrations have not been indicative of the presence of DNAPLs; (2) measurable concentrations chlorides indicative of DNAPLs were not produced during the injection program and (2) DNAPLs have not been found in site soil or groundwater.

The pH, H_2O_2, E_H, CO_2, O_2, and other chemical and physical data collected during the CO injection program suggest that the materials injected into the aquifer were dispersed laterally over distances exceeding conservatively projected estimates. Therefore, based on the chemical and physical observations, it does not appear likely that the rebounding TCE concentrations resulted from poor lateral distribution of the reaction materials.

The authors believe that TCE concentrations in groundwater rebounded because the CO technology was effective predominantly in preferential subsurface flow paths. This is consistent with multiple case studies recently completed, which indicate that soil heterogeneity plays a strong role in ultimate success or failure of the technology (ESTCP, 1999). The highly heterogeneous soils of the subject site inherently have a wide range of permeability characteristics. Under these circumstances, the injected CO reagents most likely moved predominantly through and reacted with TCE in the higher permeability horizons of the soil column (e.g., sand lenses), leaving the bulk of the lower permeability zones (e.g., silts) at least partially untreated. Groundwater samples procured from site monitoring wells following the CO injection program similarly yielded groundwater preferentially from these treated higher permeability zones and thus, initially provided the promising preliminary data. Once the CO treatment program was completed (i.e., the injections stopped and the reagents in the subsurface dissipated) adsorbed and dissolved TCE remaining in the low permeability horizons of the soil column began diffusing back into the treated higher permeability zones. Later, samples from the site monitoring wells show a remarkably uniform rate of rebound in TCE concentrations that would be consistent with this explanation.

As for the apparent re-configuration of the TCE plume, the authors believe that it is possible that the injection pressures reached at a number of the injectors could have been high enough (i.e., the soil overburden pressure was exceeded) to fracture the soil or opened/extended existing preferential flow paths. Opened or extended flow paths may have intersected soil zones that are more highly impacted with TCE due to isolation from predominate groundwater flow channels. Newly opened channels to TCE concentrated soil horizons hydraulically connected with groundwater monitoring wells would produce groundwater samples containing TCE levels more consistent with historic levels.

In summary, *in-situ* CO of a dissolved TCE plume in glacial till overburden material using Fenton's reagent chemistry did not provide permanent reductions in groundwater TCE concentrations. It is believed that the CO reactants predominantly addressed TCE in higher permeability soil horizons (e.g., sand lenses) and failed to adequately address TCE in the lower permeability dense till material (e.g., silts). Although TCE mass reduction is likely to have occurred, its magnitude cannot be readily quantified and the achieved mass reduction was insufficient to have reduced dissolved TCE levels to below pre-existing concentrations. Follow-up CO injections would likely incrementally reduce subsurface TCE mass, however, the poor results of the initial injection program has precluded this alternative from further consideration at this time.

REFERENCES

Bryant, D.J., Wilson, J.T., 1999. "In-Situ Fenton's Reagent Chemical Oxidation of Hydrocarbon Contamination in Soil and Groundwater". *Remediation.* 1999.

Environmental Security Technology Certification Program (ESTCP), 1999. *Technology Status Review, In-Situ Oxidation.*

Jerome, K.M, Riha, B., and Looney, B.B. 1997. *Final Report for Demonstration of In-Situ Oxidation of DNAPL Using the Geo-Cleanse Technology.* U.S. Department of Energy, Technical Report, WSRC-TR-97-00283, Westinghouse Savannah River Company, Savannah River Technology Center, Savannah River Site, Aiken, SC.

USEPA, 1998. Field Applications of In-Situ Remediation Technologies. EPA 542-R-98-008.

CHEMICAL OXIDATION OF PCE AT A DRY CLEANER SITE

H. Eric Nuttall (University of New Mexico, Albuqurque, NM)
Venkat M. Rao (BioManagement Services, Inc., Tinley Park, IL)
Sudhakar R. Doppalapudi (GLSEC, Lansing, IL)
William Lundy (BioManagement Services, Inc., Tinley Park, IL)

ABSTRACT: This study investigates the feasibility of soil remediation at a dry cleaner site using an in-situ oxidation technology known as The BiOx® Process. A pilot scale test was conducted at the subject site, contaminated with perchloroethylene, to determine the remedial potential of the BiOx® Process. Using modified Fenton's Reactions, the BiOx® Process yields powerful oxidizing agents which are capable of destroying chlorinated organic hydrocarbon compounds within the contaminated soil matrix. The destruction of these organic compounds by chemical oxidation yields harmless byproducts, primarily CO_2, water and $Cl_{(aq)}$.
The chemical degradation of chlorinated hydrocarbon compounds via in-situ oxidation permits site remediation in a timely, cost effective manner. The pilot test was conducted in the area of highest soil impacts of PCE at the site. Two (2) soil borings were conducted at the previously identified locations where PCE was detected at concentrations of 24.9 mg/kg and 71.11 mg/kg. The post pilot test analytical results indicated non detect PCE concentrations at both of these locations.

INTRODUCTION

Tetrachloroethylene (perchloroethylene, PCE) has been widely used as a reagent for dry cleaning, metal degreasing, and as an industrial solvent. Chlorinated aliphatic hydrocarbons are a common group of groundwater contaminants due to their widespread industrial use. In the environment, these compounds are persistent and toxic. Of this group, PCE is the only chloroethene isomer that persists throughout aerobic biodegradation[1].

The subject pilot test site is a former dry cleaner site located in a southeastern suburb of The City of Chicago, Illinois. While applying for a building permit to remodel the former dry cleaning property, it was discovered that a release incident (perchloroethylene release) had been previously reported to the Illinois Environmental Protection Agency. An environmental assessment revealed the soil to be impacted by perchloroethylene. The usual remedial techniques, including excavation and soil vapor extraction were evaluated. Site geology, consisting of primarily sandy material, supported SVE as a plausible option. However, the site was further complicated by the presence of the PCE plume beneath the building and the parking lot. Excavation was eliminated as a remedial option because of potential damage to the building structure. Similarly, soil vapor extraction would have necessitated installation of SVE system components partially inside the building which would have interrupted the normal flow of business.

Conventional Fenton chemical oxidation using hydrogen peroxide was considered, but there was a concern that heat and pressure associated with the technology might present a potential health and safety risk for on-site employees of the dry cleaner facility. The BiOx® Process was proposed as the remedial technology option because of its ease of application and the very low temperature changes (<10 °F) produced by the oxidation reaction, which greatly reduced the safety hazards for on-site employees. A pilot scale study was conducted to evaluate the feasibility of this technology at the site.

THE BiOx® PROCESS

The BiOx® Process oxidative chemistry produces a controlled release of free radicals which act as powerful oxidizing agents to destroy a variety of organic contaminants. Destruction of contaminants occurs by cleavage of the parent compound, creating less complex molecular fragments. Upon completion of the initial oxidation phase, BiOx® reagents continue to release molecular oxygen and nutrients for a period of time dependent upon contaminant flux and application rate. A complete destruction of organic contaminants results in carbon dioxide, Cl(aq) and water as the primary end products. Any remaining intermediate byproducts can also be mineralized by intrinsic biodegradation processes. This is particularly evident by the biological oxidation of vinyl chloride, in the presence of dissolved oxygen[2].

The disadvantages of the conventional oxidation processes mentioned above can be overcome by The BiOx® Process. BiOx® does not require a low pH range; reactions take place within the normal pH range of groundwater (6 to 8 standard units). The process does not produce any excessive heat. The BiOx® Process occurs at temperatures <10 °F above ambient temperature. Because the BiOx® reaction is controlled, it eliminates the heat and pressure prevalent with processes that use only liquid hydrogen peroxide[2].

One of the major limiting factor for those implementing in-situ oxidation remedial technologies has been the ability to cost effectively place reagents in contact with significant portions of the contaminant media. BiOx® Technology consists of utilizing BMS Inc.,'s Direct Injection Delivery (DID™) System to introduce BiOx® reagents into the subsurface[3]. Using a soft advance drilling technique, the DID™ System creates a high velocity fluid jet which liquefies the soil at the injector tip. This allows the injector probe to be easily advanced by hand. Because liquid accomplishes drilling, safety and protection of subsurface structures are assured. This approach should significantly reduce the dependency upon natural migration and maximize contact with the contaminated media. This approach allows to more evenly permeate the contaminant plume without the need of injection wells[2].

By placing a smaller volume of reagent into a larger number of surface access points, the required volume of reagent is applied evenly over the site. This minimizes the chance for areal expansion or migration of the contaminant plume. Because the injector is hand-held, contaminants under buildings can be reached by drilling small access holes through slabs or basement floors. If necessary, angular injection can also be made from outside the structure.

PILOT TEST METHODOLOGY

The pilot study area encompassed approximately 4,200 sq ft. in areal extent. Vertically, soil was impacted by the PCE release over an interval of eleven (11) feet, from two(2) feet to thirteen (13) feet below ground surface (BGS). The total volume of impacted soil within the pilot test area was approximately 1,900 cubic yards. Because the contaminated media was composed primarily of high permeable sand, an injection matrix of five (5) feet was selected. This equated to 168 injection points through which the BiOx® reagents were injected.

Access to the subsurface requires drilling ¾" holes through the floor or parking lot, to permit access for the injection rod. Treatment of the contaminant plume residing under the building slab could be easily accomplished. The remedial injection process was carried out by isolating a small portion of the building during injection. This allowed normal company business to proceed during the work. Once the injection is completed the injection holes are patched and surfaced with the appropriate material (asphalt/concrete etc.)

The BiOx® application was made on 2 September 1998. Prior to application, perchloroethylene concentrations over most of the site ranged from 0.38 to 5.42 mg/kg with an average level of 4.39 mg/kg. Two (2) areas with PCE concentrations as high as 24.6 mg/kg and 71.11 mg/kg, located at soil boring locations B-2 and B-7, respectively were identified within the pilot test area prior to remedial treatment.

PILOT TEST RESULTS AND DISCUSSION

For purposes of tracking the progress of the remedial treatment, locations B-2 (10' - 12') and B-7 (10'-12') were sampled seven (7) days after the injection event on September 9, 1998. Based on the results of the pilot test, it is apparent that BiOx® reagents are capable of destroying PCE in soil and groundwater. Confirmatory soil borings collected seven (7) days after the completion of the pilot test injection event revealed that PCE concentrations had dropped from 24.6 mg/kg to <0.001 mg/kg at location B-2, and from 71.1 mg/kg to <0.001 mg/kg at location B-7. This represents a contaminant reduction of greater than 99% in both locations.

The primary purpose of the pilot test was to evaluate the applicability of the treatment technology with respect to the PCE reduction and it's ability to place the remedial fluids in contact with the contaminated media underneath the parking lot and the building. The BiOx pilot test results have indicated that the technology is the best alternative for remediation of the entire plume at the subject site. The process itself is discrete and can be carried out with minimal site disturbance. The technology is cost effective in comparison with other alternatives evaluated and requires no operation and maintenance once the injection is completed. Because of the budget constraints additional sampling and analysis was not conducted which would have helped in a more detailed technical analysis of this technology.

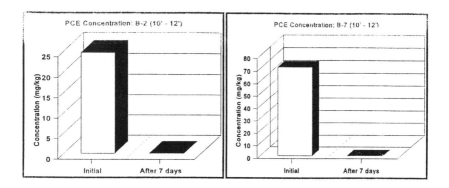

FIGURE 1: Pretretament and Post Treatment Concentreations of PCE (mg/kg)

REFERENCES

Chaudhary, G.R., and Chapalamadugu., 1991, *Biodegradation of Halogenated Organic Compounds*, Microbiol. Rev 55(1):59-79.

Nauta, Robert, and Lundy, William, *Remediation of Soils Contaminated with Tetrachloroethylene*, Wisconsin Fabricare Institute Magazine, July-August, 1999.

Product Literature on The BiOx® Process, BioManagement Services, Inc., Tinley Park, IL, ph. (888) 477-0550.

Ravikumar, Joseph X., and Gurol, Mirat D., *Chemical Oxidation of Chlorinated Organics by Hydrogen Peroxide*, Environ. Sci. & Technology, Vol 28, No. 3, 1994.

Schwarzenbach, R. P., Gschwend, P. M., and Imboden, D. M., *Environmental Organic Chemistry*, John Wiley & Sons, New York, 1993.

Tyre, Brian W., Watts, Richard J., and Miller, Glenn C., *Treatment of Four Biorefractory Contaminants in Soils Using Catalyzed Hydrogen Peroxide*, J. Environ: Qual 20: 832-838, 1991.

FENTON OXIDATION: BRINGING POLLUTANTS AND HYDROXYL RADICALS TOGETHER

Matthew A. Tarr, Michele E. Lindsey[1], Jia Lu, and Guoxiang Xu, University of New Orleans, New Orleans, LA, USA

Abstract: Degradation of aqueous aromatic compounds with hydroxyl radical produced by Fenton chemistry is inhibited by dissolved natural organic matter (NOM). The degree of inhibition is significantly greater than that expected based on a simple model in which the fraction of aromatic compound bound to (NOM) is considered to be unreactive. The reaction of several aromatic compounds (o-cresol, phenol, fluorene, pyrene, and phenanthrene) with hydroxyl radical was monitored as a function of Suwannee River fulvic acid or humic acid concentration. Steady state hydroxyl radical concentration was produced by Fenton chemistry ($H_2O_2 + Fe^{2+} \rightarrow Fe^{3+} + HO^- + HO\cdot$) with continuous addition of hydrogen peroxide to acidic Fe^{2+} solutions. Separation of the hydroxyl radical formation sites from the location of the aromatic compound is believed to be the reason for the observed reduction in rate constants, indicating that these systems are heterogeneous on a microenvironmental scale. The larger than expected inhibition of Fenton degradation by NOM indicates that natural organic matter presents a significant impediment to remediation of pollutants in natural waters and soils. In order to improve remediation efficiency better methods are needed to bring the pollutant into close proximity with the hydroxyl radical formation sites.

INTRODUCTION

Sorption of pollutants to humic substances can dramatically effect the transport, bioavailability, and chemical reactivity of the pollutant. Although partitioning to microenvironments within humics has been studied with respect to pollutant transport [Landrum *et al.*, 1984; Herbert *et al.*, 1993; Backhus and Gschwend, 1990] and bioavailability [Guha and Jaffe, 1996; Nam and Alexander, 1998; Cornelissen *et al.*, 1998], little effort has been made to understand how partitioning affects chemical reactivity of the pollutants [Lindsey and Tarr, 2000a; Sedlak and Andren, 1994]. Both physical and chemical changes upon sorption can alter pollutant reactivity. For example, a pollutant in the interior of a particle may be physically isolated from reactants in bulk solution. In addition, rate constants or mechanisms may be different for freely dissolved and sorbed compounds. Detailed information about chemical reactions within humic microenvironments is needed in order to understand both naturally occurring and engineered chemical transformations of pollutants sorbed to humics.

Polycyclic aromatic hydrocarbons (PAHs) are well studied pollutants with known carcinogenic effects [Jacob, 1996]. PAHs are ubiquitous pollutants with both natural and anthropogenic sources. In addition, PAHs are hydrophobic compounds which have been shown to partition into hydrophobic microenvironments, including hydrophobic sites of natural organic matter such as humic and fulvic acids [Backhus and Gschwend, 1990; Chin *et al.*, 1997]. One method under considerable investigation for the degradation of hydrophobic pollutants is reaction with hydroxyl radical (HO·) formed by the Fenton reaction [Walling, 1975]:

$$H_2O_2 + Fe^{2+} \rightarrow Fe^{3+} + HO^- + HO· \qquad (1)$$

The ferrous ion is regenerated through additional reactions, and therefore plays a catalytic role in hydroxyl radical production. The iron catalyzed formation of HO· from hydrogen peroxide has been widely studied [Watts *et al.*, 1990; Gau and Chang, 1996; Sun and Pignatello, 1993].

We have investigated the reaction of hydroxyl radical generated by the Fenton reaction with polycyclic aromatic hydrocarbons and chlorinated aromatic pesticides in the presence of humic substances [Lindsey and Tarr, 2000b; Lindsey and Tarr, 2000c]. Steady state concentrations of hydroxyl radical were generated by continuous addition of hydrogen peroxide to iron (II) containing solutions. The rate constant for pollutant degradation was determined as a function of added Suwannee River fulvic acid or humic acid.

EXPERIMENTAL

High purity water was obtained from a NanopureUV (Barnstead) water treatment system using a distilled water feed. Suwannee River fulvic acid and humic acid standards were purchased from the International Humic Substances Society (http://www.ihss.gatech.edu). Hydrogen peroxide (EM science, ~30%) was standardized using iodometric titration. Iron (II) perchlorate (99+%) was purchased from Alfa. Phenanthrene (99+%), pyrene (99%), fluorene (99%), benzoic acid (99.5+%), o-cresol (99+%), and p-hydroxybenzoic acid (99+%) were purchased from Aldrich. Phenol (99+%) was purchased from Fisher, and 1-propanol (99+%) was purchased from Mallinkrodt. All reagents were used as received.

Hydroxyl radical was produced using Fenton chemistry. Prior to degradation, individual compounds were dissolved in water at concentrations below their solubility, and the pH was adjusted to 2.5 with hydrochloric acid. Just prior to the use of each solution, an aliquot of $Fe(ClO_4)_2$ (e.g., 12 µL of 12.5 mM aqueous at pH 2.5) was added to yield an initial Fe^{2+} concentration of 50 µM. Hydrogen peroxide was then added continuously with a syringe pump (KD Scientific) at a low flow rate of 0.5 mL hr^{-1} so that volume changes were negligible over the course of all experiments. Continuous addition of H_2O_2 was used to establish a steady state hydroxyl radical concentration in each degradation reaction. All reactions were performed in the dark (except as noted) at 20°C with constant stirring. Steady state hydroxyl radical concentrations were determined by following the reaction of benzoic acid to form p-hydroxybenzoic acid [Lindsey and Tarr, 2000a; Lindsey and Tarr, 2000b; Zhou and Mopper, 1990].

Phenol, fluorene, and o-cresol were determined by HPLC [Lindsey and Tarr, 2000b; Lindsey and Tarr, 2000c]. Each of these compounds was allowed to react at a steady state hydroxyl radical concentration for predetermined time intervals. At these times, the reactions were quenched with 0.50 mL 1-propanol for every 10 mL of reaction solution, and the analytes were then quantitated. Each of these reactions represents one time point. Taken together, the individual points were used to reconstruct kinetic plots of analyte concentration verses time.

The degradation of phenanthrene and pyrene were monitored continuously using fluorescence detection [Lindsey and Tarr, 2000b; Lindsey and Tarr, 2000c]. Phenanthrene and pyrene were excited at 292 nm and 318 nm, and fluorescence emission was monitored at 344 nm and 370 nm, respectively. Data was collected at 1s intervals as the reaction proceeded. Dark control experiments were carried out to ensure that the excitation radiation did not affect the reaction and to verify that adsorption to glassware was not an important factor in these studies.

RESULTS AND DISCUSSION

For all compounds studied, pseudo first order degradation was observed in the absence of humics. Except for pyrene, all compounds studied maintained pseudo first order kinetics upon addition of humic or fulvic acid. For pyrene, addition of fulvic acid resulted in apparent second order kinetics. Figures 1 and 2 illustrate the observed kinetics for pyrene degradation in the absence and presence of fulvic acid. One possible explanation is that pyrene sorbed to fulvic acid in groups of two or more pyrene molecules in close proximity.

All compounds studied showed a decrease in rate constant for reaction with hydroxyl radical upon addition of dissolved humics. This effect is at least partially due

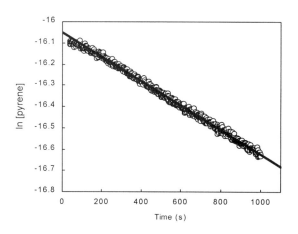

Figure 1. Pseudo first order disappearance of pyrene in the presence of a steady state hydroxyl radical concentration. Pyrene concentration monitored by fluorescence.

to sorption of the pollutant into microenvironments within humics. The more hydrophobic compounds, which had a higher degree of sorption to humics, showed a more dramatic decrease in rate constant for reaction with hydroxyl radical. Figure 3 illustrates the normalized decrease in rate constant for several pollutants.

Although the decrease in rate constant was correlated to pollutant hydrophobicity, the magnitude of these changes was too large to be explained by a simple partitioning model. For example, in the presence of 30 mg L^{-1} fulvic acid, phenol is predicted to be less than 0.1% bound. Nevertheless, the rate constant for

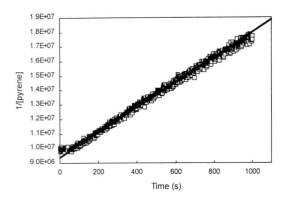

Figure 2. Pseudo second order disappearance of pyrene in the presence of a steady state hydroxyl radical concentration with added fulvic acid. Pyrene concentration monitored by fluorescence.

phenol dropped by 40% at this concentration of fulvic acid.

Since humics are known to bind metals, it is likely that the catalytic iron ions existed predominantly at hydrophilic binding sites within the humic molecules. Consequently, generation of hydroxyl radical was probably isolated to these sites and was not uniformly distributed within solution. Compounds of increasing hydrophobicity would be less and less likely to be close to the iron coordination sites. Since hydroxyl radical is very short lived in aqueous solution, collisions between the radical and hydrophobic pollutants is unlikely. A two fold mechanism in which hydroxyl radical is concentrated into polar regions and non-polar compounds partition into hydrophobic regions could explain why organic compound partition coefficients were not sufficient to explain the observed rate constant effects.

CONCLUSIONS

Dissolved natural organic matter dramatically inhibits the Fenton degradation of even slightly hydrophobic compounds. Decreased hydroxyl radical concentration at the microenvironmental site of the hydrophobic compound is proposed to explain this observation. Even low concentrations of dissolved NOM substantially hinders degradation of hydrophobic compounds. Remediation technologies that rely on Fenton chemistry will therefore be inefficient in the presence of NOM. Although this study focused on aqueous solutions, heterogeneous systems such as soil and sediment will probably exhibit even more dramatic effects.

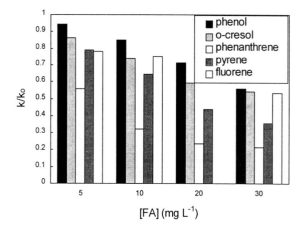

Figure 3. Normalized rate constants for reaction with hydroxyl radical as a function of added fulvic acid concentration.

Increasing the effectiveness of Fenton degradation in real systems requires the following: 1) minimizing scavenging of hydroxyl radical by non-pollutant compounds, 2) stabilizing iron in soluble forms that can participate in the catalytic Fenton cycle to form hydroxyl radical, and 3) maximizing the accessibility of hydroxyl radical to pollutants. New technologies to accomplish these three goals are currently being developed in our research group.

ACKNOWLEDGMENT

This work was supported by the Louisiana Board of Regents under grant LEQSF(1998-01)-RD-A-34.

REFERENCES

Backhus, D. A. and P. M Gschwend. 1990. "Fluorescent polycyclic aromatic hydrocarbons as probes for studying the impact of colloids on pollutant transport in groundwater." *Environ. Sci. Technol. 24:* 1214-1223.

Chin, Y. P., G. R. Aiken, and K. M. Danielsen. 1997. "Binding of pyrene to aquatic and commercial humic substances: the role of molecular weight and aromaticity." *Environ. Sci. Technol. 31:* 1630-1635.

Cornelissen, G., H. Rigterink, M. M. A. Ferdinandy, and P. C. M. van Noort. 1998. "Rapidly desorbing fractions of PAHs in contaminated sediments as a predictor of the extent of bioremediation." *Environ. Sci. Technol. 32:* 966-970.

Gau, S. H., and F. S. Chang. 1996. "Improved Fenton method to remove recalcitrant organics in landfill leachate." *Water Sci. Tech. 34:* 455-462.

Guha, S. and P. R. Jaffe. 1996. "Bioavailability of hydrophobic compounds partitioned into the micellar phase of nonionic surfactants." *Environ. Sci. Technol. 30:* 1382-1391.

Herbert, B. E., P. M. Bertsch, and J. M. Novak. 1993. "Pyrene sorption by water-soluble organic-carbon." *Environ. Sci. Technol. 27:* 398-403.

Jacob, J. 1996. "The significance of polycyclic aromatic-hydrocarbons as environmental carcinogens." *Pure Appl. Chem. 68:* 301-308.

Landrum, P. F., S. R. Nihart, B. J. Eadie, and W. S. Gardner. 1984. "Reverse-phase separation method for determining pollutant binding to Aldrich humic acid and dissolved organic carbon of natural waters." *Environ. Sci. Technol. 18:* 187-192.

Lindsey, M. E. and M. A. Tarr. 2000a. "Quantitation of hydroxyl radical during

Fenton oxidation following a single addition of iron and peroxide." *Chemosphere*, in press.

Lindsey, M. E. and M. A. Tarr. 2000b. "Inhibition of hydroxyl radical reaction with aromatics by dissolved natural organic matter." *Environ. Sci. Technol. 34*: 444-449.

Lindsey, M. E. and M. A. Tarr. 2000c. "Inhibited Hydroxyl Radical Degradation of Aromatic Hydrocarbons in the Presence of Dissolved Fulvic Acid." *Wat. Res.*, in press.

Nam, K, and M. Alexander. 1998. "Role of nanoporosity and hydrophobicity in sequestration and bioavailability: tests with model solids." *Environ. Sci. Technol. 32*: 71-74.

Sedlak, D. L. and A. W. Andren. 1994. "The effect of sorption on the oxidation of polychlorinated-biphenyls (PBCs) by hydroxyl radical." *Wat. Res. 28*: 1207-1215.

Sun, Y. and J. J. Pignatello. 1993. "Activation of hydrogen-peroxide by iron(III) chelates for abiotic degradation of herbicides and insecticides in water." *J. Agric. Food Chem. 41*: 308-312.

Walling, C. 1975. "Fenton's reagent revisited." *Acc. Chem. Res. 8*: 125-131.

Watts, R. J., M.. D. Udell, P. A. Rauch, and S. W. Leung. 1990. "Treatment of pentachlorophenol-contaminated soils using Fenton's reagent." *Haz. Waste Haz. Mat. 7*: 335-345.

Zhou, X. and K. Mopper. 1990. "Determination of photochemically produced hydroxyl radicals in seawater and freshwater." *Mar. Chem. 30*: 71-78.

OZONE MICROSPARGING FOR RAPID MTBE REMOVAL

William B. Kerfoot (K-V Associates, Inc., Mashpee, Massachusetts)

ABSTRACT: Ozone microsparging allows rapid removal of methyl tert-butyl ether (MTBE) and benzene, toluene, methyl benzene, xylenes derivatives (BTEX) encountered at gasoline spill sites. Bench-scale testing indicated applicability to MTBE removal. Field testing was performed on two types of sites: (1) a commercial automotive station which had suffered a gasoline spill, and (2) a zone upgradient of a water supply well. At the former, MTBE and BTEX compounds were treated, while at the latter only MTBE was present for treatment. The objective at the station was remediation of soil and groundwater, while the second site objective was use of a "bubble fence" to control a plume advancing towards the well. During a 20-day monitoring period at the station site, 95% reduction of BTEX and 99% reduction of MTBE was observed in a monitoring well within 18 ft (5.6 m) of the main injection well. The protective bubble fence at the second site served to reduce exiting MTBE concentrations to non-detect levels.

INTRODUCTION

MTBE presents a challenge for effective treatment in groundwater-based spills. A common fuel additive in the United States since the 1990's, MTBE is resistant to aerobic and anaerobic microbial breakdown (Suflita and Mormile, 1993). Activated carbon is not a very effective adsorber since MTBE is highly water soluble and poorly adsorbed (Reisinger, et al., 1986). Because of its relative low volatility, air stripping systems are capable of removing MTBE only if exceptionally high air to water ratios are used.

Ozone microsparging (the C-Sparge™ process)* appears capable of effecting rapid removal of MTBE from contaminated groundwaters. The use of microscopic bubbles with high surface-to-volume ratios allows efficient extraction of MTBE from aqueous to gas phase. Differing from its attack on benzene ring compounds, ozone attack on ether occurs through insertion in the C-H bond. The use of hydrogen peroxide supplementation does not appear necessary when microbubble ozone is used (Karpel vel Leitner, et al., 1994). The resultant breakdown products are normally tert-butyl formate, tert-butyl acetate, and tert-butyl alcohol. These breakdown products are either further oxidized or biodegraded rapidly under the aerobic conditions promoted by ozone decomposition to oxygen.

TECHNOLOGY DESCRIPTION

Ozone microsparging is a patented technology for *in-situ* treatment of volatile organic compounds (VOCs) in groundwater and surface water. The

---

*U.S Patent #5,855,775; other U.S. and foreign patents pending

technology combines the unit operations of air stripping and oxidative decomposition in a single process which can be catalytically accelerated. In the C-Sparge™ process, air and ozone are injected directly into groundwater through specially-designed spargers to create small "microbubbles" that have a very high surface area to volume ratio. As these microbubbles rise within a column of water, they extract or "strip" VOCs from groundwater by aqueous to gas partitioning. Upon entering the microbubbles, the VOCs are rapidly oxidized by the process of Criegee oxidation. The ozone contained within the bubbles reacts to decompose the chlorinated ethane molecule in an extremely rapid gas/liquid phase reaction whose end products are carbon dioxide, very dilute hydrochloric acid, and water. By increasing the ozone content within the bubbles, the rate of oxidation reaction is increased. Properly applied, the reaction has been used to rapidly detoxify halogenated VOCs (HVOCs), ethene and ethane derivatives and aromatic compounds in groundwater sources to below drinking water standards without producing unwanted harmful by-products (Kerfoot, 1997; Kerfoot, et al., 1998).

Gas entering a small bubble of volume $4\pi r^3$ increases until reaching an asymptotic value of saturation. If we consider the surface of the bubble to be a membrane, a first order equation can be written for the monomolecular reaction:

$$\frac{dx}{dt} = -k(Q-x) \tag{1}$$

Where: x = the time varying concentration of the substance in the bubble
  dx/dt = the rate of change of vapor concentration within the gaseous phase
  Q = the external concentration of the HVOC
  k = the absorption constant, set by Henry's partitioning coefficient

If at time $t = 0$, $x = 0$, then:

$$X = Q(1-e^{-k}) \tag{2}$$

The constant k is found to be:

$$-k = \frac{dx/dt}{Q-x} \tag{3}$$

Normally, the rate of adsorption reaches an asymptotic value because the concentration of x within the bubble becomes at equilibrium with the Henry's coefficient. However, if ozone decomposes the incoming vapor, the mass removal rate will increase.

Since MTBE occupies a partitioning region similar to the common fuel aromatics benzene, toluene, methyl benzene, xylenes (Reisinger, et al., 1986), where the volatile compounds can move from aqueous to gaseous phases and have reasonable water solubility, it appeared likely that its reaction to ozone degradation would be similar. The Henry's Constant of MTBE is about one-tenth that of the benzene derivatives. Figure 1 shows the position of MTBE relative to water solubility (atm m³/mol1) compared with the aromatics and common soluble alkanes in groundwater from fuel spills (Reisinger, et al., 1986; Kerfoot, 1994).

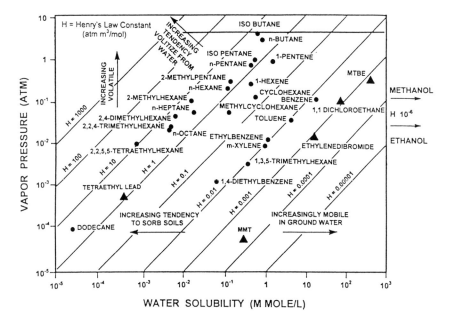

FIGURE 1. Partitioning properties of selected gasoline constituents (modified from Reisinger, et al., 1986).

BENCH-SCALE TESTING

Materials and Methods. The reaction of microbubble ozone on MTBE in dilute aqueous solution was studied under controlled experimental conditions. For batch-type experiments, ozone was bubbled under pressure in 500 ml Erlenmyer flasks. A corona-type ozone generator fed by oxygen was used for generation. Ten grams of fine glass beads (.3 mm) were added to the flask to simulate a fluidized soil mixture. A 10-micron porous cylinder (2 cm long x 2.5 cm diameter) was used to introduce an ozone-air mixture through the solution. The pH was maintained in the range of 5.0 to 6.0. Ozone concentration, pressure, and iron silicate content were experimentally varied. After the addition of ozone, the flasks were covered and stored for 60 minutes at room temperature, and then analyzed.

Ozone content in the gas phase was analyzed by Gastec tubes (#18L per ANSI/ISEA 102-1990). In the liquid phase, ozone content was analyzed by the Indigo Trisulfonate Method (Bader and Hoigne, 1982). MTBE and ozonation products were identified using a Hewlett Packard gas chromatograph (Model 6890) and mass spectrometer (Model 5973). A laboratory gas chromatograph

(HNU Model 320 with capillary column) was used for quick inspection of MTBE removal during testing.

Results. Figure 2 represents the results obtained under laboratory conditions (gas flow = 6.0 L/hr). MTBE is rapidly degraded by microbubble ozone injection. The results are presented as ozone residual in the aqueous phase, compared to ozone dosage, to enable comparison with previous literature results. The rate of decay is similar to that previously reported by Karpel vel Leitner, et al. (1994). In the bench-scale testing, ozone microbubbles appeared effective in reducing MTBE concentrations to beyond 90% of original levels. The rate of removal was sensitive to ozone concentration, pressure, and iron silicate content.

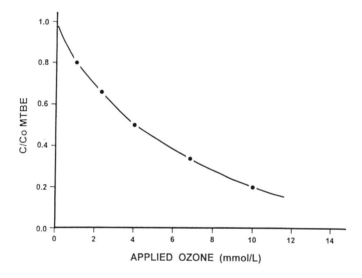

FIGURE 2. MTBE removal with exposure to microbubble ozone during bench-scale semi-batch testing.

FIELD TESTING

Field testing for MTBE removal in groundwater was performed at two different sites: (1) a source region of a gasoline spill at an automotive service station, with aromatic (BTEX) contaminated soil andground water, and (2) at the forefront of a solely MTBE plume upgradient of a water supply well. Leading portions of gasoline spill plumes often have MTBE separating from other fuel constituents because it is very soluble and tends to not be adsorbed to soils. The rate of removal could be compared with other sites where aromatic fractions have been treated (Kerfoot, 1998).

Site 1: Automotive Service Station. A C-Sparge™ unit (Model 3600) was installed at a gasoline service station. The unit was equipped with an oxygen generator, allowing ozone concentrations to be adjusted between 100 and 300 ppmv. A recirculating well installation was employed to assure adequate ozone mixing in the vicinity of the monitoring well. Dissolved oxygen content and redox potential (ORP) were monitored to define the radius of influence (ROI). Previous borings on the site showed static groundwater at depths of 3.0 to 6.0 ft (1 to 1.8 m) below grade. A brownish-gray silt extended from fill at 2.5 ft (.8 m) to about 6 ft (1.8 m) below grade. Medium sand and gravel extended to depths of 20 ft (6 m) below grade. Soil permeability ranged from 10^{-3} to 10^{-7} cm/sec. The upper soils between 3.5 to 4.0 ft (1 to 1.2 m) were measured at 9.0 x 10^{-7} cm/sec with a moisture content of 299. Fine materials (silt and clay) in smaller fractions (5% or less) were found intermixed with the sand and gravel. The pH of the groundwater was determined to fall in the range of 7.0 to 8.5. Groundwater flow was low, with a hydraulic gradient of 0.0036 ft/ft across the site. Specific conductance ranged from 850 to 1,100 µS.

Total BTEX concentrations ranged from 5000 to 24,000 ppb in groundwater samples. The vertical extent of hydrocarbon contamination extended to a maximum depth of 15 ft (4.6 m) sub-grade. No light non-aqueous phase liquid (LNAPL) was observed at the site. Dissolved oxygen contents in the groundwater were below 1 ppm in nearby monitoring wells.

Combined air/ozone flow to the recirculation well was 2.2 cfm. The system was set to run 18 minutes for the lower Spargepoint®, 11 minutes for the in-well Spargepoint®, and 5 minutes for the submersible pump. The system operated on 14 cycles of 34 minutes each during a 24-hour period.

Monitoring was performed during a 20-day test period on the site, beginning July 29, 1997. Analytical results from a monitoring well located 18 ft (5.5 m) from the main recirculation well (CS-1) are presented in Figure 3. During the period of the 20-day test, total BTEX plus MTBE concentrations dropped from 19,220 ppb to 1004 ppb. This represents a reduction of 94.8% during the test period. MTBE concentrations dropped from 520 ppb to 6 ppb during the test period, representing in excess of 99% removal during ozone treatment.

Site 2: Upgradient of Water Supply Well. An oxidative curtain, or "bubble fence", was employed to attempt to contain an advancing MTBE plume threatening a water supply well near Lake Tahoe, California. The soil consisted of silty, clayey sand. The static groundwater level at the time of installation was 5 ft (1.5 m) below grade, although levels are known to drop to 10 ft (3 m) below grade. A groundwater gradient of 5 ft (1.5 m) per 40 ft (12.2 m), or 0.125 ft/ft, was measured. The observed forward advance of the plume was between 3 ft (1 m) to 10 ft (3 m) per month. The bottom of the contaminated zone was 20 ft (6 m) below grade. The upgradient breadth of the plume was estimated to be about 50 ft (15.2 m).

After careful study of optimal placement, an oxidative curtain was installed at right angles to the advancing plume front to intercept and contain the MTBE. A bubble fence composed of 4 dual-level Spargepoints® fed by a Model 3600 C-Sparge™ unit was installed. The Spargepoints® were installed about 13 ft (4 m) on center in a line about 40 ft (12.2 m) long, 100 ft (30.5 m) upgradient of the water supply well. Two levels of Spargepoints® were installed, one at

between 21 ft and 24 ft (6.4 m and 7.3 m) below grade, and another at between 36
ft and 39 ft (11 m and 11.9 m) below grade

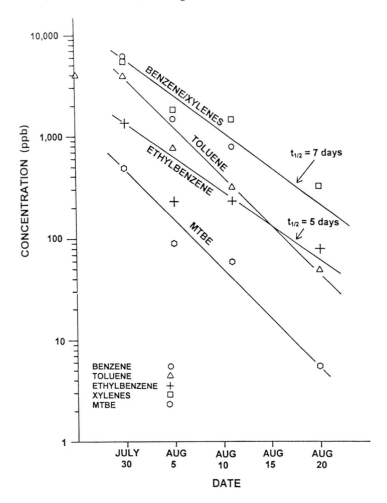

**FIGURE 3. Removal of MTBE and BTEX compounds during C-Sparge™
oxidation of ground water at a service station.**

Monitoring wells were installed upgradient and downgradient of the
oxidative curtain. Monitoring wells 9 A/B (shallow/deep) were installed 15 ft
(4.6 m) upgradient, and 17 A/B (shallow/deep) were installed be00tween 15 ft and
20 ft (4.6 m and 6 m) downgradient. Figure 4 shows the results of the MTBE
concentration change across the oxidizing zone. A level of up to 6 ppb was found
with the advancing front. The level measured in the upgradient monitoring well
did decrease somewhat with time, probably because it fell within the normal 20 ft

(6 m) radius of influence of the spargewells. The MTBE content in the downgradient well dropped to non-detect.

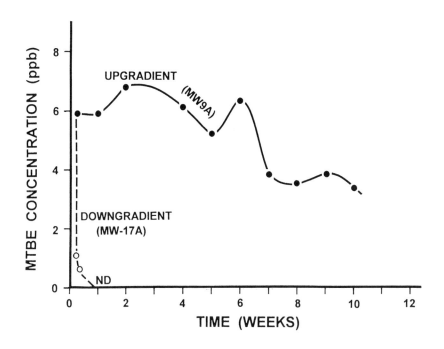

FIGURE 4. MTBE reduction across oxidative curtain.

CONCLUSIONS

The combination of fine bubbles, which increases the amount of stripping into the bubbles, and ozone decomposition of the constituents results in rapid removal of MTBE from contaminated groundwater. The knowledge that both BTEX and MTBE lie in similar partitioning regions, where volatile compounds move from aqueous to gaseous phases with reasonable water solubility, led to the assumption that MTBE could be treated with ozone microsparging as effectively as BTEX. Bench-scale tests showed this to be true. Field tests proved the usefulness for both point source treatment and bubble fence treatment. MTBE and toluene are observed to be removed with field attenuation coefficients of five days or less, for reduction to one-half concentration. Benzene, ethylbenzene, and xylenes exhibit half-lives of seven to nine days. The adsorbed soil fractions of MTBE are less than0 those of benzene fractions. MTBE removal occurs primarily from aqueous fraction.

REFERENCES

Bader, H. and J. Hoigne. 1982 "Determination of Ozone in Water by the Indigo Method: A Submitted Standard Method. *Ozone Science & Engineering.* 4:169-176.

Karpel vel Leitner, R. N. K. V., A. L. Papailhou, J. P. Crove, J. Payrot, and M. Dore. 1994. "Oxidati0on of Methyl Tert-Butyl Ether (ETBE) by Ozone and Combined Ozone/Hydrogen Peroxide." *Ozone Science & Engineering. 16:41-54.*

Kerfoot, W. B. 1994. "Soil Gas Movement – Monitoring Under Natural and Remediation Conditions." In D. I. Wise and D. J. Trantolo (Eds.), *Remediation of Hazardous Waste Contaminated Soils,* pp. 791-828. Marcel Dekker, Inc., New York, New York

Kerfoot, W. B. 1997. "PCE Removal from Groundwater with Dual Gas Microporous Treatment System." *Soil and Groundwater Cleanup,* October ed., pp. 38-41

Kerfoot, W. B., C. J. J. M. Schouten, and V. C. M. van Engen-Beukeboom. 1998. "Kinetic Analysis of Pilot Test Results of the C-Sparge™ Process." In G. B. Wickramanayake and R. E. Hinchee (Eds.), The Proceedings of the: *Physical, Chemical and Thermal Technologies, Remediation of Chlorinated and Recalcitrant Compounds,* Vol. C 1-5, pp. 271-277. Battelle Press, Columbus, Ohio.

Reisinger, H. J., D. R. Burris, L. R. Cessar, and G. D. McCleary. 1986. "Factors Affecting the Utility of S00oil Vapor Assessment Data." *Proceedings of: Petroleum Hydrocarbons and Organic Chemicals in Groundwater: Prevention, Detection, and Restoration,* Houston, Texas, November 12-14. Water Well Journal Publishing Co., Dublin Ohio.

Suflita, J. M. and M. R. Mormile. 1993. "Anaerobic Biodegradation of Known and Potential Gasoline Oxygenates in Terrestrial Subsurface." *Environmental Science & Technology.* 27:976-978.

AN IMPROVED CHEMICAL-ASSISTED ULTRASOUND TREATMENT FOR MTBE

Hung-Li Chang (University of Southern California, Los Angeles, California)
Teh Fu Yen (University of Southern California, Los Angeles, California)

ABSTRACT: Chain reactions involving ultrasound and Fenton's reagent are the epitome of this system. Both ultrasonic irradiation and Fenton's reagent can generate hydroxyl radicals in the solution. Fenton's reagent can also be continuously supplied by sonolysis of water and naturally occurring metal salts in the groundwater. The degradation strength of organic substrates is in the order: ultrasound with Fenton's reagent plus Cu^{2+} >> ultrasound with Fenton's reagent > Fenton's reagent only >> ultrasound with hydrogen peroxide oxidation > hydrogen peroxide alone > ultrasound alone > ferrous ion alone. In these systems, radical oxidation by Fe^{3+} is an electron-transfer process, but if Cu^{2+} is added, then the free radical competitors can be removed by either ligand transfer or organocopper intermediate formation. The feasibility of methyl tertiary butyl ether (MTBE) remediation by these processes is being studied. The preliminary results have shown 98% MTBE degradation. For subsurface operations in the field, a robotic self-powered mining head containing ultrasonic transducers that attached on the inner wall of the double-wall pipe can be used for site remediation as well as drinking water pretreatment. In addition, directional drilling technology can greatly increase the effective cleanup subsurface area.

INTRODUCTION

The metal-ion-catalyzed decomposition of hydrogen peroxide was discovered by H. J. H. Fenton in the 1890s who reported that iron salts strongly promoted the oxidation of malic acid by hydrogen peroxide. Hydroxyl radicals play the most important role in this process. The chain reaction can be described in simple aqueous solution as shown in Table 1. Fe3+ is the major species in the solution and hydroxyl radicals are generated. In the presence of organic substances, the hydroxyl radicals produce organic free radicals ($R\bullet$) through hydrogen abstraction, which can undergo oxidation by Fe^{3+} ($R_i\bullet$), dimerisation ($R_j\bullet$), or reduction by Fe^{2+} ($R_k\bullet$). Fe^{2+} is regenerated as a result of this process.
The chemical effects of ultrasound are mainly attributed to cavitation. Cavitation is a three-step process consisting of nucleation, growth, and collapse of a gas- or vapor- filled bubble in a body of liquid. During bubble implosion, intense heating of the bubble occurs. The localized hot spots have a temperature of roughly 5000°C, pressures of about 500 atmospheres, and lifetimes of a few microseconds. It is this high temperature and pressure that affect chemical reactions. The intensity of cavitation is affected by the following major factors:
- Temperature
- Cavitation bubble size
- Number of nuclei

- Intensity and frequency of the applied field
- Attenuation characteristics of bubbles
- Surface tension, viscosity, pH, and density of the liquid
- Vapor pressure
- Concentration and diffusion rates of dissolved gases
- Heat transfer rate in the fluid

As shown in Figure 1, sonochemistry occurs at three different regions: (1) Interiors of collapsing bubbles where extreme conditions of temperature and pressure exist transiently, which induces chemical reactions yielding products that are typical of pyrolysis/combustion reactions in the gas phase. In aqueous solutions, thermal decomposition of water vapor inside the bubbles produces hydroxyl radicals and hydrogen atoms. (2) Interfacial regions between the cavitation bubbles and bulk solution. The nonvolatile solute accumulated in this region can undergo thermal decomposition and then induce radical reactions. (3) Bulk solution where the radicals, being produced in the interior of bubble and in the interfacial region, that survive migration from the interface can undergo radical reactions with solute present in the bulk solution.

Hydroxyl radicals are generally nonselective and more powerful than oxygen or ozone alone. Upon initiation, the reactive radicals will go through a series of chain reactions until termination reactions end the reaction chain. The chain length depends on the reaction condition. The longer the chain length, the more effective the radical reaction. 2,2-Diphenyl-1-picrylhydrazyl (DPPH) was used to investigate the behavior of free radical, because of its unusual stability. When DPPH reacts with free radicals, 1,1-diphenyl-2-picrylhydrazine (DPPH$_2$) will be generated with a color change from violet to light yellow in the solution. Both DPPH and DPPH$_2$ have absorption maximum at 320 nm and 520 nm.

Cavitation bubbles

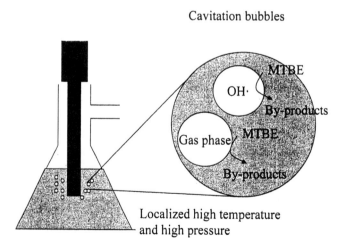

Localized high temperature and high pressure

Figure 1. Chemical reactions occur at three different regions under ultrasound.

Table 1. Chain reactions involving ultrasound and Fenton's reagent.

Ultrasound	Fenton's reagents
↓ $H_2O \xrightarrow{\text{)))}} OH\cdot + \cdot H$	
↓ $\xrightarrow{\text{)))}} H_2O_2 + H_2 \longrightarrow$	↓ $H_2O_2 + Fe^{2+} \rightarrow OH\cdot + OH^- + Fe^{3+}$
↓ $MTBE + OH\cdot$	↓ $OH\cdot + Fe^{2+} \rightarrow Fe^{3+} + OH^-$
↓ $\xrightarrow{\text{)))}} TBA$	↓ $OH\cdot + R_iH \rightarrow H_2O + R_i\cdot$
↓ $\xrightarrow{\text{)))}} acetone$	↓ $OH\cdot + R_jH \rightarrow H_2O + R_j\cdot$
↓ $\xrightarrow{\text{)))}} acetic\ acid$	↓ $OH\cdot + R_kH \rightarrow H_2O + R_k\cdot$
↓ $\xrightarrow{\text{)))}} carbon\ dioxide$	↓ $R_i\cdot + Fe^{3+} \rightarrow Fe^{2+} + product$
	↓ $2\ R_j\cdot \rightarrow product\ (dimer)$
	↓ $R_k\cdot + Fe^{2+} \rightarrow Fe^{3+} + R_k^- \rightarrow R_kH$
	↓ $R\cdot + Cu^{2+} \rightarrow product + Cu^+$
	↓ $Cu^+ + Fe^{3+} \rightarrow Cu^{2+} + Fe^{2+}$

In order to evaluate the degradation efficiency of the chemical-assisted ultrasound processes, we choose methyl tertiary butyl ether (MTBE) as our target chemical. MTBE has been used initially as an octane enhancer. Now MTBE is one of the major oxygenates blending into gasoline to meet Clean Air Act goals. At the same time, this environmental recalcitrant compound has been found both in groundwater, particularly shallow urban wells, and surface water, because it is highly soluble in the water and it migrates quickly. Beside chronicle health effect, MTBE has a distinctive taste and odor like ether that can be smelled at relatively low concentration. Therefore, California's secondary maximum contaminant level (MCL) is 5 ppb that is close to the odor threshold. The EPA method detection limit (MDL) is about 2 ppb with other organic substrate present. At this stage, scientists are still assessing the health risks of MTBE. All groundwater needs to be treated to ppb level, which is a big challenge for current available remediation methods.

Current researches had shown that MTBE can be degraded rapidly in atmosphere, but really slow in biodegradation, even with many different enhancements. Due to the fact that MTBE has high solubility and low Henry's constant, extraction processes including air stripping and air sparging can only have lower efficiency and require off-gas treatment. Advanced oxidation processes, such as ultrasound with ozone system still has problems with bromate occurrence in coastal area. Ultrasonic irradiation initiated a free radical reaction at the cavitation centers, which eventually can degrade MTBE. The combination of hydrogen peroxide and a ferrous salt, "Fenton's reagent," is an effective oxidant of a wide variety of organic substrates. Without ultrasound involved, the optimal hydrogen peroxide concentration has a molar ratio of H_2O_2: organic substrates as high as 50~500 : 1. However, in this system, hydrogen peroxide can be produced from sonolysis of water. The addition of little amount of hydrogen peroxide in this process is considered an initiator in the solution. Table 1 shows the chain reactions involving ultrasound and Fenton's reagent. Hydroxyl radicals play the most important role in attacking groundwater contaminants. Fenton's reagent can be continuously supplied by sonolysis of water and naturally occurring metal salts in the groundwater. In these systems radical oxidation by Fe^{3+} is an electron-transfer process. Due to Cu^{2+} was involved, MTBE degradation by-products such as tertiary butyl alcohol (TBA), acetone, and acetic acid can by removed by either ligand transfer or organocopper intermediate precipitation. The preliminary results have shown 98% MTBE degradation in 60 minutes through the chemical-assisted ultrasound processes.

In-situ ultrasound treatment has been used for enhancement of oil recovery. For subsurface operations in the field, a robotic self-powered mining head containing in excess of 200 ultrasonic transducers can be attached on the inner wall of the double-wall pipe. The drilling bit combines mechanical, hydraulic, and pneumatic systems to move through the contaminated zone. In addition, this new technology of horizontal drilling can greatly increase the effective area of the chemical-assisted ultrasound treatment. Many soil microorganisms possess enzymes that catalyze the destruction of hydrogen peroxide by converting it to oxygen. Therefore, chemical-assisted ultrasound processes can also enhance in-situ bioremediation.

MATERIALS AND METHODS

A Branson's ultraonic cleaning tank was used. The reaction temperature was kept at room temperature. The reaction solution was prepared by mixing a fixed volume of saturated solution of DPPH in methanol with 500 ml methanol-water (60:40 in volume) binary solvent mixture to obtain an initial DPPH concentration close to 2×10^{-5} M.

Ferrous ion was determined by the phenanthroline method. Ferrous ion test solution was prepared by mixing HCl, phenanthroline solution, ammonium acetate buffer, and distilled water with a ratio of 1:40:20:39. The ferrous ion concentration was determined photometrically at 510 nm by an UV-VIS spectrophotometer (Hewlett Packard 8452A). This method is good for the measurement of total ferrous ion amounts up to 50 μg.

MTBE and its degradation by-products were analyzed by "Standard Method 524.2" that is measurement of purgeable organic compounds in water by capillary column gas chromatography/mass spectrometry.

RESULTS AND DISCUSSION

The conversion of DPPH to $DPPH_2$ occurs in all seven conditions. The experiments conducted with DPPH that gave proof of free radical formation in the solution. The potential of radical generation are in the order: ultrasound with Fenton's reagent plus Cu^{2+} >> ultrasound with Fenton's reagent > Fenton's reagent only >> ultrasound with hydrogen peroxide oxidation > hydrogen peroxide alone > ultrasound alone > ferrous ion alone.

As shown in Figure 2, Fenton's reagent reacts with MTBE very rapidly, usually in less than ten seconds, 80% MTBE already been converted to TBA and acetone. The preliminary results have shown 98% MTBE degradation through 60 minutes ultrasonication.

CONCLUSIONS

When the cavitation bubbles explode, it generates local high temperatures and high pressures. These extreme conditions lead to the formation of free radicals in the interior of bubbles and in the interfacial region between bubbles and the bulk solution through pyrolysis processes. If radicals can survive the reactive pathway in the interior of a bubble, they could diffuse out into the bulk solution and react with non-volatile solute. The overall effect of ultrasound may accelerate the original chemical reaction. Under optimal conditions, the efficiency of MTBE removal can reach 98 percent removal. The experimental results indicate ultrasound and Fenton's reagent were equally important. The overall reaction appears to be a pseudo-first-order reaction.

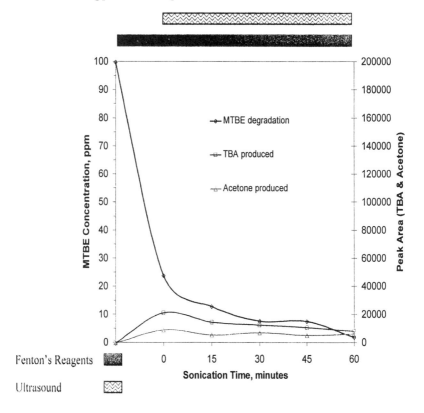

Figure 2. MTBE degradation and its by-products generation.

By Fenton's reagent alone, the degradation rate is dependent on H_2O_2 to organic substrates ratios, because several organic substrates will retard H_2O_2 decomposition. So, huge amount of hydrogen peroxide was needed to reach high removal rate. However, once Cu^{2+} is introduced into the system, those inhibitors, TBA and acetone, will form organocopper intermediate and precipitate out. Therefore, only very little amount of hydrogen peroxide used as free radical initiator is required in the Fenton's reagent/ultrasound/Cu^{2+} system.

Directional drilling can highly enhance the efficiency of site remediation by tracking contaminants plume. Liquid transducers can be installed in the inner wall of the double-wall pipe. This equipment will benefit to small water treatment plant at remote area, because easy maintenance and less expensive. The chemicals used in the processes are environmental benign. And these processes can also be used for other application such as color removal, iron and manganese removal, taste and odor control (oxidation of sulfides), and supplying supplemental dissolved oxygen for biodegradation. The intensity of ultrasound affects the degree of chemical reaction. Liquid transducers use either a simple "edge tone" principle or a more complicated valve oscillator. The use of vibrating blade transducers in underground double-wall pipes still need to be studied.

REFERENCES

Kang, J., and Michael R. Hoffmann. 1998. "Kinetics and Mechanism of the Sonolytic Destruction of Methyl tert-Butyl Ether by Ultrasonic Irradiation in the Presence of Ozone." Environmental Science & Technology. 32(20): 3194-3199.

Suslick, K. S. 1990. "Sonochemistry." Science. 247:1439-1445

Walling, C. 1975. "Fenton's Reagent Revisited." Accounts of Chemical Research. 8:125-131.

Walling, C., and Shin'ichi Kato. 1971. "The oxidation of Alcohols by Fenton's Reagent. The Effect of Copper Ion." Journal of the American Chemical Society. 93(17): 4275-4281.

SONOCHEMICAL DEGRADATION OF AZO DYES IN AQUEOUS SOLUTION

Hugo Destaillats, Jiju M. Joseph and Michael R. Hoffmann
(California Institute of Technology, Pasadena, 91125 CA.)

ABSTRACT: The sonochemical degradation of aqueous solutions of azobenzene (AB) and the related azo dyes methyl orange (MO), *o*-methyl red (*o*-MR) and *p*-methyl red (*p*-MR) was performed at 500 kHz and 50 W, under air, O_2 or Ar saturation at 288 K. Reaction products and intermediates were identified by HPLC-ES-MS. Total Organic Carbon was also determined as a function of reaction time. We propose a reaction mechanism based on the observed species and the extent and rate of TOC depletion. The effect of the dye structures and of the background gas on the sonochemical bleaching rates were investigated. The addition of Fe(II) increased the sonochemical bleaching rates, while O_3 sparging increased considerably the mineralization extent.

INTRODUCTION

Aromatic azo derivatives constitute a major part of all commercial dyes employed in a wide range of processes in the textile, paper, food, cosmetics and pharmaceutical industries. They are characterized by the presence of the azo group (-N=N-) attached to two substituents, mainly benzene or naphthalene derivatives, containing electron withdrawing and/or donating groups. Most of these highly soluble dyes are found to be resistant to normal waste water treatment processes, thus several methods have been developed to achieve their degradation, such as treatment with UV light in the presence of H_2O_2, photocatalysis in aqueous TiO_2 suspensions, gamma radiolysis techniques, and ultrasonic irradiation. Eliminating the strong color of the effluents implies lowering of the concentration of the dye to under the ppm range. Azo compounds often become toxic to organisms after reduction and cleavage of the azo bond .

Under ultrasonic irradiation, the target molecules can undergo degradation by two different main pathways depending on their chemical nature: pyrolytic reactions inside the cavitation bubbles or oxidation by hydroxyl radical in the bulk medium. Non-volatile compounds degrade mainly by reaction with hydroxyl radicals in the bulk solution, while those with higher vapor pressure can partition into the bubble and also undergo thermal degradation in the gas phase. The rate of sonochemical reactions is influenced by the nature of the background gas, since the temperature attained within the cavitation bubble upon collapse depends on physicochemical properties such as the polytropic ratio (C_p/C_v), thermal conductivity, and the solubility of a gas in water. The yield of the free radicals formed depends essentially on the temperature reached within the bubble in the final collapse stages (Leighton, 1992).

Many efforts have been devoted to improve the efficiency of sonochemical reactions, considering that a substantial amount of the energy employed in cavitational collisions is not effectively converted into an optimum yield of the desired products. The recombination of •OH to yield H_2O_2 both in the gas phase within the bubbles and in solution, are two of the major processes that limit the amount of reactive radicals accessible to the target molecules. In the present work (Joseph et al., 2000), the addition of Fe(II) ions provides a secondary source of •OH radicals through their reaction with the sonochemically generated H_2O_2 (Fenton's reaction). The use of O_3 as background gas (in mixtures with O_2) facilitates the oxidation of persistent sonolysis products, while providing more reactive species in solution.

MATERIALS AND METHODS

10 μM aqueous solutions containing AB or the different dyes were prepared before each measurement, and filtered through Millipore GS 0.22 μm discs. The concentrations were determined using the absorbance at λ_{max} (ε_{max} = 22000 at 319 nm for AB, ε_{max} = 26900 at 464 nm for MO, ε_{max} = 20900 at 430 nm for o-MR and ε_{max} = 26300 at 464 nm for p-MR) . The high molar absorptivities values typical of these compounds allowed for experiments to be carried out at very low concentrations, in order to avoid the formation of dimers and also neglect the formation of complexes with the metal ions employed as catalysts. The pH of the solutions was measured using an Altex 71 pH meter.

The sonochemical reactor consisted of a 650 mL glass chamber surrounded by a self-contained water jacket and a piezoelectric transducer. Sonication at 500 kHz was performed with an ultrasonic transducer (Undatim Ultrasonics) operating at 50 W (2 W/cm^2). The ultrasonic power input to the reactor was determined using a standard calorimetric method. The solution was stirred magnetically during each experiment and the temperature was kept constant at 15.0 $\pm$ 0.5 ^{O}C with a VWR Scientific thermostat. When the solutions were saturated with oxygen or argon, a stream of the gas was sparged into the reactor at a flow rate of 80 mL/min for one hour before the reaction. For experiments conducted in presence of Fe(II), the corresponding volume of a stock $FeSO_4$ solution (0.1 M or 0.01 M) was injected into the solution before saturation with the respective gases. pH measurements were also performed under gas sparging. When ozone was employed, different O_3/O_2 ratios were obtained by applying different voltages in an Orec ozone generator. The ozone concentration in the liquid was measured spectrophotometrically at λ_{max} = 260 nm.

Samples were collected for analysis at different times through a septum port by means of a syringe. The UV-VIS absorption spectra were recorded using a Hewlett Packard 8452 A diode array spectrophotometer. The analytical quantification of the products was carried out using a Hewlett Packard 1100 HPLC system with UV detection and coupled to a mass spectrometer through an electrospray interface (ES-MS). The mass spectra were recorded for both, positive and negative ions in the range 50 to 1000 m/z units, at 3 s/scan, and a skimmer cone voltage of 3.5 kV. A 3 μm, 100 x 4 mm Hypersil BDS-C18 column (Hewlett Packard) was used for MO and a 5 μm, 100 x 2.1 mm Hypersil MOS-C8 column was used in the case of o-MR, p-MR and AB. The eluent solution consisted of acetonitrile/water mixtures in the proportion 15/85 for MO, 20/80 for *o*-MR, 10/90 for *p*-MR, and 50/50 for AB. The flow rate was 0.3 mL/min in all cases. The wavelength used to follow the reduction in the dye concentrations was that corresponding to the maximum of the visible band for each dye. Total Organic Carbon analyses were carried out with a Shimadzu 5000A TOC analyzer operating in the non-purgable organic carbon (NPOC) mode. Calculations were performed with *Mathematica* 3.0 ®.

MO (Baker, >95 %), *o*-MR (Sigma, >95 %), *p*-MR (Sigma, >97 %), AB (Aldrich, > 99 %) and $FeSO_4$ (Fisher, 99 %) were used without further purification. The solutions were prepared using water purified by a Millipore Milli-Q UV Plus system (R = 18.2 $M\Omega$.cm). The sparged gases (O_2 and Ar) were provided by Air Liquide.

RESULTS AND DISCUSSION

The degradation of the azo compounds was studied following both the disappearance of the parent molecule (bleaching) and the degree of mineralization, as measured by the reduction of the total organic carbon present in the sample. Bleaching is a fast process, which is completed within the first 40 min of sonication, whereas mineralization is attained (only partially) after a longer period of time. The non-volatile nature of the azo dyes suggests that pyrolytic reactions within the collapsing bubbles can be negligible, and that degradation occurs exclusively *via* the attack of hydroxyl radicals in the liquid phase.

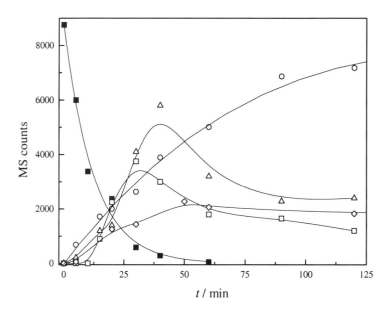

FIGURE 1. Depletion of *o*-methyl red (■) and variations in the observed concentrations of NO_3^- (O), benzoquinone (◇), nitrophenol (△) and dinitrobenzene (□) vs time during sonolyses at 500 kHz with an applied power of 50 W under Ar saturation.

The fast bleaching, *i.e.*, the decrease of the absorption band in the visible region, suggests that the addition of hydroxyl radicals to the azo double bond is one of the first steps in the degradation process, producing the loss of the chromophoric characteristic in the products and intermediates that are generated. Figure 1 illustrates the depletion of o-MR together with the production of some byproducts and intermediates, as followed by HPLC-ESMS. Based on these observations, the scheme on Figure 2 summarizes some common degradative patways for the studied substrates, including the most observed byproducts.

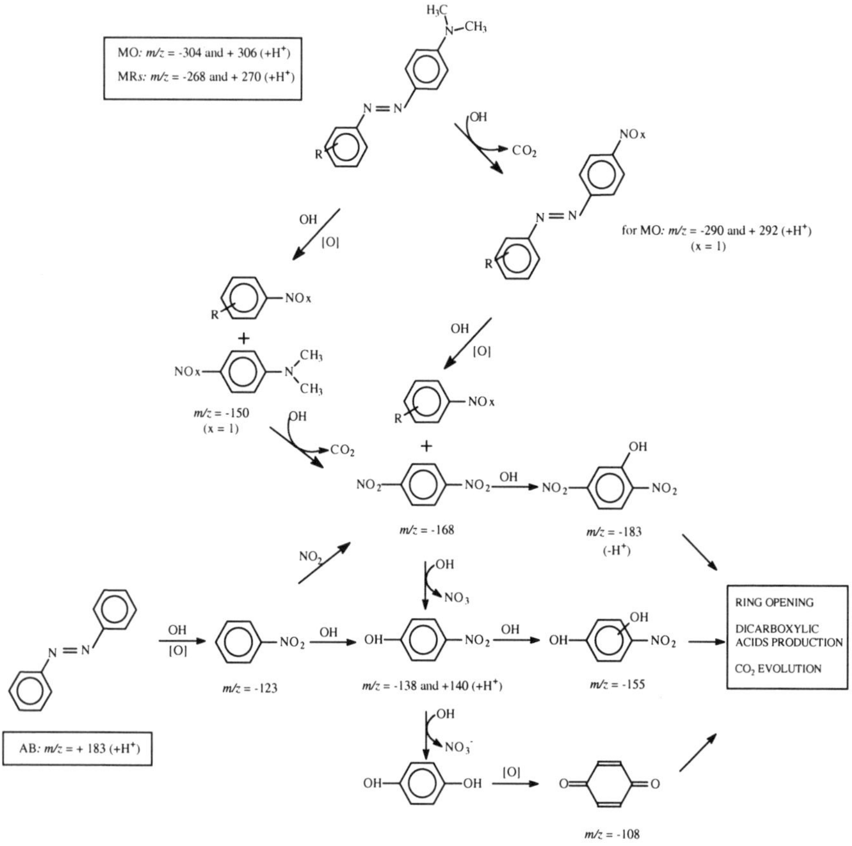

FIGURE 2: Main degradative pathways for AB, MO and the MRs.

The depletion of the target molecules followed pseudo-first order kinetics with respect to the dye concentration. This suggests that a steady-state $\bullet$OH concentration exists in the liquid phase under continuous irradiation. The resulting steady-state concentration depends in principle only on the nature of the saturation gas and the temperature. The observed pseudo first order rate constants,

k_{gas} determined for the bleaching of the different compounds under air, O_2 and Ar are presented in Table 1. In each case, the use of Ar as a background gas enhances the value of k_{Ar} by 10 %, compared with k_{O2} and k_{air}. This enhancement correlates linearly with an increment in the steady state concentration of •OH radicals (Hua and Hoffmann, 1997). The faster process observed in the case of o-MR can be attributed to a positive substituent effect of the carboxylic group in the *ortho* position, near the azo group.

TABLE 1. Pseudo first order rate constants for the decolorization of AB, MO and MR under different gas saturation.

Dye	k_{air} (min^{-1})	k_{O2} (min^{-1})	k_{Ar} (min^{-1})
AB	0.042 ± 0.002	0.043 ± 0.002	0.049 ± 0.002
MO	0.040 ± 0.002	0.042 ± 0.002	0.046 ± 0.002
p-MR	0.035 ± 0.002	0.033 ± 0.002	0.043 ± 0.002
o-MR	0.054 ± 0.002	0.053 ± 0.002	0.059 ± 0.003

Effect of Fe(II) addition. The enhancement of the bleaching reaction rate in the presence of Fe(II) ions was investigated as a function of the concentration of $FeSO_4$ in the range 10 μM to 5 mM. The lower Fe(II) concentrations had a minor effect, while the upper concentration limit was established by the precipitation threshold of iron oxides during the reaction. All the kinetic measurements for this system were carried out with 10 μM MO solutions, in absence of iron oxides precipitation, and under Ar saturation. Figure 3 illustrates the changes observed in the apparent first order rate constants k_{Ar} when the experiments were carried out in presence of different Fe(II) concentrations. A maximum 3-fold increase in the measured rate constant was observed when Fe(II) concentration was between 0.1 mM and 0.5 mM. This increment was due to the higher •OH radicals concentration produced through the Fenton's reaction

$$Fe^{2+} (aq) + H_2O_2 (aq) \rightarrow Fe^{3+}(aq) + •OH (aq) + OH^- (aq) \qquad (1)$$

Further increases in [Fe(II)] showed no further catalytic activity, due to the direct reduction of •OH radicals by the metal ions,

$$Fe^{2+} (aq) + •OH (aq) \rightarrow Fe^{3+} (aq) + OH^- (aq) \qquad (2)$$

A kinetic model based on a simple reaction scheme in the bulk liquid phase, considering only eqs. 1 and 2, together with the sonochemical production of •OH radicals and H_2O_2, the recombination of the radicals and their reaction with the dyes, is able to reproduce our experimental observations. Figure 3 shows also the calculated curve for the rate constant and the concentration of H_2O_2 when the dye concentration was reduced to half of its initial value, $(H_2O_2)_{1/2}$.

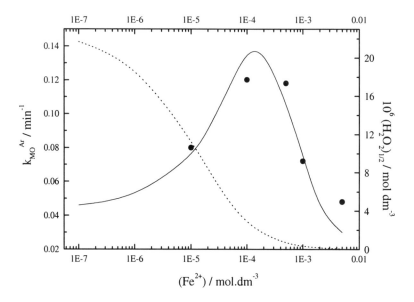

FIGURE 3. Effect of added Fe(II) on the pseudo first-order rate constant for the bleaching of methyl orange under Ar saturation (k_{MO}^{Ar}). Experimental values (●) were reproduced by the adjusted model (solid line). The amount of H_2O_2 produced when half of the MO is consumed, $(H_2O_2)_{1/2}$, is also shown (dashed line, scale on the right).

Mineralization. The degree of mineralization was also studied as a function of the reaction time. In the case of MO and the MRs, after 90 to 120 minutes of continuous irradiation, the organic load remained constant in every case, reaching a limiting value of approximately 50% of the initial concentration. This result is in agreement with the observed persistence of a variety of stable organic acids, quinones and other species described above. Within the experimental error of the measurements (~10 %, estimated from repeating calibrations in the concentration range of this work) no difference was observed in the mineralization process under air, O_2, or Ar. While mineralization of MO and the MRs presented similar characteristics, the process was much slower for AB. After 300 min, only a 30 % of the total organic matter present in the sample was destroyed. The reason for a slower mineralization is that the intermediates produced after the oxidative cleavage of the azo group of AB are essentially monosubstituted aromatic compounds (nitrobenzene). In the case of the other compounds, MO and the

MRs, the oxidation of the disubstituted aromatic intermediates was easier and the amount of CO_2 produced was higher.

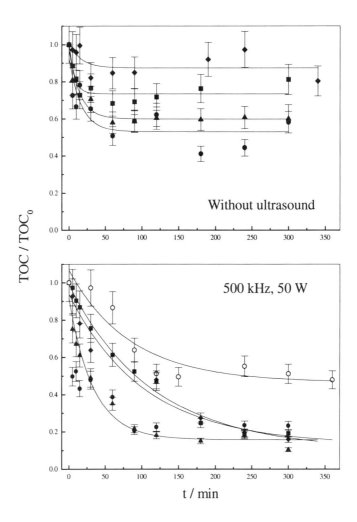

FIGURE 4: Mineralization of MO under ozonation (upper plot), ultrasonic irradiation (○) and the combined sonolysis/ozonolysis under 24 (●), 12 (▲), 7 (■) and 2 (♦) mbar of O_3 (lower plot).

The effect of Fe(II) on the yields of mineralization was also analyzed. In the presence of 0.1 mM Fe(II), the final limiting value for total organic carbon depletion was the same as the observed in the absence of the metal ions (50 %), but this final state was reached much faster (only 20 minutes, versus 120 in the absence of iron). Thus, the extent of mineralization was not increased. However, the velocity at which the limit level is reached is increased in the presence of Fe(II). When the concentration of Fe(II) was lower or higher than that optimum value, the mineralization curves were similar to those recorded in the absence of the metal ions.

In order to achieve a more complete mineralization, ultrasonic irradiation must be combined with the addition of a strong oxidant such as O_3, which is able to attack some of the remaining stable organic compounds. Figure 4 shows the effect of saturating the solution under four different O_3/O_2 ratios in the extent of mineralization of MO, measured as TOC depletion. When ultrasonic irradiation is not applied, ozonation alone accounts for less than a 50 % of the mineralization degree.

An effective chemical synergism exists for the combined technique. When sonolysis and ozonolysis are applied together, the extent of mineralization exceeds the 80 % of the original organic carbon concentration, at all O_3 (aq) concentrations. When any of the two methods are applied independently, the extent of mineralization is never higher than 50 %. The same effect was observed for AB: the mineralization extent with sonolysis or ozonolysis alone are close to the 30 %, while the combined technique yields also more than an 80 %. The limiting value $TOC_{final} = 0.2\ TOC_0$ was reached earlier for higher O_3/O_2 ratio mixtures, both for MO and AB. That limiting value is determined by the formation of persistent byproducts, such as carboxylic acids.

REFERENCES

Hua, I., Hoffmann, M.R. 1997. "Optimization of Ultrasonic Irradiation as an Advanced Oxidation Technology". *Environ. Sci. Technol. 31*(8): 2237-2243.

Joseph, J.M., Destaillats, H., Hung, H-M., and Hoffmann, M.R. 2000. "The Sonochemical Degradation of Azobenzene and Related Azo Dyes: Rate Enhancements via Fenton's Reactions". *J. Phys. Chem. 104*(2): 301-307.

Leighton, T. G. 1992. *The acoustic bubble*. Academic Press, London.

MTBE AND PRIORITY CONTAMINANT TREATMENT
WITH HIGH ENERGY ELECTRON BEAM INJECTION

William J. Cooper (UNC Wilmington, Wilmington, NC)
Greg Leslie (Orange County Water District, Fountain Valley, CA)
Paul M. Tornatore and Wayne Hardison (Haley & Aldrich, Rochester, NY)
Paris A. Hajali (Haley & Aldrich, Brea, CA)

ABSTRACT: A pilot scale mobile electron beam system was used to examine the removal efficiency of MTBE (methyl *tert*-butyl ether) in single solute systems and in complex (multi-solute) systems, including perchlorate ion. Ground water and RO permeate were used to prepare mixtures of the compound(s). Using several different waters it was possible to examine the factors effecting removal efficiency and reaction by-product formation and destruction. Fifteen additional organic compounds and perchlorate ion were also studied. Factors affecting the removal and efficiency of removal of the organic compounds were determined. Surprisingly, in one water the removal of perchlorate was also observed.

INTRODUCTION

Irradiation of water can be thought of as the injection of electrons into a solution. The resulting products of that electron injection are described by equation 1:

$$H_2O -\wedge\wedge\wedge-> (2.7) \ \cdot OH, (0.6) \ H\cdot, (2.6) \ e_{aq}^-, (0.45) \ H_2, (0.7) \ H_2O_2, (2.6) \ H_3O^+ \quad [1]$$

Where the numbers denote the yield (G-value) of each species per 100 eV deposited energy. This can be thought of as an efficiency estimate, the relative concentration of a radical, excited species or molecule per unit absorbed energy. The species that are of principal importance with respect to the electron beam process are three reactive species, the reducing species, e_{aq}^- and $H\cdot$, and the oxidizing radical, $\cdot OH$.

The electron beam process is the only advanced oxidation technology (AOT) that has been studied at large scale (> 100 gallons per minute) that produces both the oxidizing and reducing radicals (Bolton et al., 1998). Therefore, in complex mixtures of organic compounds it is possible to have both processes occurring simultaneously.

Objective. The objective of this study was to evaluate the potential of the electron beam process for destroying MTBE and other organic compounds in waters of varying quality. A secondary objective was to evaluate the potential for the destruction of perchlorate ion.

MATERIALS AND METHODS

All of the experiments were conducted in a mobile unit, 20 kilowatt, that is housed in a 48 foot by 8 foot semi trailer, located at Water Factory 21 in Orange County, CA (Nickelsen et al., 1998). The trailer is divided into three rooms - the pump room, the process room and the control room. The pump room contains all ancillary equipment required for water and air handling processes. The process room contains the electron beam and a 500 keV power supply. The control room contains the two control consoles where the system parameters are monitored, maintained and adjusted.

All of the analytical methods used were the standard methods for the analysis of volatile organic compounds. Ion chromatography was used for the perchlorate ion determinations.

RESULTS AND DISCUSSION

The results are reported for samples to which only MTBE was added and for samples where MTBE was added in the presence of 15 other organic compounds. In addition to the removal of organic compounds three studies were conducted where perchlorate ion was also added (the complex mixture samples).

MTBE. Sample 1, a deep well water, was characterized by relatively low TOC, the average was 4.09 mg/L, an alkalinity of 158 and a hardness of 19.1 mg/L as $CaCO_3$, and a pH of between 7.7 and 8.8. Two experiments with only MTBE were conducted using this source water. The data obtained for the destruction of MTBE are summarized in Tables 1 and 2. Only two reaction by-products, *tert*-butyl alcohol (TBA) and *tert*-butyl formate (TBF), were determined and the concentrations for those are also summarized. The data are tabulated as both weight and molar concentrations. The reason for the conversion to molarity is to account for the differences in molecular weight of the three different compounds. That is, to analyze data for removal efficiencies and reaction by-product formation and loss for a variety of compounds, it is necessary to convert all of the concentrations into molar units.

A summary of the results for the destruction of MTBE in all of the experiments is shown in Table 3. Sample 1 was the only sample where only MTBE was spiked. Samples 2-4 had 15 other organic compounds present.

Note the difference in the D_{90} value, essentially an estimate of the energy required to remove 90% of the MTBE, between sample 1 and 4. The alkalinity, pH and hardness were essentially the same. However, the removal of the MTBE in sample 4 was five times less efficient that in sample 1. This demonstrates the effect of other organic compounds present in the mixture on the removal of MTBE. This mixture sets up a complex competition for all of the solutes that results in reduced removal of the MTBE. This observation has significant implications in assessing this or any destruction technology when it is to be applied to real world samples that are often contaminated with multiple solutes.

TABLE 1. MTBE removal data in sample 1, replicate 1.					
Dose (krads)	Concentration (ug/L)			Concentration (uM)		
	MTBE	TBA	TBF	MTBE	TBA	TBF
0	170	1.9	1.3	1.93	0.026	0.013
125	NA	NA	NA	NA	NA	NA
250	14	29.8	33.9	0.16	0.4	0.33
500	1.4	9.3	13.4	0.016	0.12	0.13

Conversion factors for weight to molarity: NA = not applicable, no data for this dose
MTBE, mol weight = 88;   1 uM = 88 ug/L
TBA,   mol weight = 74;   1 uM = 74 ug/L
TBF,   mol weight = 102;   1 uM = 102 ug/L
To obtain uM, divide ug/L by the above factor

TABLE 2. MTBE removal data in sample 1, replicate 2.					
Dose (krads)	Concentration (ug/L)			Concentration (uM)		
	MTBE	TBA	TBF	MTBE	TBA	TBF
0	196	1.4	0.64	2.23	0.019	0.006
125	62.9	59.6	43.9	0.71	0.81	0.43
250	15.9	31.8	21.7	0.18	0.43	0.21
500	1.8	10.6	9.5	0.02	0.14	0.093

TABLE 3. Summary of MTBE removal in waters of different quality.								
Water Sample	MTBE Conc. (ug/L)	MTBE Dose Constant	D_{90} (krads)	MTBE Or Mixture	TOC (mg/L)	Alkalinity (mg/L as $CaCO_3$)	Hardness (mg/L as $CaCO_3$)	PH
1	170	-0.00960	229	MTBE	4.09	158	19.1	7.7 - 8.8
1	196	-0.00944	233					
2	219	-0.00232	947	Mixture	13.5	14.8	2.7	7.7
2	257	-0.00198	1110					
3	198	-0.00191	1150	Mixture	14.1	147	153	8.1
3	223	-0.00136	1620					
4	224	-0.00179	1230	Mixture	16.0	156	20.0	8.9
4	260	-0.00147	1500					

Complex Mixtures. The 15 organic compounds that were studied in addition to MTBE in the complex mixture were: trichloroethylene (TCE) and tetrachloroethylene (PCE); benzene, toluene, ethyl benzene *o*-xylene (BTEX);

chloroform, $CHBrCl_2$, $CHClBr_2$ and $CHBr_3$ (THMs); Ethylenedibromide (EDB); dibromochloropropane (DBCP); nitrosodimethyl amine (NDMA); and the herbicides, atrazine and simazine. These compounds represent many different classes of organic compounds and compounds that are destroyed by very different mechanisms.

All of the organic compounds that were spiked into the water were removed. The exact removal of each compound varied depending on the mechanism of removal. However, the removal efficiencies were predictable based on an understanding of the of the underlying radiation chemistry. A summary of the results are presented in Tables 4, 5 and 6, where we have chosen to compare the removal of seven compounds that are representative of all of the compounds.

TABLE 4. Organic compound concentration (ug/L) at increasing dose using the electron beam process, RO permeate water (water sample 2).

Compound (ug/L)	Experi-ment	Dose (krads)				% Removal
		0	125	250	500	
		Solute Concentration (ug/L)				
MTBE	A	219	180	135	70	68.1
	B	257	225	167	98.4	61.7
TCE	A	38.6	17.2	6.37	0.9	97.7
	B	42.9	21.2	7	1.2	97.2
Benzene	A	55	20	5.8	0.36	99.3
	B	62.9	25.1	7.1	0.8	98.7
$CHCl_3$	A	45.7	33	23.7	10.4	77.2
	B	57.8	40.9	26.7	10.8	81.3
EDB	A	1.54	1.13	0.62	0.13	91.7
	B	1.76	1.24	0.63	0.15	91.8
NDMA	A	4.19	2.19	1.66	0.5	88.1
	B	4.08	3.12	1.28	0.73	82.1
Atrazine	A	8.82	4	2.9	0.78	91.2
	B	10.4	6.55	2.14	1.08	89.6

Table 7 summarizes the water quality of the four samples that were studied. The major differences in water quality were in sample TOC and alkalinity.

In general, the natural organic matter, characterized by TOC, tends to be a good hydroxyl radical, ·OH, scavenger. Therefore the higher the concentration of the TOC the more competition for the ·OH exists in the solution. And, the more inefficient an AOT is for destroying targeted organic compounds that are removed primarily by the ·OH. In the case of the three spiked samples, 2, 3 and 4, the TOC was very similar in concentration. However, it is not possible to provide a detailed description of the differences in MTBE removal at the different TOC's, as the water with low TOC had only MTBE while the high TOC waters had the complete mixture of organic compounds.

TABLE 5. Organic compound concentration (ug/L) at increasing dose using the electron beam process, monitoring well M-14/1 (water sample 3).

Compound (ug/L)	Experi-ment	Dose (krads)				% Removal
		0	125	250	500	
		Solute Concentration (ug/L)				
MTBE	A	198	157	139	75.3	61.8
	B	223	202	165	115	48.6
TCE	A	28.3	11.5	4	0.8	97.2
	B	38.5	18.9	6.2	1.6	95.8
Benzene	A	39.8	16.7	6.1	1.2	97.0
	B	52	26.9	10.3	2.4	95.4
$CHCl_3$	A	37.9	23.6	10.4	1.5	96.0
	B	51.6	33.2	13.6	2.1	95.6
EDB	A	1.17	0.73	0.25	0.036	96.6
	B	1.54	0.91	0.34	0.059	96.2
NDMA	A	2.18	1.05	0.6	0.06	97.2
	B	3.04	1.54	0.6	0.09	97.0
Atrazine	A	6.82	4.46	1.88	0.39	94.3
	B	8.72	4.36	2.39	0.66	92.4

TABLE 6. Organic compound concentration (ug/L) at increasing dose using the electron beam process, deep well water (water sample 4).

Compound (ug/L)	Experi-ment	Dose (krads)				% Removal
		0	125	250	500	
		Solute Concentration (ug/L)				
MTBE	A	224	185	159	91.4	59.2
	B	260	228	192	119	54.2
TCE	A	32	12.2	4.2	1	96.9
	B	40	17.4	5.7	1.3	96.8
Benzene	A	44.7	21.3	8.7	1.8	95.9
	B	53.3	28.9	12.2	2.6	95.1
$CHCl_3$	A	44.8	24.3	9.1	1	97.8
	B	55.9	28.9	11	1.2	97.9
EDB	A	1.19	0.62	0.21	0.039	96.7
	B	1.48	0.68	0.24	0.05	96.6
NDMA	A	2.87	1.55	0.49	0.06	97.9
	B	3.22	1.23	0.44	0.07	97.8
Atrazine	A	6.44	4.15	1.76	0.27	95.8
	B	8.22	4.77	1.98	0.28	96.6

TABLE 7. Carbonate speciation in the sample waters.					
Water Sample	TOC (mg/L)	Alkalinity (mg/L)	PH	CO_3^{2-} (mM)	HCO_3- (mM)
1	4.09	158	8.8	0.104	2.94
1B	4.09	158	7.7	0.00846	3.01
2	13.5	14.8	7.7	0.00079	0.281
3	14.1	147	8.1	0.0201	2.85
4	16.0	156	8.9	0.127	2.85

The carbonate system, as expressed by alkalinity, is present in different concentrations in the waters that were tested. A simplified description of the carbonate system in natural waters is described by the following equations:

$$H_2CO_3 + H_2O \longleftrightarrow HCO_3^- + H_3O^+ \qquad pK_a = 6.35 \quad [2]$$

$$HCO_3^- + H_2O \longleftrightarrow CO_3^{2-} + H_3O^+ \qquad pK_a = 10.25 \quad [3]$$

The pK_a is the pH at which the concentration of the acid and conjugate base are equal. For example, at a pH = 10.25 the concentration of the HCO_3^- = CO_3^{2-}. Table 7 summarizes the calculated values for the concentration of the CO_3^{2-} and HCO_3- for the sample waters that were used in this series of tests. It is obvious from the table that the concentration of the two species varied considerably from sample to sample.

In the three waters studied where the complex mixture was used, the TOC was essentially the same, 13.5 - 16.0, the major variable was alkalinity and the carbonate ion concentration. The presence of the carbonate ion affects the removal of organic compounds differently depending upon whether the organic compounds are removed primarily via hydroxyl radical attack or hydrated electron attack. For example, if we look at benzene and chloroform, as primarily removed via hydroxyl radical and hydrated electron reactions, respectively, then with increasing carbonate, 4>3>2, the removal efficiency of benzene decreased and the removal of chloroform increased.

The concentration of the CO_3^{2-} in the waters that were used for this test varied from 0.127 to 0.00079 mM. The effect of the carbonate/bicarbonate system can essentially be ignored in the RO permeate water (sample 2). However, in the two other waters that were used, the CO_3^{2-} increased in sample 3 from 0.0201 to 0.127 mM in sample 4.

A concentration of 0.127 mM (or 12.7 uM) represents a considerable ·OH radical scavenger especially considering that the typical concentration of MTBE in the starting solutions were in the concentration range of 2-3 uM. For the other organic compounds $CHCl_3$ was typically in the range of 0.5 uM, TCE was 0.2 - 0.4 uM and benzene was just under 1 uM.

Although the reaction rate constants for e^-_{aq} and NDMA and atrazine have not been reported, it is apparent by comparing the removal efficiencies in water samples 2, 3 and 4, and by analogy with $CHCl_3$, that this reaction is significant in the destruction of these compounds.

Perchlorate ion. Perchlorate ion was spiked into water samples for three different experiments. The waters used for the perchlorate ion studies were, 2 (RO permeate water), 3 (Monitoring Well M-14/1) and 4 (Deep Well). In general the TOC for all three waters was similar, i.e. approximately 12-16 mg/L. The major difference in the waters appears to be the alkalinity/hardness concentrations. In waters 3 and 4, the alkalinity was approximately 150 mg/L as $CaCO_3$, with an approximate hardness of 150 and 20 for 3 and 4, respectively. However, in the RO permeate water (sample 2) the alkalinity was 15 and the hardness approximately 4.

For the two water samples having high alkalinity (3 and 4) no removal of the perchlorate ion was observed in the four experiments (each water was run in duplicate). It was the low alkalinity/hardness water in which the perchlorate ion concentrations appear to be significantly reduced at the highest dose. The removal appears to be greater than 50% in both samples. If this result could be duplicated, this may have significant implications as a strategy to remove perchlorate ion from contaminated ground and surface waters.

CONCLUSIONS

In all of the experiments a control, zero dose, was run. There were no differences in concentration between the influent and effluent samples for MTBE. Therefore, the observed loses are due to the treatment process and not other physical phenomenon. In solutions where MTBE was the only organic solute to be removed, it was efficiently destroyed. The removal of MTBE was affected by the presence of carbonate alkalinity. That is as alkalinity increased the removal efficiency decreased, as it would with all AOTs. The presence of natural organic matter (NOM), acting as a OH radical scavenger, reduced the removal efficiency, as well. In the presence of other organic compounds where competition for the radicals exists the efficiency was also affected (Zele et al., 1998).

Probably the most significant finding was that in the presence of 15 other organic solutes and perchlorate ion, MTBE was one of most difficult compounds to be destroyed. This has implications in the remediation of contaminated ground waters where it is likely that complex mixtures will exist. However, the removal of MTBE and 15 other organic compounds in the spiked matrix was consistent with the chemistry of the system. That is, the removal of each compound was predictable based on the fundamental radiation chemistry of the compounds. Overall the removals were very good (up to 99.9%) in one pass. Three different waters were studied for the removal of the mixture of organic compounds. In the three waters studied the TOC was essentially the same, 13.5 - 16.0, the major variable was alkalinity and the carbonate ion concentration. The presence of the carbonate ion affects the removal of organic compounds depending upon whether the organic compounds are removed primarily via hydroxyl radical attack or hydrated electron attack. If we take two examples, benzene and chloroform, as compounds primarily removed via hydroxyl radical and hydrated electron reactions, respectively, then, with increasing carbonate, the removal efficiency of benzene decreased and the removal of chloroform increased.

Based on the results obtained in two experiments perchlorate ion removal should be studied in more detail in waters with low alkalinity and hardness. We observed 50% removal in several of the experiments and if this trend is confirmed it is important to develop a better understanding of the potential for the removal of perchlorate ion using the electron beam process.

ACKNOWLEDGEMENTS

This research was funded by the Orange County Water District, the California Energy Commission, and, Haley and Aldrich. WJC would like to acknowledge the assistance of a NSF grant, BES 97-29965 for partial support.

REFERENCES

Bolton, J.R., J.E. Valladares, W.J. Cooper, T.D. Waite, C.N. Kurucz, M.G. Nickelsen and D.C. Kajdi. 1998. "Figures-of-Merit for Advanced Oxidation Processes - A Comparison of Homogeneous UV/H$_2$O$_2$, Heterogeneous TiO$_2$ and Electron Beam Processes " *J. Adv. Oxid.. Technol.* 3: 174-181.

Nickelsen, M.G., D.C. Kajdi, W.J. Cooper, C.N. Kurucz, T.D. Waite, F. Gensel, H. Lorenzl and U. Sparka. 1998. "Field Application of a Mobile 20-kW Electron Beam Treatment System on Contaminated Groundwater and Industrial Wastes" In, *Environmental Applications of Ionizing Radiation*, W.J. Cooper, R.D. Curry and K.E. O'Shea, Eds., John Wiley and Sons, Inc. N.Y. 451-466.

Zele, S. M.G. Nickelsen, W.J. Cooper, C.N. Kurucz and T.D. Waite. 1998. "Modeling Kinetics of Benzene, Phenol and Toluene Irradiation in Water using the High Energy Electron-Beam Process" In, *Environmental Applications of Ionizing Radiation*, W.J. Cooper, R.D. Curry and K.E. O'Shea, Eds., John Wiley and Sons, Inc. N.Y. 395-415.

CHLORINE DIOXIDE TREATMENT OF TCE IN FRACTURED BEDROCK

Timothy V. Adams, ENSR Consulting and Engineering, Warrenville, IL USA
Gregory J. Smith, Radian International, Rolling Meadows, IL USA
Gregory P. Vierkant, Lucent Technologies, Inc., Springfield, MO USA

ABSTRACT: Chlorinated aliphatic hydrocarbons (CAHs) are present in groundwater in fractured shale/limestone bedrock at a former electronics manufacturing facility in Missouri. Trichloroethene (TCE; up to 71 mg/l) is the predominant CAH detected in three identified source areas. The remediation program is being implemented under the Missouri Department of Natural Resources (MDNR) Voluntary Remediation Program following protocols developed by the Air Force Center for Environmental Excellence (AFCEE) (Wiedemeier, et. al., 1996) to evaluate natural attenuation of CAHs.

In former source areas, TCE concentrations in groundwater exceed 1% of its aqueous solubility (1,100 mg/L), indicating the presence of DNAPL (Cherry and Feenstra, 1991). The presence of DNAPL is considered toxic to microorganisms believed responsible for creating redox conditions to promote dehalogenation of TCE. Source reduction measures including soil vapor extraction and chemical oxidant flushing are being used at the site. Chemical oxidant flushing using chlorine dioxide was performed to promote partitioning of residual CAHs from the aquifer matrix into groundwater. This partitioning phenomenon was observed by Smith, et. al., (1998a).

To evaluate the degree of natural attenuation occurring, groundwater was sampled for major ions, Eh, dissolved oxygen and iron, to provide evidence of natural attenuation. Eh measurements across the site average -222.47 millivolts (mV), indicating highly reducing conditions. Bouwer (1994) suggests the optimal Eh range for reductive dehalogenation is –220 to –240 mV, indicative of sulfate reducing conditions. Background sulfate concentrations are above the inhibitory levels presented in Wiedemeier, et. al., (1996) and were reduced downgradient of the source areas. Correspondingly, chloride concentrations increased relative to background downgradient of former source areas, reflecting the dehalogenation of TCE. This would indicate that high concentrations of sulfate at this location do not have the inhibitory effects described in Wiedemeier, et. al., (1996) Methane was locally detected downgradient of former source areas at concentrations as high as 2.3 mg/l and 7.5 mg/l, indicating areas of highly reducing conditions.

BACKGROUND

Work activities described herein are being conducted at a former electronics manufacturing facility in the Kansas City, Missouri area. The facility produced microelectronic parts and coatings for use in telecommunications and data processing equipment. The site was recently sold for redevelopment into a mixed use, light industrial/office/commercial complex in a campus-type setting. The property encompasses approximately 132 hectares (328 acres; 72.8 hectares of natural vegetation and 60 hectares used for the former manufacturing operation).

CAHs, mainly in the form of TCE, were discovered in groundwater beneath the site during initial due diligence investigations for the sale of the property conducted in 1992. Subsequent investigations determined CAHs were likely released to groundwater in the vicinity of three main use areas including:

- Former Underground Storage Tank (UST) Farm and Chemical Storage Area - (South Plume)
- Former Waste Solvent UST – (East Plume)
- Unknown Process Area – (West Plume)

Remedial activities are being conducted at this site under the Missouri Voluntary Cleanup Program following the protocols established under the CALM (Cleanup Action Levels for Missouri) guidance manual (Missouri Department of Natural Resources, 1998). Soil and groundwater cleanup objectives for the site have been determined using a Tier 2 evaluation under CALM. Since the groundwater is not being utilized and exposures can be limited, the cleanup criteria are to provide against offsite trespass, the cleanup criteria are based on attenuation to drinking water maximum contaminant limits at the property line.

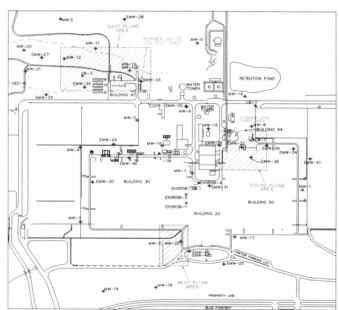

FIGURE 1. Site Plan and Areas of Remediation

SITE HYDROGEOLOGY

The bedrock geology at the site is sedimentary in origin, deposited during the Pennsylvanian time period. The bedrock investigated at the site is part of the Kansas City Group and consists of the Westerville Limestone, Wea Member, Block Limestone, Fontana Shale and Winterset Limestone. The Westerville Limestone is medium to thick bedded, microcrystalline, and fossiliferous, with a 0.15 m (6-inch) shale interbed. The Wea Member consists of thin to thick bedded, fissile shale, weathered with a 0.3 - 0. 6 m (1 - 2 foot) thick cherty limestone (known as the Bull Ledge) layer. The Block Limestone is thin to thick bedded, fossiliferous, with argillaceous shale interbeds. The Fontana Shale is highly fossiliferous, thin bedded, with metamorphosed coal stringers. The Winterset Limestone is thick bedded, massive, with wavy argillaceous shale and cross beds. The Winterset contains a continuous chert layer 0.3 - 0.9 m (1 - 3 feet) thick in its upper sequence. The presence of the Winterset chert layer separates two distinct groundwater flow regimes beneath the site, as defined by unique hydrogeochemical differences. The Winterset chert layer has been observed to retard vertical migration of residual solvent in the upper bedrock.

The bedrock is overlain by 0 - 9.1 m (0 - 30 feet) of Pleistocene and Holocene age loess deposits, and residuum derived by the in-situ weathering of the bedrock. Surface geophysical survey results (seismic, resistivity, and induced polarization) indicate 3 primary fault zones trending in a northwest-southeast direction with secondary fault zones trending perpendicular (northeast-southwest).

Depth to groundwater varies from 0.6 - 12 m (2 to >40 feet) below grade. Monitoring well depths in bedrock range from 15 to 63 feet below grade. Groundwater flow above the Winterset chert is to the northeast on the east side of the site and northwest on the west side at a gradient of approximately 0.037.

SOURCE REDUCTION TECHNIQUES

Source removal/reduction has been proven as a means of promoting the natural attenuation of residual chlorinated solvents in a groundwater flow system under reducing conditions (Smith, et. al., 1998b). Due to the large site area, the remediation strategy at this location was to address the source areas while letting the aqueous phase naturally attenuate. Further, as stated previously the cleanup goals are based on reducing source area concentrations to levels where attenuation results in achieving drinking water maximum contaminant levels at the property line.

Concentrations of TCE in groundwater at in the source area wells range from 16 mg/l to 71 mg/l. The aqueous solubility of TCE is 1,100 mg/l. From Cherry and Feenstra (1991), concentrations in groundwater exceeding 10% of the aqueous solubility (110 mg/l) suggest the presence of DNAPL, while concentrations exceeding 1% (11 mg/l) indicate that DNAPL is located proximal to the monitoring well. The measured concentrations (1998) in wells MW-2 (55 mg/l), MW-10 (16 mg/l), and EMW-33 (71 mg/l) exceed 1% of the aqueous solubility. As such, it is expected that TCE in these 3 areas have the potential to represent a continuing source of groundwater contamination.

In situ chemical treatment can involve the addition of an oxidant or a bulk reductant. Typically, this requires delivery of the oxidant to where the DNAPL is located to directly oxidize and destroy the compounds. Delivery of oxidants to react with source material can be problematic in fractured media. Chlorine dioxide has been used to a limited extent to oxidize chlorinated solvents (Smith, et. al., 1998a), forming trichloroacetic acid, which in-turn breaks down in the groundwater eventually forming carbon dioxide, water and chloride. However, for purposes of treatment at this site chlorine dioxide also interacts with soil minerals surfaces. That interaction improves the partitioning of TCE from the soils, and modifies groundwater redox conditions. It has been observed at other sites (Smith, et. al., 1998a), that the addition of chlorine dioxide forms Fe III from Fe II, which provides the mineral source of oxygen for iron reducing bacteria (in-turn converting Fe III back to Fe II). This biological reduction of iron in solution results in a reduction in redox potential, which is beneficial for bacteria-mediated reductive dehalogenation of TCE.

Chlorine dioxide (ClO_2) was selected for use at the east plume area (near the former waste solvent UST). ClO_2 flushing was performed at the east plume area from September 15 through 25, 1999. ClO_2 was generated using a Rio Linda® generator, consisting of a mixing venturi and chamber equipped with adjustable rotometers. ClO_2 solution was prepared by combining sodium hypochlorite, sodium chlorite, and 15% hydrochloric acid. An electric feed pump transferred ClO_2 solution through a PVC distribution manifold to up to 4 flushing well points at 1 to 1.5 gallons per minute (gpm). Prior to conducting ClO_2 flushing, groundwater samples were collected from 2 flushing well points (IR1-2 and IR4-1) along with monitoring wells MW-10 and EMW-33. The results are presented in Table 1.

TABLE 1. VOCs (mg/L) in Groundwater Pre and Post ClO_2 Injection

Well Location	TCE		cis 1,2-DCE		1,1-DCE		Vinyl Chloride	
	9/99	10/99	9/99	10/99	9/99	10/99	9/99	10/99
IR1-2	2.4	NA	0.24	NA	0.001	NA	ND	NA
IR4-1	3.2	NA	0.27	NA	ND	NA	ND	NA
MW-10	18	12	2.2	2.8	0.013	0.009	0.037	ND
EMW-33	56	5.6	0.83	1.5	0.008	0.012	0.003	0.019

Note: Pre ClO_2 injection sampling occurred 9-14-99; Post injection sampling occurred 10/19/99

NA = not analysed
ND = not detected

Groundwater samples from MW-10 and EMW-33 after ClO_2 flushing were collected in October 1999 as part of semi-annual groundwater monitoring. Well MW-10 is located downgradient from the former waste solvent tank, whereas well EMW-33 is located immediately adjacent and upgradient from the former waste solvent tank. It can be seen that significant reductions in concentration of TCE at well EMW-33 were observed, with corresponding increases in cis-1,2-DCE and vinyl chloride. This was the desired response, for the ClO_2 was injected to increase the partitioning of the TCE from the soil or rock mineral surfaces into the groundwater for bio-attenuation. The detection of increased concentrations of cis-1,2-DCE and vinyl chloride may be indicative of the bio-attenuation occurring with partitioning into the dissolved phase. The cleanup criteria was developed from the following equation:

$$Cw_{attn} = \frac{GTARC}{\left(\dfrac{C_x}{C_{source}}\right)} \bullet DF$$

Where: $GTARC$ = groundwater target concentration (mg/l)
DF = dilution factor (unitless)
C_x = concentration at some point "x" (mg/l)
C_{source} = concentration at source (mg/l)
Cw_{attn} = is the allowable concentration in groundwater at the source

From this, and the distance to the property line, the source area cleanup concentration for TCE was determined to be 4.02 mg/l. The concentration reduction observed in the October 1999 sampling, would result in an estimated concentration of TCE at the property line of 0.018 mg/l. This compares to the drinking water maximum contaminant level of 0.005 mg/l. Further reductions may be realized through bio-attenuation at the source area.

The ClO_2 resulted in a concentration reduction, which also results in reduction in organic carbon demand. From Wiedemeier, et. al. (1996), each mg of dissolved organic carbon that is oxidized via reductive dehalogenation results in the consumption of 5.65 mg of organic chloride. Table 2 presents the determination of organic chloride, organic carbon demand.

TABLE 2: Change in Organic Carbon Demand

Well	Organic Carbon Demand	
	Sampled 9/14/99	Sampled 10/19/99
MW-10	2.87 mg/l	2.08 mg/l
EMW-33	45.97 mg/l	5.65 mg/l

Organic carbon was not measured as part of this sampling, but was 0 mg/l in MW-10 and 4.5 mg/l in EMW-33 when measured in 1998. This would indicate that in the order of 80% of the organic carbon demand may now be satisfied at well EMW-33. Reductive dehalogenation may show increases at this location in the future. Further monitoring will help determine if the organic carbon demand is being met, whether additional ClO_2 treatment would be advantageous, or whether amendments are required.

Soil vapor extraction (SVE) is being conducted as a means of source reduction in the South Plume area in the vicinity of a former chemical sump and UST tank farm. Horizontal drilling techniques were used to install 341 m (1,120 feet) of 5.1 cm (2-inch) diameter slotted HDPE pipe underneath Building 50.

The SVE pipe was installed in unconsolidated silt and clay above the bedrock from 1.8 – 3.05 m (6 - 10 feet) below grade. The soils contained TCE as high as 240 mg/kg. Over 46.4 kg (102 pounds) of hydrocarbon were removed from the soil during the first 4 months of system operation, and the system is scheduled to continue operation through July 2000. Because this area is beneath a building, the vapor extraction in the vadose zone is not expected to have an immediate impact on the groundwater.

NATURAL ATTENUATION EVALUATION

To evaluate where natural attenuation is taking place, concentrations of chloride and sulfate were contoured (Figures 2 and 3). The data indicate that there are area of high chloride in monitoring wells that are downgradient of former solvent use areas.

The chloride distribution would suggest that as the solvents migrated downgradient with the groundwater, the compounds are being dehalogenated. The sulfate distribution shows reduction of sulfate in association with the releases and downgradient therefrom. Sulfate reducing conditions are interpreted to be the result of the microbial activity and result in optimal redox conditions for reductive dehalogenation. Background sulfate concentrations were measured as high as 379 mg/l, which is above the inhibitory level presented in Wiedemeier, et. al., (1996). The data do not suggest inhibition of the reductive dehalogenation process.

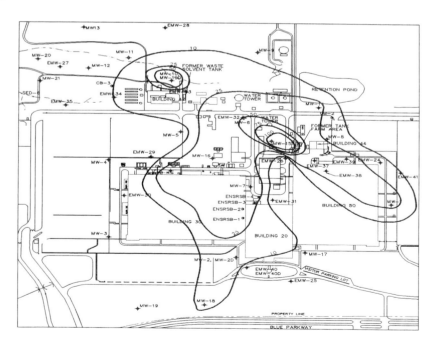

FIGURE 2. Chloride Distribution in Groundwater (mg/L)

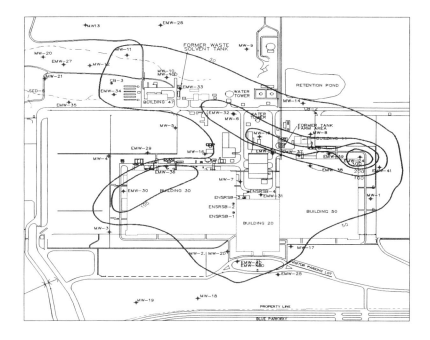

FIGURE 3. Sulfate Distribution in Groundwater (mg/l)

SUMMARY AND CONCLUSIONS

Natural bio-attenuation of TCE is occurring at the site in naturally high sulfate groundwater. This is contrary to the inhibitory effects described in Wiedemeier, et. al., (1996).

Source reduction techniques using oxidants in this testing have shown encouraging results. The organic carbon demand at the source area of the east plume have bee reduced to a level were previous samplings have shown that available organic carbon would satisfy approximately 80% of the organic carbon demand. Additional monitoring will determine if additional treatments or amendments are required to achieve the risk based cleanup goals.

The source area reductions will result in reduction in the concentrations downgradient. It is estimated (based on site specific parameters) the reductions observed will now result in an estimated TCE concentration of 0.018 mg/l, at the property line. This compares to a drinking water maximum contaminant level of 0.005 mg/l.

There is insufficient data to determine if vapor extraction being conducted at the south plume is resulting in reductions in groundwater concentrations.

REFERENCES

Bouwer, E. J. (1994) "Bioremediation of chlorinated solvents using alternate electron acceptors". In Norris, R. D., R. E. Hinchee, R. Brown, P. L. McCarty, L. Semprini, J. T. Wilson, D. H. Kampbell, M. Reinhard, E. J. Bouwer, R. C. Borden, T. M. Vogel, J. M. Thomas, and C. H. Ward, eds, *Handbook of Bioremediation*: Lewis Publishers. P 149-175

Cherry, J. A. and S. Feenstra (1991). *Identification of DNAPL Sites: An eleven point approach,* draft document in Dense Immiscible Phase Liquid Contaminants in Porous and Fractured Media, short course notes, Waterloo Centre for Ground Water Research, Kitchener, Ontario.

Smith, G., B. Dumdei, and V. Jurka (1998a). "Modification of Sorption Characteristics in Aquifers to Improve Groundwater Remediation". *Proceedings of the First International Conference on Remediation of Chlorinated and Recalcitrant Compounds, Monterey, California, May 18 – 21, 1998.*

Smith, G. J., T. V. Adams, and V. Jurka (1998b) "Closing a DNAPL Site Through Source Removal and Natural Attenuation". ." *Proceedings of the First International Conference on Remediation of Chlorinated and Recalcitrant Compounds, Monterey, California, May 18 – 21, 1998.*

Wiedemeier, T. H., M. A. Swanson, D. E. Moutoux, E. K. Gordon, J. T. Wilson, B. H. Wilson, D. H. Kampbell, J. E. Hansen, P. Haas, and F. H. Chapelle, (1996). *Technical Protocol for Evaluating Natural Attenuation of Chlorinated Solvents in Groundwater, Draft – Revision 1.* Air Force Center for Environmental Excellence, Brooks Air Force Base, San Antonio, TX

OZONE-ENHANCED REMEDIATION
OF PETROLEUM HYDROCARBON-CONTAMINATED SOIL

Heechul Choi and Hyung-Nam Lim (Kwangju Institute of
Science and Technology, Kwangju, Korea)
Jeongkon Kim (HydroGeoLogic, Inc., Herndon, VA 20170)

ABSTRACT: Experiments were conducted by using a soil column to investigate
the feasibility and efficiency of *in-situ* remediation of PAH-contaminated soil by
ozone. Column tests using gaseous ozone also revealed the enhanced
decomposition of ozone molecule due to the catalytic reaction as evidenced by as
much as 25 times shorter half-life of ozone in a sand packed column than in a
glass bead packed column. Ozone was readily transported through columns
packed with field soil and sand and oxidized more than 75% of commercial diesel
initially present as non-aqueous phase liquid 12 hours after the initial injection of
ozone gas. Chromatogram analysis of contaminants before and after ozonation
clearly indicated that carbon numbers shifted from higher numbers to lower ones
as ozone injection continued. The effects of other various conditions such as soil
media, ozone dosage, bicarbonate ion, and humic acid on the ozone oxidation of
Phs in slurry phase were also investigated.

INTRODUCTION

Petroleum hydrocarbons (Phs) are often found in many places as main
contaminants via leaking of fuel storage tanks and refinery factories, oil spills,
and dumping of waste, etc (Edward J. Calabrese, *et al.*, 1991). Phs contaminated
sites have a great concern as a source of ground water pollution due to containing
polycyclic aromatic hydrocarbons (PAHs) known as carcinogenic compounds.
PAHs are generally semi- or non-volatile and hardly biodegradable compounds.
Therefore soil vapor extraction (SVE) and bioventing process may not be effective
for attenuating of PAHs in soil. Currently, as an innovative remediation
technology, *in situ* ozone enhanced remediation process (*in situ* OERP), was
proposed (Susan J. Masten *et al.*,1997; H.C.Choi *et al.*, 1998). As a combination
of SVE and chemical oxidation processes, *in situ* OERP can remove both volatile
hydrocarbons (VHs) and semi- or non-volatile PAHs. Moreover injected ozone on
sites generates more powerful oxidant, e.g., hydroxyl radical ($\bullet OH$, E^o=3.06V) via
it's catalytic decomposition. Although the mechanism of $\bullet OH$-generation is not
clear, it may be considered to be the catalytic decomposition of ozone on reactive
sites of soil where $\bullet OH$-generation takes place through heterogeneous reaction
between oxidants and natural inorganic (e.g., metal oxides) and organic matters.
And the fate and reaction mechanisms of ozone and $\bullet OH$ in porous media are
neither well understood nor examined. Therefore, the objectives of this research
are to investigate the fate of ozone and $\bullet OH$, as well as feasibility for the removal

of diesel and PAHs as model compounds that are hard to be removed by the conventional processes.

MATERIALS AND METHODS

Ozone was generated by PCI[®] ozone generator (Model GL-1, U.S.) and measured by the PCI[®] ozone monitor which was calibrated by potassium iodide (KI) method. A mass flow controller (MFC) was employed for regulating the gas flow at the rate of 200ml/min. Column tests for the removal of contaminants and the decomposition of ozone were performed under various conditions. The soil samples were extracted with dimethyl chloride using modified EPA sonication method. PAHs and diesel were quantified by HPLC (WATERS[TM] 717) and GC (HP-5890) analysis. The transport and demand of ozone in the soil column were conducted by using a modified continuous gas-flow detection method.

Materials and Reagents. Phenanthrene (Wako, 99.9%), methylene chloride (Fisher, HPLC grade), diesel (commercial product), and other chemicals were used without further purification. Sand (Jumunjin Sand, Korea) was passed through 400-600μm sieved to provide a consistent surface area for the sorption of model compounds. To eliminate organic matter effects, sand was washed and baked at 500 °C for 24 hr . Glass beads was purchased (400-600μm) and also washed and dried at 105°C for 24hr. Adding PAHs (in methylene chloride) and diesel, 10mg/kg, 1400mg/kg as initial concentrations prepared two types of spiked media, respectively. Characteristics of soils are shown in Tables 1 and 2.

TABLE 1. Characteristics of Sand (Jumunjin, Korea) and Soil (Kyunggi, Korea)

Soil Property	Test Method	Soil	Jumunjin Sand
Organic Matter Content (%)	Miller et al., 1992	4.37	0.12
Porosity (%)	Miller et al., 1992	45.6	40.8
Bulk Density (g/cm^3)	KS F 2308	1.34	1.48
Specific Weight	ASTM D2216-92	2.66	2.5
Water Content (%)	ASTM D422-63, D421-85	8.39	6.3
Particle Size (μm)	Standard Method 209D	Silt Sand[a]	0.6-0.8

[a] Coarse sand (4.75-0.25mm) = 50%, fine sand (0.25-0.05mm) =25%, silt (0.05-0.005mm) =21%, clay (<0.005mm) =4%.

Ozonation Experiments. To reduce analysis error, which may come from sampling and the heterogeneity of contaminant distribution in soil matrix and perform kinetic study, replaceable unit reactors were used. Ozone was generated from dried oxygen by electric discharge, using a PCI ozone generator (model GL-1). All oxidation runs were made in the batch mode. Temperature was maintained constantly at 25 °C throughout all experiments. The gaseous ozone was monitored by PCI [®]ozone monitor (HC-400). Off-gas ozone was trapped in the potasium iodide (KI) solution to calculate the amount of ozone that was consumed in the reactor.

TABLE 2.   Result of X-ray fluorescence (XRF) analysis of sand (Jumunjin , Korea) and soil (Kyunggi, Korea).

Contents	Analysis Data (%-wt)	
	Jumunjin Sand	Soil
SiO_2	90.41	66.25
Al_2O_3	5.48	15.26
Fe_2O_3 [a]	0.12	6.96
TiO_2	0.02	0.97
MnO	<74ppm	0.08
CaO	0.07	0.47
MgO	<85ppm	1.90
K_2O	3.45	2.76
Na_2O	0.36	0.19
P_2O_5	0.01	0.09
L.O.I [b]	0.09	5.03
TOTAL	100	99.95

[a] Fe_2O_3: Total of Fe   [b] L.O.I. : Loss on Ignition

Analysis. Quantification of   phenanthrene was carried out on a WATERS[TM] HPLC system equipped with a autosampler (WATERS[TM] 717) and a Youngin UV absorbency detector(λ = 254 *nm*). With a Nova-Pak C_{18} column(3.9×150 *mm*, Waters) at a flow rate of 1.0 *ml/min* after extraction using a modifed EPA sonication method for 1hr. The temperature of column box was 40 ^oC. The eluent was consisted of a   65:35(water/acetonitrile) isocratic mixture. GC-FID (HP-5890) was utilized to analyze diesel compounds.

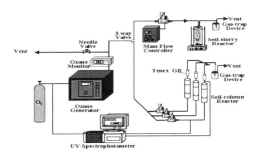

FIGURE 1.   Experimental set-up.

RESULTS AND DISCUSSION

Ozone Decomposition in Porous Media. Gaseous ozone was pulse-input into a column packed with dry sand and glass beads to investigate the characteristics of ozone decay and to confirm the enhanced decomposition of ozone. Ozone was injected in the column for several pore volumes until the equilibrium ozone concentration is reached. And then the column is closed on both sides and the rate of ozone decay was calculated by sacrificing each column to measure ozone concentration. The ozone-decay rate constants in glass beads (GB) and sand (S) packed columns were $4.3 \times 10^{-4} s^{-1}$ and $9.9 \times 10^{-3} s^{-1}$, respectively (Figure 2). Even though radius sizes of S and GB were similar (400-600μm), the half-life of ozone in S packed column was almost 25 times shorter than in GB packed column. It is apparent that sand (Jumunjin sand, Korea) containing soil organic matter (SOM) and metal oxides (related data shown in Table 1, 2) can accelerate the decomposition of ozone. But, it is not clear that whether the accelerated ozone decomposition on sand will also enhance the removal of contaminants by generating more powerful oxidants such as hydroxyl radical (•OH).

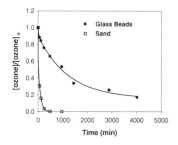 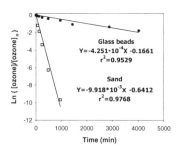

FIGURE 2. Decomposition of ozone in soil column.

Removal of Phenanthrene in Soil Column.   To investigate the effects of inorganic and soil organic matter (SOM) on the phenanthrene removal in soil column, three types of media such as sand (S), baked sand (BS) and glass beads (GB) were used for ozonation experiments. A modified UV-detection method continuously monitored breakthrough curves (BTCs) of ozone. The effects of soil types on BTCs of ozone were also investigated. Figure 3 shows BTCs of ozone of sand columns with and without phenanthrene for various types of soil. The flow rate of gaseous ozone was 200ml/min, the initial concentration of phenanthrene, 100mg/kg and water content, 0%. BTCs of clean and PAH-contaminated sands showed that there was no significant retardation to transport ozone in both columns. But in the case of PAH-contaminated sand, it was noticed that the consumption of ozone substantially increased in phenanthrne contaminated sand due to the ozone demand to oxidize the contaminant. However, the amount of ozone consumption reduced gradually as phenanthrene was removed in the column. After ozonation, the residual concentration of phenanthrene on baked sand was lower than on glass beads (Figure 3 (b)). This is considered that reactive sites such as metal oxides in the sand accelerated the decomposition of ozone to enhance •OH generation.   There were many related reports about enhancement effects of sand, metal oxides and artificial catalysts on the decomposition of ozone. The species of oxygen, such as $O_2^{•-}$ (superoxide anion) known as an intermediate of the ozone decomposition, producing hydroxyl radicals (•OH), and O_2^{2-} (peroxide anion), O (oxygen atom), and $O^{•-}$(oxygen radical anion) were observed via gaseous ozone decomposition on a silver (Ag) catalyst (Immamura, *et al.*, 1991; Li *et al.*, 1998).

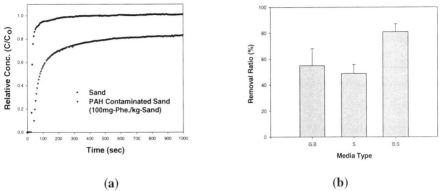

(a)   (b)

FIGURE 3. (a) Breakthrough curves (BTCs) in soil columns and (b) removal ratios of phenanthrene contaminated soils.

Removal of Diesel in Soil Column. Figures 4 and 5 show the breakthrough curves (BTCs) of ozone and the removal trends of diesel in 1-dimensional columns packed with various types of soil. Gaseous ozone was continuously injected to respective columns packed with DRO-contaminated sand and natural soil. Compared with sand packed columns, the natural soil column containing 4.37(%) of soil organic matter (SOM) revealed a considerable retardation and consumption of ozone. It took more than 6 hours for the gaseous ozone to breakthrough the natural soil column while 1 hour for the sand column. The effect of types of soil on the removal of diesel range organic (DRO) by ozone was also investigated simultaneously. The results from soil and sand column experiments revealed that removal pattern in early stage of ozone injection depends on the properties of soil such as media size, SOM and inorganic matter. However, after 14hours of ozonation, the overall removal efficiencies of DRO in sand and soil columns were 83.0 % and 78.7 %, respectively. Figures 4 (b, c) and 5(b, c) present relative concentrations of DRO, ranging from dodecane (C_{10}) to tetracosane (C_{24}), according to the ozonation time. Within 2 hours, more than 90% of C_{10} in the inlet of both sand and soil columns were removed via volatilization and oxidation. Similarly, most DRO ranging C_{12}-C_{24} in the column inlet were oxidized within approximately 4 hours and then gradually removed. However, in the outlet of both cases, it is observed that the normalized DRO concentration of smaller carbon numbers, i.e., C_{10}-C_{18} slightly increased higher than the initial concentration in the beginning. This is simply because the higher carbon numbers of DRO shifted to the lower ones as ozone injection continued. It is interesting to notice that ozone molecule reacts with most DRO compounds to oxidize in relatively general pattern in multi component chemicals like diesel. Close observation of the removal pattern of both cases indicates that the removal trend of each DRO shows similar pattern in the inlet of both columns, whereas the outlet shows the apparent difference. Sand columns show higher removal for lower carbon number compounds and lower for higher carbon numbers at the end of ozone injection. On the contrary, the column packed with natural soil revealed much higher removal rate for higher carbon numbers and gradual increase pattern of the lower carbon numbers. This may be considered that oxidation byproducts of higher molecules are trapped in upper part of the soil column. Whereas they escaped in the sand column.

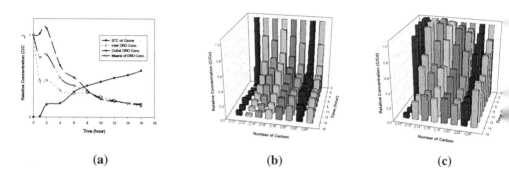

(a)   (b)   (c)

FIGURE 4. (a) Removal of diesel (DRO) and breakthrough curve (BTC) of ozone in sand packed column. (b) Residual of DRO profile in inlet. (c) Residual of DRO profile in outlet. Flow rate was 50mL/min, conc. of ozone was 119mg/L, initial conc. of diesel was 1,485mg/kg-sand (as DRO), organic matter content was 0.12(%), and moisture content was 6.3(%-wt) at 25°C.

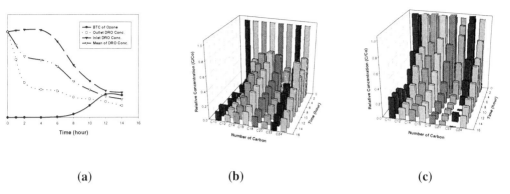

(a)   (b)   (c)

FIGURE 5. (a) Removal of diesel (DRO) and breakthrough curve (BTC) of ozone in natural soil packed column. (b) Residual of DRO profile in inlet. (c) Residual of DRO profile in outlet. Flow rate was 50mL/min, conc. of ozone was 119mg/L, initial conc. of diesel was 1,485mg/kg-sand (as DRO), organic matter content was 4. 37 (%) and moisture content was 8.4%-wt at 25°C.

ACKNOWLEDGEMENT

This work was supported in part by the Korea Science and Engineering Foundation(KOSEF) through the Advanced Environmental Monitoring Research Center (ADEMRC) and by the Brain Korea 21 of the Ministry of Education through The Graduate Program for Chemical and Environmental Engineering at Kwangju Institute of Science and Technology (KJIST).

REFERENCES

Edward J. Calabrese and Paul T. Kostecki. (1991). *Hydrocarbon Contaminated Soils*. volume I, Lewis Publishers.

Susan J. Masten and Simon H.R. Davies. (1997). Efficacy of *in-situ* ozonation for the remediation of PAH contaminated soils. *Journal of Contaminants Hydrology*, 28, 327-335.

Susan J. Masten. (1991). Use of *in-situ* ozonation for the removal of VOCs and PAHs from unsaturated soils. Proceedings of the symposium on soil venting. Houston, Texas. 29-53.

H.C.Choi, K. S. Kim, D. Y. Yoo and M. K. Jung. (1998). Development of a *in-situ* remediation technology for petroleum hydrocarbons-contaminated soil. KICT, pp. 47-48.

Li, W. and S. T. Oyama, Mechanism of ozone decomposition on a manganese oxide catalyst. 2. Steady-state and transient kinetic studies, *Journal of the American Chemistry Society*, Vol. 120, No. 35, pp. 9047-9052, 1998b.

Imamura, S., M. Ikebata, T. Ito, and T. Ogita, Decomposition of ozone on a silver catalyst, *Industrial & Engineering Chemistry Research*, Vol. 30, No. 1, pp. 217-221, 1991.

PHOTOCATALYTIC OXIDATIVE DEGRADATION OF ACRIDINE ORANGE SOLUTION WITH METALLOPORPHYRINS

Jinzhang Gao, Xinglong Jin, Yanjun Fang, Ruiqiang Yang, and *Hui Chen*[*]
(Northwest Normal University, Lanzhou, Gansu, P. R. China)

ABSTRACT: In the presence of suspension of polymeric M [meso-tetra (4,4'-biphenybisulfony) phenylporphyrin] (PMTBPBSOPP M = Co, Mn, Cu, Zn, Fe), the photocatalytic oxidative degradation of acridine orange (AO) was investigated in the conditions of irradiation with high pressure mercury lamp (HPML), iodine tungster lamp (ITL) and natural sunlight (NSL) respectively. The results indicated PCoTBPBSOPP had better effects on photodegradation of AO under mild condition. When lower concentration of hydrogen peroxide was presented in the system, PCoTBPBSOPP showed stronger photocatalytic activity and AO aqueous solution was degraded in shorter time. The degradation of AO obeys to the first order kinetics with the equation as $lnC_t = 2.52052 - 0.56546t$, $R = -0.99704$. When the initial concentration of AO was 43.792mg/L, the degradation rate and the decolorization rate of AO was up to 97.76% and 97.16% respectively under the irradiation with HPML. Polymeric of metalloporphyrins could decompose AO by using artificial light source such as HPML and ITL. Furthermore, degradation could be completed in three hours with solar light. It was proved that polymeric of metalloporphyrins had a potential effective application in the wastewater treatment in the heterogeneous photocatalytic oxidation system under natural conditions.

INTRODUCTION

The conventional techniques of removing organic pollutants from water resources and water supply system are usually focused on sedimentation, flocculation, steam stripping, adsorption with activated carbon and filtration etc. However, some pollutants are poorly or even not absorbed by activated carbon while adsorption may transfer some pollutants from the aqueous solution to the solid phase without discernible reduction of the total quantity or hazardous potentials. As early as the 1960s, much work has been undertaken on the photocatalytic oxidation for removal of various organic compounds, priority attention is paid to the identification of intermediate degradation product.

---

[*] Corresponding author. **Email: kun@mail.lzri.edu.cn**

Titanium dioxide (TiO_2) is the most commonly used photocatalyst because certain forms of it have reasonable photo-activity (Matthews, Ralph W., 1991 and David F. Ollis et al., 1991). Besides, it has many advantages, such as non-toxic, insoluble and comparatively inexpensive. But Titanium dioxide has broader forbidden bend (Eg=3.2eV), which means that it is not easy to be excited by visible light since the available utilization of solar energy is only about 3%. In recent years, some metalloporphyrin complexes have been found to photocatalyze the organic substance oxidation by O_2 under mild conditions (Maldotti, A. et al., 1993). Moreover, these photocatalysts are extensively used in homogeneous phases rather than in heterogeneous phases although the latter was reported. Recently metalloporphyrin as a new photocatalyst started to attract environmental chemists' interest (Carlo Bartocci et al., 1996). A portion of investigations aimed to the discovery of new catalytic systems to induce selective oxygenation reactions of inactivated C-H bonds in hydrocarbons through molecular oxygen. A photochemical method to oxidize dye with O_2 under mild conditions was conducted in our laboratory using high molecular polymers modified by metalloporphyrins as catalyst (Hui Chen et al., 1999).

In this paper, the polymers of metalloporphyrins were used to catalyze and accelerate the degradation reaction of recalcitrant acridine orange (AO). It was proved that the polymers of metalloporphyrins had higher photoactivity than single porphyrins and were ready to recovery and recycle, even easy to be excited by ultra or visible light. Due to the characteristics mentioned above, it is possible to realize the scale-up industrialization because it can float or suspend in ambiet waters.

MATERIALS AND METHODS

Polymeric M [meso - tetra (4,4'-biphenybisulfony) phenylporphyrin] (PMTBPBSOPP M = Co, Mn, Cu, Zn, Fe) were prepared, purified and characterized. AO was obtained from importing in bulk and used without further purification. Other chemicals were commercial products of analytical grade or reagent-grade and were used without further purification. All the solutions were prepared with double-distilled water.

The process of AO solution degradation was scanned by UV-3400 spectrophotometer (Shimadzu, Japan) at stationary time and the maximum absorption wavelength of AO solution was identified at 492nm. UV-vis spectra data were recorded with UV-754 spectrophotometer (Shanghai, China) at every 20nm range from 420nm to 620nm. The pH of solution determined by phS-4C digital presentation acidometer with composite electrodes (Chengdu, China). The temperature of reacting solution was kept at 30°C by super thermostatic recycling water bath, air was provided by air pump (Hong Kong) and the solutions stirred by ET-1 magnetic stirrer (Hubei, China). Irradiation was carried out with 450 w and 125 w HPML (Beijing bulb factory) or 300 w Iodine Tungster Lamp (ITL, China). A cylindrical, double layer quartz photochemical reactor (internal diameter: 35mm; external diameter: 50 mm; height: 50 mm; volume: 40 ml) was used for photocatalytic reaction.

Photocatalytic oxidative degradation of AO aqueous solutions was studied using polymer of metalloporphyrins as catalyst under mild conditions. Appropriate amount polymeric metalloporphyrins and 35 ml of 13.34 mg/L AO aqueous solution were added into the quartz reactor, then stirred with a magnetic stirrer prior to irradiated with HPML, ITL or NSL and kept the temperature at 30 ℃ by recycling water with super thermostatic bath and simultaneous air sparged. The absorbances of the sample solution were determined at 492nm with glass cells (10mm optical path length) every hour after standing for 5 minutes.

RESULTS AND DISCUSSION

Kinetics of photocatalytic degradation of AO. AO, an orange solution, has the maximum absorption at 492nm. The absorbances of AO versus its concentrations gave a calibration curve. It was found to be linear in the range of 4.379~21.879 mg/L with the regression equation as the following:

$$A = 0.02884 + 0.06203 \ C \qquad R = 0.9991$$

Where A is the absorbance value of AO solution, C is the concentration of AO solution (mg/L) and R is the coefficiency for the straight line.

The photocatalytic degradation of AO aqueous solution was investigated in various conditions and the results were shown in FIGURE 1. AO solution could be store stable in the dark and coexist a long time with PCoTBPBSOPP (see curve a and b). The absorbance decreased with increasing irradiation time after irradiating the solution and sparging air, but the degradation rate wasn't high, only 49.58% (see curve c). When the catalyst was added into the solution above, it was apparent that AO solution degraded dramatically (see curve d), and the degradation rate reached 85.86% after eight hours. When 1.26×10^{-2} mol/L hydrogen peroxide was presented in the solution, the efficacious reaction was achieved in three hours (see curve e), and the degradation rate was up to 92.41%. That indicated photocatalytic activity of catalyst was improved remarkably when the catalyst and H_2O_2 presented in solution together.

The primary degradation in curve c, d and e obeyed to the first order kinetic law according to the following equation:

$$-dC_t / dt = k_a \ C_t$$

Where C_t is the residual concentration of AO solution at time t and k_a is the apparent first-order decay constant.

According to the above equation, the slopes of lines that were plotted with lnC_t versus irradiation time were determined by regression analysis and given in Table 1. AO solution could be photodegraded completely in PCoTBPBSOPP-H_2O_2 system. Not only was degradation rate improved apparently but also decolorization rate was increased accordingly.

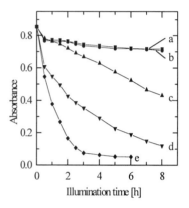

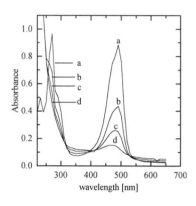

FIGURE 1. The dynamic curves in different conditions a - No catalyst and in dark; b - No illumination c - No catalyst in HPML; d -3 mg catalyst HPML and air sparged; e - 3 mg catalyst, 1.26×10^{-2}mol/L mol/L in HPML and air sparged;

FIGURE 2. The changes of visual and violet spectrum of AO aqueous solution with illumination;   a – no illumination; b -- illumination 2 h;   c -- illumination 4 h; d -- illumination 6 h.

The Process of AO Degradation. The absorption spectral changes of AO aqueous solution with irradiation and catalysis were measured and shown in FIGURE 2. The visual spectra of AO aqueous solution could be degraded completely within 8h, and ultra-violet absorption between 250~360nm were decreased too. But the absorbance of UV below 250nm was increased and the maximum wavelength was shifted to short wave with prolonged irradiation time. It was reasonably regarded that AO was decomposed into smaller molecular after photocatalytic degradation. When the concentration of H_2O_2 was 1.26×10^{-2}mol/L, and the pH of solution was 8.80, the photodegradation of AO were more rapid and complete within three hours.

Selection of Photocatalyst. Five polymeric M [meso-tetra (4,4'-biphenybisulfony) phenylporphyrin] (PMTBPBSOPP M = Co, Mn, Cu, Zn, Fe) were prepared and their ability of photocatalytic degradation AO was investigated (FIGURE 3). The experiments showed that the catalyst PCoTBPBSOPP had stronger photocatalytic ability than the others. It was possible that molecule oxygen can be activated easily through the electron transformation from cobalt (II) ion abstracted the hydrogen atom of the substrate and induced the photochemical reaction. The mechanism of photodegradation of AO may be the following:

$$Co^{2+}PTBPBSOPP + O_2 \rightarrow [Co^{3+}PTBPBSOPP]\cdots O_2^-$$

$$[Co^{3+}PTBPBSOPP]\cdots O_2^- + AO \rightarrow [Co^{3+}PTBPBSOPP]\cdots O_2H^- + products$$

$$[Co^{3+}PTBPBSOPP]\cdots O_2H^- \rightarrow Co^{2+}PTBPBSOPP + H_2O \cdot$$

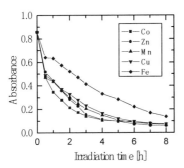

FIGURE 3. The effect of different metal complexes

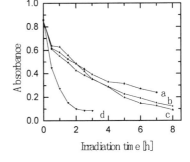

FIGURE 4. The degradation curves in various light source   a -- 125w HPML; b -- 450w HPML; c -- 300 ITL;   d -- NSL.

The function of H_2O_2 in system. It is well known that $\cdot$ OH radicals which play an important role to initiate the chain reaction are easy to generate from under illumination. Hydrogen peroxide can be decomposed under light of wavelength less than 380nm and produced HO $\cdot$ radicals.

In this photocatalytic system, the substrate AO performed photodegradation in correspondence with radical chain reaction in the presence of $\cdot$ OH radicals, and then AO was removed rapidly. From curve e in FIGURE 1, the H_2O_2 not only oxidized AO but also induced the catalyst to photolysis of AO. When the concentration of H_2O_2 was reached to the certain level, the degradation rate couldn't be changed any longer. The hydroxyl radicals were possibly consumed by excessive H_2O_2 and began to restrain reaction. The amount of H_2O_2 in $5.0\sim12.5 \times 10^{-3}$mol/L was feasible, it was selected in the photocatalytic degradation of AO reaction.

Effect of light source. The irradiated AO solution with different light sources showed different degradation rates (see FIGURE 4). The dynamics regression equations with HPML, ITL and NSL were shown in Table 1. The degradation rate was increased with the strengthened power and the decolorization rate was improved with addition of long wavelength. In order to investigate the practical application of this technology in ambient water supply, the NSL was chosen to be used as ideal source of irradiation.

TABLE 1. The kinetic parameters of various light source effect

No.	Light source	Conc. of H_2O_2	Kinetic equation	k_a (h^{-1})	R	Half time (h)	Degrada tion rate(%)	Decolor ization rate(%)
FIG.1 c	HPML 450w		lnCt=2.58281−0.06864t	0.06864	-0.99649	10.098	49.53	46.81
FIG.1 d	HPML 450w		lnCt=2.43977−0.22477t	0.22477	-0.99389	3.804	85.86	77.36
FIG.1 e	HPML 450w	1.26×10^{-2}mol/L	lnCt=2.52052−0.56546t	0.56546	-0.99704	1.226	91.94	81.12
FIG.4 a	HPML 125w		lnCt=2.34823−0.14863t	0.14863	-0.99241	4.664	72.20	72.00
FIG.4 c	ITL 300w		lnCt=2.57765−0.2758t	0.2758	-0.99755	2.513	89.49	83.33
FIG.4 d	NSL		lnCt=2.5654−1.0738t	1.0738	-0.99783	0.646	90.19	77.29

Effect of pH. As shown in FIGURE 5, AO solutions were photodegraded with PCoTBPBSOPP in the pH range from 1.67 to 10.00 which adjusted previously with diluted NaOH or HCl. It indicated higher pH of solution was beneficial to the photocatalytic oxidation degradation of AO aqueous solution. It was evident that degradation of AO solution was easy in alkalescent but difficult in acid. The degradation rate and decolorization rate were increased to 94.74% and 93.43% respectively with pH 8.80 and 1.26×10^{-2}mol/L H_2O_2 in three hours.

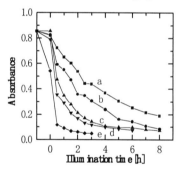

FIGURE 5. The effect of pH for photodegradation of AO solution
a - pH = 1.67;   b- pH = 3.40;   c- pH = 6.85;   d- pH = 8.80;   e- pH = 10.00

CONCLUSIONS

AO aqueous solution can be degraded under irradiation with HPML, ITL and NSL in the presence of polymers of metalloporphyrins, which showed higher activity of photocatalyst in solar light and made full use of long wavelength of light. PCoTBPBSOPP is a beneficial and environmental friendly catalyst. Simultaneous, illustrated polymers of metalloporphyrins have potential prospects of application in photocatalytic oxidation degradation of organic pollutants and broad usage of sunlight in water treatment.

PCoTBPBSOPP can be deposited without the secondary pollution. The added photocatalyts was micro-amount with dosage of 85.71 mg per liter. The degradation of AO in water completely obeys to first-order kinetics. The photocatalytic degradation of AO in water could be greatly accelerated with alkalescent condition and in the presence of diluted hydrogen peroxide. After photocatalytic degradation, the recalcitrant compounds (AO) could be completely decolorized and finally the solution became colorless.

ACKNOWLEDGMENTS

We are grateful to the Natural Scientific Foundation of Gansu (Project ZR-96-066) and the Natural Scientific Foundation of China (29977015) for supporting the research work.

REFERENCES

Carlo Bartocci, Andrea Maldotti and Graziano Varani. 1996. "Photoexcited Iron Porphyrin As Biomimetic Catalysts" *Science and Technology, 78-La Chimica e l'Industria* 1097-1104.

David F. Ollis, Ezio Pelizzetti, Nick Serpone. 1991. "Photocatalyzed Destruction of Water Contaminants" *Environ. Sci. Technol.*, 25 (9): 1522~1529.

Hui Chen, Taicheng An, Yanjun Fang and Kun Zhu. 1999(b). "Photocatalytic Oxidation of Aromatic Aldehydes with Co(II)tetra-(Benzoyloxyphenyl)Porphyrin and Molecular Oxygen" *J. Mol. Catal.* A: Chemistry, 147, 165-167.

Maldotti , A., Amadelli, R., Bartocci, C.,Carassiti, V., Polo, E., and Varani, G. 1993. "Photochemistry of Iron-porphyrin complexes. Biomimetics and Catalysis" *Coordination Chemistry Reviews.* 125, 143-154.

Matthews, Ralph W. 1991. "Photooxidative Degradation of Coloured Organics in Water Using Supported Catalysts TiO$_2$ on Sands" *Wat. Res.* 25(10): 1169-1176.

PHOTOLYTIC AND PHOTOCATALYTIC OXIDATION OF PCE IN AIR

Steven Y.C. Liu, *KUYEN LI*, and Daniel H. Chen
Lamar University, Beaumont, Texas

ABSTRACT: Concentric glass reactor (55 mm I.D.) was used in this study with a 4-watt UV-C (for photolytic reaction) or black light (for photocatalytic reaction) lamp located at the center of the reactor. TiO_2 (Degussa P25) was coated on the inner surface of the glass reactor for photocatalytic reaction. Three PCE concentrations (12, 110, and 430 ppmv), were used in this study. For photolytic oxidation of PCE, the conversion increased with both PCE inlet concentration and space time. Tri-chloro acetyl chloride (TCAC) was detected as the major by-product, except at low concentration (12 ppm) and short space time (1.8 sec) at which no TCAC could be detected. Phosgene was also detected but in minor amount. For photocatalytic oxidation of PCE, the conversion increased with the space time. Phosgene was found to be the major by-product at low PCE inlet concentration and slightly increased when the space time increased. TCAC was not detected at low PCE inlet concentration (12 ppm) but increased steadily with PCE inlet concentration and space time. A reaction pathway using an initiation of chlorine free radicals was proposed to explain the by-products detected in this experiment.

INTRODUCTION

Perchloroethylene (PCE) is a widely used organic solvent in dry-cleaning business and vapor-degreasing industry. Air, groundwater, and soil have been contaminated as a result of careless or improper disposable practices. Animal studies have clearly demonstrated PCE's hepatic and renal toxicity. For humans, PCE is toxic to the liver, kidney, and central nervous system and may be a human carcinogen (Ulm et al., 1996).

Current treatment methods like air stripping, activated carbon adsorption, and soil vapor extraction all have their unavoidable shortages. They could be either higher treatment cost (McGregor et al., 1988) or merely transfer pollutant from one phase to another (Bhowmick and Semmens, 1994). Two developing processes are introduced here as the alternative air remediation technologies: radical oxidation process (ROP) and photocatalytic oxidation (PCO). These processes can be used for converting the undesirable chemical compounds in air and water to environment benign or more easily removed products. They are cost-effective, energy-efficient, and capable of oxidizing a wide range of organic pollutants into innocuous substances in dilute system. For PCE, in addition to the expected final oxidation products, carbon dioxide and hydrogen chloride /chlorine, there are still some undesired by-products such as trichloroacetyl chloride (TCAC) and phosgene. These by-products need to be further treated, by process such as wet scrubber, before discharging into the air (Haag & Johnson, 1996).

The ROP uses UV energy to cleave the bond between atoms and produce the reactive radicals to proceed the photolytic reaction. Typical low pressure mercury light bulb offers UV peak output at 254 nm. This photon energy is strong enough to cause PCE photolytic reaction. The oxygen source of the oxidation process is from the air. Some research groups add oxidants, such as O_3 and H_2O_2, to enhance the oxidation reaction (Kang & Lee, 1997; Shen & Ku, 1998).

The PCO technique uses near UV or near UV (black light main peak at 365 nm) as the energy source to excite semiconductor catalyst (titanium dioxide) and produce electron-hole pair on the catalyst surface. The electron-hole pair could be recombined and go back to the ground state, then release the heat energy to the nearby or the excited electrons could associate with the nearby oxygen molecules to form oxygen ion, O_2^-. On the other hand, the holes could be reacted with water molecules to produce hydroxyl free radicals, $\cdot OH$. Both O_2^- and $\cdot OH$ are capable to initiate a rapid chain reaction to degrade organic contaminants (Fujishima et al., 1999).

Recently, it has been reported that chlorine free radicals could be responsible for the chain initiation instead of water molecules (Blystone et al., 1993; Yamazaki-Nishida et al., 1996). It is the purpose of this paper to study the effects of humidity, space time, and inlet PCE concentration on photolytic (without TiO_2) and photocatalytic (with TiO_2) oxidation of PCE. By-products will be identified by an on-line GC-MS and the reactive pathways for both photolytic and photocatalytic reactions will be proposed.

MATERIALS AND METHODS

The experimental setup is a continuous flow photoreactor system as shown in Figure 1.

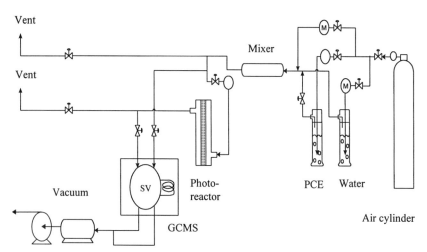

FIGURE 1. Schematic diagram of the experimental setup

Air source is from a gas cylinder with water content less than 3 ppm. One air stream picks up the PCE vapor while the other one picks up water vapor from two wash bottles, respectively. Both air streams are mixed with the main air stream to obtain the appropriate PCE concentration and the humidity in the test air. A static in-line mixer is used right before the reactor.

The concentric cylindrical photo-reactor is made of Pyrex glass with inside diameter of 55 mm (the reactor volume is 253 cm^3 and the inner surface area is 198.7 cm^2). The black light (FL4BL) is purchased from NEC Co. and is mainly near UV (300 to 400 nm). The germicidal UV lamp (GL4T5) is purchased from Sankyo Denkei Co. and is emitting light spectrum mainly at 254 nm. The germicidal UV lamp is a low pressure mercury lamp with a screen off the spectrum below 200nm. The light bulb is located at the center of the Pyrex glass reactor. The outside of the reactor is wrapped with aluminum foil and the two ends are sealed with viton septum which is also wrapped with aluminum foil.

Chemical Analysis. The inlet and outlet gas samples were analyzed by the Varian star 3400 gas chromatograph with a flame ionization detector (FID) and a Supelco column (Bentone 34/DNDP 30 m× 0.53 mm). The HP G1800B GC/MS equipped with electron ionization detector (EID) and HP-5 capillary column (30 m× 0.25 mm× 0.25 μm film thickness) was used for by-product identification. Two six-port sampling valves were used for the on-line GC and GC/MS analyses. A 250 μl sampling loop was adopted to guarantee a constant injection volume for each sample. The vacuum pressure was kept at 15 ±1 psig to ensure a continuous steady flow through the sampling system. This on-line GC & GC/MS sampling system can avoid a possible transfer error and sample contamination. A blank test indicated that the standard error of this sampling system is within 2%.

Chemicals. The P-25 TiO_2 was obtained from Degussa Corp. It is about 75% anatase and 25% rutile structure with surface area of 50-70 m^2/g. The other characteristics of P-25 can be found from Degussa product information (Degussa, 1996). PCE was purchased from Fisher Scientific and was listed as reagent grade, which contained 0.5% ethanol as inhibitor. The ethanol impurity was removed by evaporation for two days and was inspected by the GC/MS before the experiment.

RESULTS AND DISCUSSION

Photolytic Reaction. The photolytic oxidation of PCE was performed under the UV-C light to breakdown the PCE molecules directly. For PCE molecules, the C=C bonding energy is 611 kJ/mole and the C-Cl bonding energy is 330 kJ/mole. While the theoretical energy of 254 nm UV is 473 kJ/mole, the UV-C light is theoretically strong enough to cleave C-Cl bond and induce a photolytic reaction.

The humidity effect on the photolytic oxidation of PCE is shown in Figure 2. The PCE conversion was not affected by the relative humidity, as can be seen from Figure 2, for all the relative humidity, from 0 to 100%, experimented.

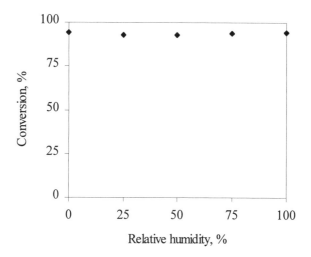

FIGURE 2. Humidity effect on photolytic oxidation of PCE

The concentration and space time effects on the photolytic oxidation of PCE are shown in Table 1. When the PCE concentrations increased from 12 to 430 ppm, as can be seen from the table, the PCE reaction rate is proportional to

TABLE 1. Experimental results of photolytic oxidation of PCE at different PCE inlet concentrations and space time

Space time	[PCE]o	[PCE]/[PCE]o	Phosgene/[PCE]o	TCAC/[PCE]o
1.8 sec	12 ppm	0.882	0.000	0.000
1.8 sec	110 ppm	0.623	0.000	0.128
1.8 sec	430 ppm	0.431	0.040	0.222
6.3 sec	12 ppm	0.197	0.090	0.273
6.3 sec	110 ppm	0.043	0.066	0.368
6.3 sec	430 ppm	0.000	0.058	0.426
18.0 sec	12 ppm	0.172	0.079	0.302
18.0 sec	110 ppm	0.042	0.059	0.363
18.0 sec	430 ppm	0.000	0.061	0.423

PCE concentration. This indicated a first-order reaction model for the photolytic oxidation of PCE. The amount of trichloroacetyl chloride (TCAC) by-product increased with the PCE concentration. While the by-product phosgene is only a minor amount for all the three PCE concentrations. The chain reactions initiated by the Cl· radicals probably lead to the formation of TCAC by-product instead of phosgene.

Table 1 also shows the effect of space time at 12, 110, and 430 ppm of PCE inlet concentrations. The conversion of PCE increased when the space time increased for all the three concentrations. The concentration of TCAC by-product increased significantly from the space time of 1.8 to 6.3 sec but insignificantly from 6.3 to 18 sec. While for the by-product of phosgene, the amount is very small and is not sensitive to either PCE concentration or space time.

Photo Reaction Pathway. Based on the discussions in the above and the intermediates identified, the photolytic oxidation reaction mechanism of PCE could be started with a photolysis of PCE molecules as shown in Equation (1):

$$Cl_2C=CCl_2 \xrightarrow{h\nu} Cl_2C=CCl\bullet + Cl\bullet \tag{1}$$

After the formation of chlorine atom, Cl•, the following chain reactions could be followed,

$$Cl\bullet + Cl_2C=CCl_2 \longrightarrow Cl_3C\text{-}CCl_2\bullet \tag{2}$$

$$Cl_3C\text{-}CCl_2\bullet + O_2 \longrightarrow Cl_3C\text{-}CCl_2O\text{-}O\bullet \tag{3}$$

Since the TCAC has been detected as the major by-product during the reaction, it could be formed through reaction steps of (4) and (5) or through Equation (6),

$$2\ Cl_3C\text{-}CCl_2O\text{-}O\bullet \longrightarrow 2\ Cl_3C\text{-}CCl_2O\bullet + O_2 \tag{4}$$

$$Cl_3C\text{-}CCl_2O\bullet \longrightarrow Cl_3C\text{-}COCl + Cl\bullet \tag{5}$$

$$Cl_3C\text{-}CCl_2O\text{-}O\bullet \longrightarrow Cl_3C\text{-}COCl + ClO\bullet \tag{6}$$

The by-product phosgene which is detected as a minor amount could be formed from Equation (7),

$$Cl_3C\text{-}CCl_2O\bullet \longrightarrow Cl_3C\bullet + COCl_2 \tag{7}$$

The termination reactions could be,

$$Cl_3C\bullet + Cl\bullet \longrightarrow CCl_4 \tag{8}$$

$$Cl\bullet + Cl\bullet \longrightarrow Cl_2 \qquad\qquad (9)$$

While chlorine was not detected from the reactor, the formation of carbon tetrachloride was detected by the mass spectrum which was hidden in the same peak as TCAC. When the PCE concentration increased the carbon tetrachloride mass spectrum matching percentage increased also.

Photocatalytic reaction. The photocatalytic oxidation of PCE was conducted by using black light (UV-A) to illuminate onto the TiO_2 (Degussa P25) thin film. The black light is strong enough to make electron-hole pairs on the surface of TiO_2 but not strong enough to induce a photolytic oxidation of PCE. This has been confirmed in this study by a blank test of black light without TiO_2.

The effect of humidity on the photocatalytic oxidation reaction of PCE is shown in Figure 3. The PCE conversion decreased from 60 to 35% when the relative humidity increased from 0 to 100%. This effect is in contradictory with

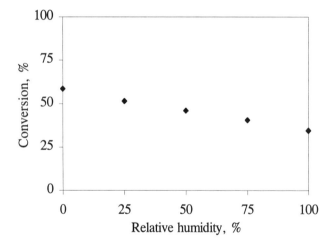

FIGURE 3. Humidity effect on potocatalytic oxidation of PCE

those observed for photocatalytic oxidation of non-chlorinated contaminants (Li et al, 1998; 1999). We have observed this humidity effect in our former study and have confirmed in this study again. It seems the chlorine free radicals produced from a chlorinated compound reacts more effectively than the hydroxyl free radicals from water molecules. At higher humidity, more water molecules block the active sites on the TiO_2 surface and therefore the active sites available for chlorine free radicals formation are reduced.

The effects of PCE concentration and space time were shown in Table 2. The conversion of PCE was found not sensitive to the PCE concentration for the

TABLE 2. Experimental results of photocatalytic oxidation of PCE at different PCE inlet concentrations and space time

Space time	[PCE]o	[PCE]/[PCE]o	Phosgene/[PCE]o	TCAC/[PCE]o
1.8 sec	12 ppm	0.507	0.142	0.000
1.8 sec	110 ppm	0.576	0.081	0.045
1.8 sec	430 ppm	0.675	0.030	0.073
6.3 sec	12 ppm	0.352	0.172	0.000
6.3 sec	110 ppm	0.328	0.122	0.058
6.3 sec	430 ppm	0.396	0.081	0.116
18.0 sec	12 ppm	0.169	0.165	0.000
18.0 sec	110 ppm	0.138	0.157	0.070
18.0 sec	430 ppm	0.163	0.122	0.125

space time of 6.3 and 18 sec. However, at a short space time of 1.8 sec, the PCE conversion increased with the PCE concentration. The amount of by-products TCAC increased with the inlet PCE concentration while the amount of phosgene decreased with the PCE inlet concentration.

The effect of space time was also shown in Table 2. The PCE conversion increased with the space time for all the three PCE inlet concentrations tested in this study.

Due to a limitation of the space of this publication, the reaction pathway of the photocatalytic oxidation of PCE will be discussed in our future publication.

CONCLUSIONS

Both the photolytic oxidation (by UV-C) and the photocatalytic oxidation (by UV-A) of PCE proceeded rapidly. The by-product identification indicated that TCAC was the predominant product while phosgene was a minor in the photolytic oxidation of PCE. For photocatalytic oxidation of PCE, both TCAC and phosgene by-products were about the same amount at higher PCE inlet concentration.

A photolysis of PCE to produce chlorine free radicals is proposed to initiate the photolytic oxidation of PCE in the air under the illumination of UV-C. The chlorine free radicals then went chain reactions to produce TCAC as the major by-product and phosgene as the minor by-product.

ACKNOWLEDGMENT

This project has been funded partially with funds from the United States Environmental Protection Agency (USEPA) as part of the program of the Gulf Coast Hazardous Substance Research Center. The contents do not necessarily reflect the views and policies of the USEPA nor does the mention of trade names or commercial products constitute endorsement or recommendation for use.

REFERENCES

Bhowmick M., and M. J. Semmens. 1994. "Ultraviolet Photooxidation for the Destruction of VOCs in Air." *Water Research* 28(11): 2407-2415.

Blystone P. G., M. D. Johnson, W. R. Faag, and P. F. Daley. 1993. "Advanced Ultraviolet Flash Lamps for the Destruction of Organic Contaminants in Air." *Emerging Technologies in Hazardous Waste Management III*, American Chemical Society, Chapter 18: 381-392.

Fujishima A., K. Hashimoto, and T. Watanabe. 1999. *TiO₂ Photocatalysis Fundamentals and Applications*: 126-156.

Haag W. R. and M. D. Johnson. 1996. "Direct Photolysis of Trichloroethene in Air: Effect of Cocontaminants, Toxicity of Products, and Hydrothermal Treatment of Products." *Environmental Science & Technology* 30(2): 414-421.

Kang, J. and K. Lee. 1997. "A Kinetic Model of the Hydrogen Peroxide/UV Process for the Treatment of Hazardous Waste Chemicals." *Environmental Engineering Science* 14(3): 183-192.

Li, K., S. Y. C. Liu, S. Khetarpal, and D. H. Chen. 1998. "TiO2 Photocatalytic Oxidation of Toluene and PCE Vapor in the Air." *J. Adv. Oxid. Technol.* 3(3): 311-314.

Li, K., S. Y. C. Liu, C. Huang, S. Esariyaumpai, and D. H. Chen. 1999. "TiO₂ Photocatalytic Oxidation of Butyraldehyde, Ethylbenzene and PCE in the Air through Concentric Reactors," paper presented in *The Fourth International Conference on TiO2 Photocatalytic Purification and Treatment of Water and Air*, Albuquerque, New Mexico, May 24-28

McGregor F. R., P. J. Piscaer, and E. M. Aieta. 1988. "Economics of treating Waste Gases from an Air Stripping Tower Using Photochemically Generated Ozone." *Ozone Scientific Engineering* 10: 339-352.

Shen, Y. and Y. Ku. 1998. "Decomposition of Gas-Phase Chloroethenes by UV/O₃ Process." *Water Research* 32(9): 2669-2679.

Ulm, K., D. Hehchler, and S. Vamvakas. 1996. "Occupational Exposure to Perchloroethylene." *Cancer Causes and Control* 7: 284-285.

Yamazaki-Nishida, S., X. Fu, M. A. Anderson, and K. Hori. 1996. "Chlorinated Byproducts from the Photoassisted Catalytic Oxidation of Trichloroethylene and Tetrachloroethylene in the Gas Phase Using Porous TiO₂ Pellets." *Journal of Photochemistry and PhotobiologyA:Chemistry* 97: 175-179.

TRANSFORMATION OF MTBE OVER A SOLID ACID CATALYST

Sarah A. Richards and **Wei-xian Zhang**
*Department of Civil and Environmental Engineering, Lehigh University,
Bethlehem, PA 18015*

ABSTRACT: An innovative technology for the removal of MTBE is studied using a solid acid catalyst to degrade MTBE through hydrolysis and dehydrogenation reactions to produce tert-butyl alcohol, acetone and possibly methanol and isobutene. The solid acid catalyst used is Nafion, consisting of a perfluorinated backbone and sulfonic acid functional groups. Nafion has surface acidity similar to 100% sulfuric acid and therefore could serve as a strong-acid catalyst for hydrolytic and oxidative reactions. A form of Nafion entrapped in silica has been tested for MTBE transformation. This form of Nafion, Nafion SAC 13 has a high surface area (100-300 m^2/g) and reactivity. Degradation of MTBE to form tert-butyl alcohol and isobutene has been observed, and the reaction can take place under both oxic and anoxic conditions. In batch experiments, equilibria were achieved and observed equilibrium constants have been found close to the theoretical equilibrium calculated from thermodynamic data. Through improved understanding of MTBE transformation by Nafion, the prospect appears good for applying Nafion as an effective catalyst for the removal of MTBE and many other recalcitrant organic contaminants in water.

INTRODUCTION

Methyl tert-butyl ether (MTBE) is used as an oxygenated fuel additive in gasoline to reduce air pollution. The 1990 Clean Air Act Amemdments mandate that oxygenated fuel be used in 10 areas where the National Ambient Air Quality Standards are exceeded for carbon monoxide and ozone emissions (U.S. EPA, 1998a).

MTBE is recently becoming a major concern for drinking water quality primarily due to leaking underground storage tanks. A survey conducted in 1993 and 1994 found that MTBE was the second most commonly detected contaminant in ground water aquifers (Squillace, 1996). MTBE is highly soluble in water and does not partition into soil easily. Therefore, it moves rapidly through aquifers with a high degree of dispersion. Furthermore, due to MTBE's persistence and high solubility in water, conventional separation techniques for the removal of organic contaminants such as granular activated carbon and air stripping are not efficient at removing MTBE (Speth and Miltner, 1990; Malley et al., 1993).

Currently, the toxicity of MTBE has not been quantified but the EPA has labeled as a potential carcinogen. Due to the potential toxicity as well as taste and odor concerns, the EPA has set a drinking water advisory of 20-40 µg/L (US EPA, 1997). However, contaminated water can rarely be treated to this level due to the lack of treatment technologies.

Ethers, in general, are unreactive compounds and are stable towards bases, oxidizing, and reducing agents. Literature of organic chemistry suggests that ethers undergo only one reaction, cleavage by strong acids (McMurry, 1996; Morrison and Boyd, 1992). Therefore, it would be logical to attempt this technique in the removal of MTBE in water. However, the use of a water-soluble strong acid would not be applicable in drinking water due to obvious obstacles such as a strongly acidic effluent and the potential corrosion of pipes.

It would be beneficial to use a strongly acidic water-insoluble solid material which may serve as a catalyst for MTBE transformation. The most important advantage of a solid catalyst is the easy separation of treated water from the catalyst. An extensive literature search suggests that Nafion may prove promising for this purpose. Nafion is a perfluorosulfonic acid resin with terminal sulfonic acid groups. Nafion has been utilized in a number of synthesis reactions including the synthesis of MTBE (Nunan, et at, 1993). This reaction was found to be a reversible reaction.

Previous research on the transformation of MTBE by a solid acid catalyst was conducted at high temperatures and pressures. Also, MTBE feedstocks were at high concentrations mixed in organic solvents. To our knowledge, this is the first research to explore the use of a heterogeneous catalyst for MTBE transformation in dilute aqueous solutions.

METHODS

Chemicals. Nafion SAC-13 was available from Aldrich. Methyl tert-butyl ether, tert-butyl alcohol, acetone, methanol, and isobutene stock solutions were all of HPLC grade and were available from Sigma-Aldrich.

Batch reaction bottles consisted of 40 ml EPA vials with standard open top flat silicone/PTFE septums and 100 ml serum bottles with aluminum crimp tops and PTFE-faced red rubbler liners. Samples were collected and stored in 2 ml glass sample vials with screw caps until analyzed.

Batch Experiments. For these experiments, 3-5 grams of Nafion SAC 13 were added to 40 ml of 72.76 mg/L MTBE in water. The tests were prepared in 100 ml serum vials with a crimp top septum. Tests were put in a batch rotator and sampled at regular intervals. Samples were stored in 2 ml vials and refrigerated until analysis. For the anoxic experiments, the water was purged with nitrogen for 15 minutes in order to remove all oxygen from the system. For the oxic experiments, water was purged with pure oxygen for 5 minutes. This water was purged previous to the addition of MTBE so that no MTBE would be purged from the system.

Analysis. Analysis was performed using a Tekmar 3000 Purge and Trap Concentrator connected to a Shimadzu GCMS-QP500 gas chromatograph mass spectrometer fitted with a DB-624 capillary column with a length of 30 m and inside diameter of .25 mm. Analysis involved a temperature program under constant pressure.

It should be noted that the analysis was not sensitive to the detection of methanol or isobutene. A large water peak masked the peaks of both these compounds. This water peak also decreased the sensitivity of detection of MTBE, TBA, and acetone. In order to increase the sensitivity of these three compounds, the pressure was increased to 50 kPa in the detector and the elution time of water was cut out of the mass spectrometer detection. Therefore, the detection of MTBE, TBA, and acetone increased, but detection of methanol and isobutene had to be eliminated. Sensitivity of TBA was low due to its nonvolatility and could only be detected at a range higher than 500 μg/L. Acetone and MTBE could be detected a range as low as 50-70 μg/L.

RESULTS AND DISCUSSION

The degradation of MTBE was first analyzed under ambient conditions. The results show that 58% of the MTBE was removed within the first hour. The degradation product, acetone was detected within the first hour. TBA was not detected until 6 hours. The concentrations of acetone and MTBE remained constant for 103 hours. This indicates that an equilibrium was achieved. The concentration of TBA continued to rise slowly until hour 103. Figure 1 shows the degradation of MTBE with time as well as the concentrations of TBA and acetone with time.

The above observations suggests that there are two reactions occurring simultaneously. The first reaction is a hydrolysis reaction of MTBE to yield TBA and methanol. The second possible reaction is a dehydrogenation reaction of MTBE to yield isobutene and methanol. These reactions are in accordance to well-documented reactions of the acid cleavage of MTBE (Morrison and Boyd, 1992) as well as MTBE synthesis reactions available in the literature (Feeley et al., 1995). Although acid cleavage of MTBE in organic solvent should produce isobutene and methanol, it is also possible that TBA can be produced due to the presence of water. For example, TBA was formed over a solid acid from

Figure 1: Transformation of MTBE with Nafion SAC 13 in a batch reactor

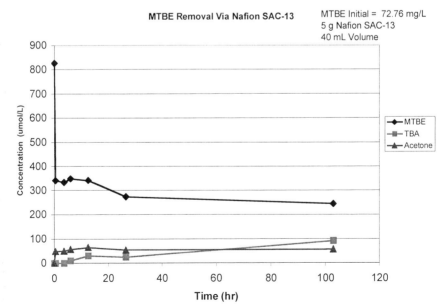

isobutene and water (Cunill et al., 1993). These byproducts are further reacted through dehydration and oxidation reactions. TBA is thought to be dehydrated to form isobutene and water. This dehydration reaction has been proven in the presence of a solid acid with TBA in the liquid phase (Matouq and Goto, 1993). It should be noted that this reaction might not be highly favored due to the presence of water as the solvent for this reaction although it has been reported (Ych and Novak, 1995; Barreto et al., 1995). Isobutene formed by dehydrogenation of MTBE and dehydration of TBA can then be oxidized to acetone and methanol.

Although isobutene was not detected due to limitations in the analysis, the presence of acetone indicates that isobutene is the most probable intermediate. Acetone is a product of isobutene oxidation (Barreto et al., 1995). The proposed reaction scheme for the degradation of MTBE and intermediate reactions is shown in Figure 2.

Figure 2: Proposed Reaction Scheme for Acid Catalyzed Destruction of MTBE. Solid boxes indicate positive confirmation of compound.

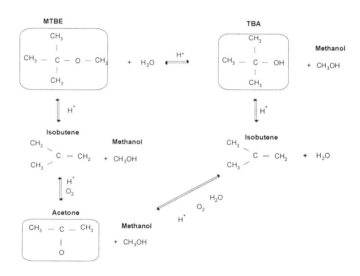

As mentioned previously, an equilibrium was achieved within the batch system. A theoretical equilibrium constant can be found using the following equation:

$$\Delta G_{rxn}^{o} = RT \ln K_{eq}$$

(1)

ΔG^{o}_{rxn} is the Standard Gibbs Free Energy Change where

$$\Delta G_{rxn}^{o} = \Delta H^{o} + T\Delta S^{o}$$

(2)

Figure 3: MTBE Degradation by Nafion SAC 13 under Anoxic Conditions

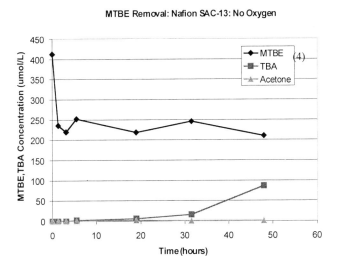

MTBE Removal: Nafion SAC-13: No Oxygen

$$\Delta H^{O} = \Delta H^{O}_{f}(products) - \Delta H^{O}_{f}(reactants) \qquad (3)$$

$$\Delta S^{O} = S^{O}(products) - S^{O}(reactants) \qquad (4)$$

$$K_{eq} = \frac{[TBA][MeOH]}{[MTBE]} \qquad (5)$$

The values for $\Delta H^{o}{}_{f}$ and S^{o} can be found in the literature (Lide, 1974; Mallard, 1998). The theoretical K_{eq} for the hydrolysis of MTBE to form TBA and methanol was found to be 1.81×10^{-4}. However, using the data from the ambient conditions, finding an equilibrium constant would be complicated due to the presence of multiple reactions taking place simultaneously. Therefore, it was important to limit the system to only one or two reactions. In oxidizing conditions, isobutene undergoes oxidation to form acetone and TBA is dehydrated to form isobutene. Therefore, an experiment was conducted under anoxic conditions to limit the reaction to the hydrolysis reaction to form TBA and the dehydrogenation reaction to form isobutene. The results are shown in Figure 3. There was very little acetone produced, which is to be expected with the absence of oxygen. A large amount of TBA was produced. This is most likely due to the lack of reaction pathways available in the anoxic condition.

Due to the inability to detect isobutene and methanol, the only reaction where K_{eq} could be calculated is for that of the hydrolysis of MTBE to form TBA and methanol. K_{eq} was calculated using the concentration of MTBE and TBA at 48 hours. The concentration of methanol was assumed to be equal to TBA. K_{eq} was found using the following equation:

The observed K_{eq} was found to be 5.32×10^{-4}. This value is slightly higher than that of the theoretical K_{eq} of 1.80×10^{-4}. This could be due to experimental error or due to the degradation of MTBE due to other reactions. This would decrease the concentration of MTBE to less than what it would be if only the hydrolysis reaction were present.

An experiment was also conducted to see whether the reaction changed when the system was saturated with oxygen. The results of the oxygen excess tests (not shown) were similar to the tests conducted in ambient conditions. Therefore, the addition of oxygen did not improve the reaction.

CONCLUSIONS

Transformation of MTBE over a solid acid catalyst was studied. The solid acid catalyst used was Nafion SAC 13. Tests were conducted under ambient, anoxic, and oxic. From the batch tests, the following conclusions were drawn:

1. The degradation of MTBE over Nafion SAC 13 was is probably due to two competing reactions. The first reaction is the hydrolysis of MTBE to form TBA and methanol. The second reaction is the dehydrogenation of MTBE to form isobutene and methanol. Isobutene is further oxidized to form acetone.

2. Tests conducted under anoxic conditions show that acetone is the product of an oxidation reaction. This reaction is most likely the oxidation of isobutene to acetone. Tests conducted with excess oxygen showed no improvement in the reactivity over tests conducted in ambient conditions.

3. An equilibrium was achieved using the Nafion SAC 13. The observed equilibrium for the hydrolysis of MTBE to form TBA and methanol is close the theoretical equilibrium found using Gibbs Free Energy Theory.

LIST OF SYMBOLS

K_{eq} = Equilibrium Constant
R = Universal Gas Constant
S^{o} = Standard Entropy Change
T = Temperature
ΔG^{o}_{rxn} = Standard Gibbs Free Energy Change
ΔH^{o} = Standard Enthalpy Change
ΔH^{o}_{f} = Heat of Formation
ΔS^{o} = Standard Entropy Change

REFERENCES

Barreto, R.D., Gray, K.A., Anders, K. (1995). "Photocatalytic Degradation of Methyl-*tert*-Butyl Ether in TiO$_2$ Slurries: A Proposed Reaction Scheme," *Water Research*, Vol. 29, No. 5, pp. 1243-1248.

Cunill, F., Meritxell, V., Izquierdo, J.F., Iborra, M., Tejero, J. (1993). "Effect of Water Presence on Methyl tert-Butyl Ether and Ethyl tert-Butyl Ether Liquid-Phase Synthesis," *Industrial Engineering Chemistry Research*, Vol. 32, pp. 564-569.

Lide, D.R. ed. (1974). *CRC Handbook of Chemistry and Physics 79th Edition*. CRC Press, Boca Raton.

Mallard, W.G. ed. (1998). *NIST Standard Reference Database Number 69*. Available on the web at webbook.nist.gov.

Malley, J.P., Eliason, P.A., Wagler, J.L. (1993). "Point-of-entry Treatment of Petroleum Contaminated Water Supplies," *Water Environment Research*, Vol. 65, No. 2, pp. 119-128.

Matouq, M.H., Goto, S. (1993). "Kinetics of Liquid Phase Synthesis of Methyl tert-Butyl Ether from tert-Butyl Alcohol and Methanol Catalyzed by Ion Exchange Resin," *International Journal of Chemical Kinetics,* Vol. 25, pp. 825-831.

McMurry, J. (1996). *Organic Chemistry: 4th Edition*. Brooks/Cole Publishing Company, Pacific Grove, CA.

Morrison, R.T., Boyd, R.N. (1992). *Organic Chemistry: 6th Edition*. Prentic Hall, Englewood Cliffs.

Nunan, J.G., Klier, K., Herman, R.G. (1993). "Methanol and 2-Methyl-1-Propanol (Isobutanol) Coupling to Ethers and Dehydration over Nafion H: Selectivity, Kinetics, and Mechanism," *Journal of Catalysis*, Vol. 139, pp. 406-420.

Speth, T.F., Miltner, R.J. (1990). "Technical Note: Adsorption Capacity of GAC for Synthetic Organics," Journal of the American Water Works, Vol. 82, pp. 72-75.

Squillace, P.J., Zogorski, J.S., Wilber, W.G., Price, C.V. (1996). "Preliminary Assessment of the Occurrence and Possible Sources of MTBE in Groundwater in the United States, 1993-1994," *Environmental Science and Technology*, Vol. 30, No. 5, pp. 1721 – 1730.

Yeh, C.K., Novak, J.T. (1995). "The Effect of Hydrogen Peroxide on the Degradation of Methyl and Ethyl tert-Butyl Ether in Soils," *Water Environment Research*, Vol. 67, No. , pp. 828-834.

CONSTRUCTION OF A FUNNEL-AND-GATE TREATMENT SYSTEM FOR PESTICIDE-CONTAMINATED GROUNDWATER

Dean Williamson, CH2M HILL, Gainesville, FL;
Karl Hoenke and Jeff Wyatt, Chevron Chemical Company, Richmond CA; and Andy Davis and Jeff Anderson, Geomega, Boulder CO

ABSTRACT: Over thirty years of handling and blending pesticides at the Marzone Superfund site, Tifton, Georgia, resulted in soil and groundwater contamination. Contaminants detected at the site include DDT, Toxaphene, Atrazine, Dieldrin, Chlordane, Lindane, alpha-BHC, Endosulfan, methyl parathion, xylene, and ethyl benzene. Faced with implementing a Superfund Record of Decision (ROD) specifying a pump-and-treat remedy for contaminated groundwater, Chevron Chemical Company worked with CH2M HILL to evaluate passive, in-situ groundwater treatment alternatives. A Funnel-and-Gate (F&G)-type passive system was selected as a viable alternative to pump-and-treat.

The Marzone project team successfully designed, built, and started-up a full-scale F&G groundwater treatment system at the Marzone Superfund site to test its efficacy as a permanent groundwater remedy. Results from the first 1.5 years of operations indicate the system is successfully operating as intended.

The F&G system has operated since August 1998, treating approximately 1 to 2 gallons of hydrocarbon- and pesticide-contaminated groundwater per minute. Total xylene concentrations are reduced from approximately 12,000 to 20,000 ug/L in the influent to less than 1 ug/L in the effluent. Total BHC concentrations are reduced from approximately 1 to 4 ug/L to less than 0.05 ug/L in the effluent. Other contaminants are similarly removed. EPA has developed a ROD Amendment to make this system the permanent groundwater remedy.

The reactive media in the treatment gates is granular activated carbon (GAC). The gates are constructed of three subsurface, precast-concrete vessels which provide significant process flexibility. The system can operate in series, parallel, upflow, or downflow mode. The system is designed to operate in a low-flow, low-head aquifer. A process modification of the standard F&G approach, using a trench for contaminated groundwater collection, was selected to maximize hydraulic efficiency. A similar trench was used to distribute the treated groundwater downgradient of the cut-off wall.

The "funnel" portion of the system was constructed using the vibrated beam method. A proprietary technology was used to install an impermeable 4-inch-thick cut-off wall similar to a conventional soil-bentonite (SB) slurry wall, but without the excessive site disturbance associated with SB wall construction.

The biopolymer slurry method was used for construction of the collection and distribution trenches. After startup, gas accumulation within the system piping resulted in periodic flow stoppages. Vents were installed in the piping to allow the gas to escape. This enabled the system to operate more effectively and significantly decreased the frequency of flow stoppage incidents. The source of the gas was believed to be generated by microbiological activity in the groundwater system. The amount of gas accumulating in the system decreased significantly over the first 6 to 9 months. A significant excess of TOC (on the order of 100 mg/L), above that attributable to known contamination, was also noted initially and decreased with time. The large amount of biogas in the system and initial presence of high TOC were both eventually attributed to the use of guar as part of the biopolymer slurry construction method.

INTRODUCTION

The Marzone site in Tifton, GA was formerly a regional pesticide formulation plant that operated from 1950 through 1983. Chevron Chemical Company (Chevron) owned and operated the site through 1970 followed by a succession of smaller companies including Tifton Chemical Company, Tifchem Products, Inc., Marzone Chemical Company, and Kova Fertilizer.

The 1.5 acre site is located on the edge of the city of Tifton, a small agricultural community in southern Georgia. The facility formulated chlorinated and organophosphate pesticides (e.g., DDT and toxaphene) used to control pests on peanuts and cotton as well as the herbicide atrazine and provided large quantities of DDT to U.S. Department of Agriculture to control fire ants during the 1950's and 1960's.. During the early 1960's liquid blending equipment was added. Xylene was received, stored in bulk containers and used as the liquid carrier solvent. The pesticide formulation slate expanded over the years to include chlordane, parathion, methyl parathion, lindane, and endosulfan.

Chevron sold the facility in 1970. A series of owners continued pesticide formulating operations. Over the course of thirty three years of operation, incidental leaks, spills and actual waste disposal resulted in significant pesticide and solvent contamination of soils and groundwater at the site. In 1989, the site was placed on the National Priorities List (NPL). Chevron and two other Potentially Responsible Parties (PRPs) entered into a Consent Decree with EPA Region IV in 1990 to perform a Remedial Investigation/ Feasibility Study (RI/FS) for the site. The RI and FS were completed in 1992 and 1993 respectively. EPA issued a Record of Decision (ROD) for the Marzone Superfund Site in 1994, specifying low temperature thermal desorption (LTTD) as the remedy for contaminated soil and a pump-and-treat (P&T) approach for contaminated groundwater.

In mid 1995, Chevron, EPA, the State of Georgia and other key stakeholders formed an innovative High Performance Team(HPT) to accelerate remediation activities. The HPT successfully completed the soil remediation at the Marzone site in early 1999.

GROUNDWATER CONTAMINATION AND SITE HYDROGEOLOGY

A general site plan, showing the original source area and general hydraulic gradients, is presented in Figure 1. Only the shallow surficial aquifer (to a depth of approximately 25 to 30 feet below land surface) has been affected by releases of contamination at the site. The shallow aquifer is comprised of primarily of interbedded silts and clays, with some sand lenses that may not be continuous.

Groundwater flow velocities in this aquifer unit are typically on the order of 10 to 20 feet per year, although within some thin sand layers the flow velocity could be somewhat higher than this. The primary groundwater contaminants include xylene and several OCl pesticides, particularly alpha BHC, beta BHC and gamma BHC (lindane). Ethylbenzene and ethoprop are also present at elevated levels. The groundwater source area has an anaerobic core zone, characterized by low ORP, low dissolved oxygen, and high dissolved iron. Groundwater becomes more oxic downgradient of the former source area.

The Marzone team was concerned that the P&T approach specified in the ROD would be ineffective based on site hydrogeologic conditions. In 1997, the team prepared a FS addendum to evaluate in-situ technologies including permeable reactive barriers. As a result of the FS Addendum, the team selected a Funnel-and-Gate approach as an effective alternative to P&T. EPA agreed to allow Chevron to construct a field–scale system as a pilot test to assess its efficacy.

F&G HYDRAULIC DESIGN

Groundwater flow modeling was used to identify the optimal configuration of the F&G system. The well-known groundwater flow model MODFLOW was used to simulate the groundwater flow regime. Potential gate configurations, including one and two gate systems, were modeled to assess the number of gates required and potential size. A single gate system was predicted to be adequate by the model.

The aquifer unit of interest is a relatively low productivity aquifer, comprised of various interbedded clay, silt and sand layers. The total groundwater flow through the target treatment area was estimated by the model to be approximately 1 gallon per minute (gpm). Because of the heterogeneous nature of the aquifer, a process modification of the F&G system involving a collection trench for collecting groundwater was developed. The purpose of the collection trench was to ensure uniform collection of groundwater flowing through the system. The collection trench was designed to fully penetrate the target aquifer recovery zone. The trench was designed to be backfilled with an inert gravel media (granite) to ensure that undesirable geochemical reactions did not occur.

Similarly, a discharge trench was selected for reinfiltration of treated groundwater into the shallow aquifer downgradient of the F&G system. The location of the discharge trench was placed as far downgradient as feasible to obtain as much hydraulic driving force (system head) as possible.

Various potential locations for the F&G system were evaluated by the project team. The decision regarding where to locate the system required balancing two issues. One the one hand, the system needed to be located hydraulically downgradient of the source area far enough to capture as much of the plume as practicable. One the other hand, the further the system was placed downgradient of the source area, the longer it would take contaminated groundwater to flow though the system for treatment. Because the groundwater flow velocity at the site is on the order of 10 to 20 ft per year, placing the system

100 feet downgradient of a particular source area location would require the groundwater to flow towards it for 5 to 10 years before the groundwater received treatment. Based on a review of potential system locations, the team selected a location relatively near the downgradient edge of the most contaminated groundwater locations, but sufficiently downgradient of the source area to capture an estimated 93% of the contaminant plume mass.

F&G SYSTEM DESIGN

Several reactive media were considered for use in the treatment gate. A medium was required that was capable of removing or degrading hydrocarbons, such as xylene and ethylbenzene, as well as the OCl pesticides present in the groundwater. Granular activated carbon (GAC) was one obvious choice. Dynamic (flowthrough) bench-scale testing of GAC using contaminated site groundwater was performed by Calgon Corporation to assess GAC loading rates performance and breakthough curves. Xylene was found to be the contaminant that broke through first, prior to other hydrocarbons and pesticides.

In addition, bench-scale testing of zero valent iron (ZVI) was conducted by EnviroMetals Technologies, Inc. (ETI) on site groundwater spiked with lindane to assess the potential effectiveness of this media. Although effective for lindane degradation, the use of ZVI was concluded to be not well suited for this application because of its inability to degrade xylene and ethyl benzene. GAC was selected as the reactive medium for this system.

Having decided on a reactive medium, the next task was to design the appropriate gate system. Key design criteria for GAC systems include residence time within the GAC (typically expressed as Empty Bed Contact Time [EBCT]) and surface loading rate. Additional design criteria related to issues such as constructability, accessibility, carbon change-out requirements, life cycle costs, corrosion resistance, and safety. Based on these considerations, a gate design based on the use of pre-cast concrete vaults was selected.

The number of vaults and configuration of the interconnecting piping was developed based on desired process operational flexibility. Because the system was intended to be in place and operating for up to several decades, it was considered likely that the operational modes could change over time and that a different reactive medium or combination of media might be used at some future time. For these reasons, three vessels were included in the gate system with interconnecting piping that would allow operation of two vessels in series or two vessels in parallel, and both upflow and downflow modes. Thus, the system had nearly the same process flexibility that an above-grade two-vessel GAC system would have.

The selection of the design and construction approach for the cutoff wall (typically the funnel portion of a F&G system) considered various available methods, including a soil-bentonite (SB) slurry wall, use of driven sheets of various synthetic materials (e.g., Waterloo barrier) and the vibrated beam (VB) method. Each of these methods produces a cutoff wall of different thickness. Other factors considered in selecting a method for cutoff wall installation included the need to construct a portion of the cutoff wall adjacent to an active railroad, community concern about the visual condition of the site, and the nature of subsurface conditions. The VB method was chosen based on its ability to install a cutoff wall of suitably-low permeability quickly, adjacent to railroad tracks, and without the typical surface disturbance associated with SB slurry walls. In the United States, the VB method is implemented using proprietary technology owned by Slurry Systems, Inc. (SSI).

For construction of the collection and discharge trenches, conventional excavation methods were evaluated and compared to other methods such as the biopolymer (BP) slurry method. The BP slurry method employs a gum derived from the guar bean to increase the viscosity of the water inside the trench, which holds up the trench side walls during excavation. This construction method eliminates the need for trench boxes and ensures that no personnel have to enter down into an earthen trench during construction, thus eliminating a significant safety hazard in trench construction. The BP slurry method was selected for construction of the collection and distribution trenches.

A schematic, 3-D diagram of the Marzone F&G system design is presented in Figure 3. The general locations of the system components at the site are shown in Figure 1.

F&G CONSTRUCTION AND STARTUP

Construction of the system began in July 1998 and was completed in August 1998. The construction was implemented largely as anticipated with no major system changes. The trenches were installed using the biopolymer slurry method, the cutoff wall was installed with the vibrated beam method and the GAC reactor vaults were install using a trench box.

Functional startup testing was initiated in August and September 1998. During the initial start of the system, it was observed that although a significant head difference (approximately 4 feet) was present across the system (as measured from the water elevations in GAC reactors and a piezometer in the discharge trench), no flow was observed to be occurring. Various activities were implemented to assess the cause of the lack of flow, including evaluating the settings on the valves (on/off), purging the lines to eliminate potential blockages, and backflushing the lines to the GAC reactors.

By the process of elimination and evaluation of monitoring and hydraulic data, it was concluded that the presence of gases and vapors in some of the interconnecting piping was inducing a "vapor lock" in some of the piping such that flow was being blocked. In particular, the piping leading from the GAC vaults to the discharge trench was found to be susceptible to build up of gases and vapor lock. By purging this line, it was possible to induce system flow.

The next task was to identify the source of the gases in the lines. After reviewing the data, it was concluded that the most likely source was biodegradation of guar used in the construction of the trenches, which were constructed using the biopolymer slurry method. During the construction of the trenches, significant amounts of guar were added to the water in the trench because the relatively high ambient temperatures at the site during construction made the guar biodegrade quickly, with a resultant loss of viscosity of the slurry.

Guar is a polysaccharide comprised primarily of D-mannose and D-galactose. These sugars are readily biodegradable. The influent TOC data indicated that a significant amount of TOC (over 100 mg/L), other than was attributable to site COCs, was present initially in the influent but dropped off exponentially over the ensuing months. The source of much of this TOC was considered likely attributable to the guar used during construction. Some biogas may also have been produced by biodegradation of xylenes and other site COCs.

In order to eliminate the accumulation of gases in the pipes, vents were installed at strategic locations to allow gases to vent to the atmosphere. The installation of the vents significantly reduced the problems of flow stoppage caused by vapor accumulation.

Performance of the GAC for COC Removal. Performance of the treatment system for removal of COCs from groundwater was assessed by analyzing influent and effluent groundwater samples. The system was found to remove all target COCs to below the applicable cleanup standard and typically to below the applicable method detection limit.

Water Quality Observations. Generally, influent water quality entering the GAC reactors was not considered optimal for most treatment systems. Total suspended solids were in the range of 12 to 50 mg/L. Dissolved iron concentrations of up to 60 mg/L were also measured in the influent. The iron was found to generally pass through the GAC system, presumably as ferrous iron, and has not noticeably interfered with system effectiveness. Over time, however, the concentrations of several parameters, such as TSS, have decreased. Plugging of the GAC beds due to the presence of TSS has not yet been a problem in this system.

Hydraulic Performance. Groundwater elevations collected after startup of the system were compared to groundwater contours predicted by the modeling during the design stage. Generally, the actual contours were found to compare well with the modeled contours. From a hydrogeologic perspective, the data indicate the system is working as intended.

SUMMARY
Overall, the Marzone site F&G system has performed largely as intended. The system is monitored for hydraulic performance on a monthly basis to ensure it is operating. If necessary, lines are purged to eliminate trapped gases within the piping. Samples are collected on a quarterly basis to assess the performance of the GAC.

Based on the effectiveness of the system, EPA has developed a ROD Amendment that changed the recommended groundwater remedy for the site from a pump-and-treat system to the funnel and gate system.

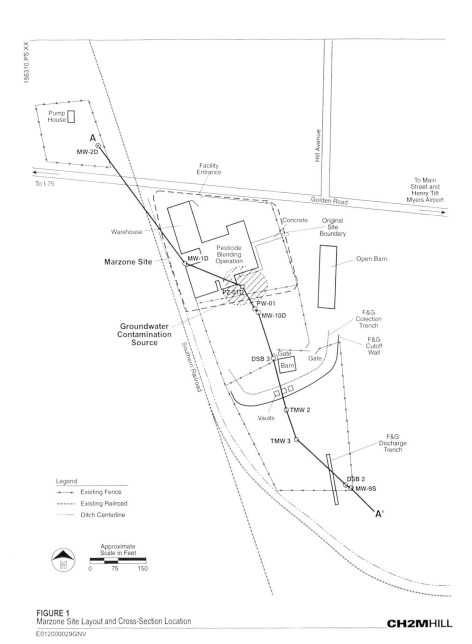

FIGURE 1
Marzone Site Layout and Cross-Section Location

E012000029GNV

CH2MHILL

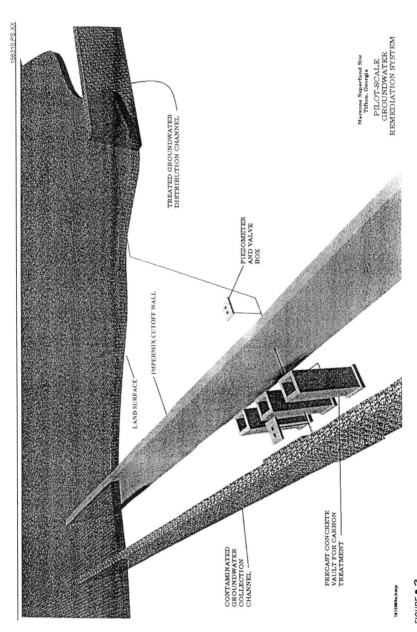

FIGURE # 2
Marzone Funnel-and-Gate Groundwater Treatment System

CONTAMINATED
GROUNDWATER
COLLECTION
CHANNEL

PRECAST CONCRETE
VAULT FOR CARBON
TREATMENT

LAND SURFACE

IMPERMIX CUTOFF WALL

PIEZOMETER
AND VALVE
BOX

TREATED GROUNDWATER
DISTRIBUTION CHANNEL

Marzone Superfund Site
Tifton, Georgia
PILOT-SCALE
GROUNDWATER
REMEDIATION SYSTEM

CH2MHILL

INVESTIGATION, TREATABILITY TESTING, AND DESIGN OF A PERMEABLE REACTIVE-IRON BARRIER

Zane H. Tuta, Kevin B. Finder, and Robert M. Yeates
(IT Corporation, Englewood, Colorado)
Robert Focht (EnviroMetal Technologies Inc., Waterloo, Ontario, Canada)

Abstract: A permeable reactive-iron barrier (PRB) is planned to contain and treat groundwater contaminated with trichloroethene, 1,2-dichloroethene, and tetrachloroethene at the Lake City Army Ammunition Plant in Independence, Missouri. The plume occurs in unconsolidated, fine-grained sediments overlying a gently sloping paleochannel wall carved in Pennsylvanian claystone and interbedded limestone. The plume extends into a broad valley, where a thick alluvial sequence containing sand and gravel constitutes a local aquifer. The barrier will intercept and treat contaminants migrating toward the aquifer. Field investigation conducted in 1999 expanded on previous site characterization with the objective of quantifying the rate of groundwater flow and the degree of heterogeneity within the flow system. Bench-scale treatability tests were conducted to assess the reactivity of granular iron under conditions simulating several possible construction alternatives. Test results showed relatively small differences in the rate of VOC degradation between a column of granular iron originally mixed with guar slurry, simulating construction by biopolymer slurry trenching, and a standard iron column. A third column containing a 1:1 mixture (by weight) of fine-mesh iron and soil with guar slurry, simulating a jetted PRB, yielded a markedly lower rate of VOC degradation. The design consists of biopolymer slurry trenching to install a continuous PRB, 380 feet (116 m) long, and keyed into bedrock at depths ranging from 20 to 60 feet (6 to 18 m).

INTRODUCTION

The Lake City Army Ammunition Plant (LCAAP), located in Independence, Missouri, is a U.S. Army Industrial Operations Command installation where small arms ammunition has been manufactured almost continuously since 1941. Several sources of contamination by volatile organic compounds (VOCs) have been identified in a portion of the facility used for various waste disposal purposes, the Northeast Corner Operable Unit (NECOU). A VOC groundwater plume extends from the source zone into a local aquifer system. A record of decision between LCAAP and the regulatory agencies calls for an interim remedial action to include elements of both plume containment and source containment. A PRB is planned to meet the objective of reducing further migration of VOCs in the groundwater plume to the aquifer. This paper summarizes investigation, treatability testing, and design of the PRB.

The LCAAP is characterized by rolling uplands traversed by broad valleys. The NECOU is principally situated in an upland area. The northwestern portion of the NECOU is located at the transition between the upland and the abandoned Lake City Valley, a former path of the Missouri River. A gently sloping paleochannel wall carved into the underlying bedrock marks the transition. The upland area to the southeast of the transition is comprised of Pennsylvanian Pleasanton Group, which is predominantly claystone with interlayered limestone, covered by a veneer of residual and colluvial clay and silt. To the northwest, alluvial sediments of clay, silt, sand, and gravel mark the paleochannel sediments in a sequence that generally becomes coarse downward.

Groundwater flow within the unconsolidated sediments of the NECOU upland is to the northwest, with a pronounced downward gradient from the water table to the top of bedrock. In the valley, flow is more northerly within the more permeable alluvial paleochannel sediments, which comprise a local aquifer. Bedrock is considered to transmit a relatively small quantity of water compared to the overlying, unconsolidated sediments, although there is evidence of limited local fracturing. The plume axis generally parallels the horizontal hydraulic gradient in the sediments (Figure 1).

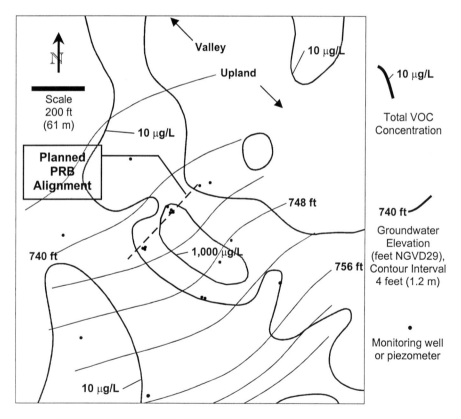

FIGURE 1. Map of total VOCs plume and groundwater elevation in unconsolidated sediments.

INVESTIGATION

Previous investigations had characterized the site hydrogeology and plume extent, principally by means of soil borings, monitoring wells, and cone penetration tests. Additional investigation in 1999 was chiefly directed toward quantification of the rate of flow within the plume and assessment of the degree of heterogeneity within the unconsolidated sediments of the upland area, in the area of transition to the alluvial sediments of the valley. Several piezometer clusters were installed to provide a three-dimensional network of hydraulic and chemical data. Aquifer testing, including slug tests and a pumping test, were conducted.

Aquifer testing confirmed a high degree of heterogeneity with respect to hydraulic conductivity of the sediments. Slug tests indicated hydraulic conductivities from 7×10^{-8} cm/sec to 9×10^{-3} cm/sec. Drawdown in response to pumping varied greatly among closely spaced observation piezometers. The piezometers exhibiting a significant response provided a geometric mean hydraulic conductivity of 7×10^{-4} cm/sec. The heterogeneity is attributed to lithologic variations, such as the presence of sandy and gravelly clay lenses, especially at greater depths, within generally fine-grained colluvial material (Figure 2), as well as features of secondary permeability.

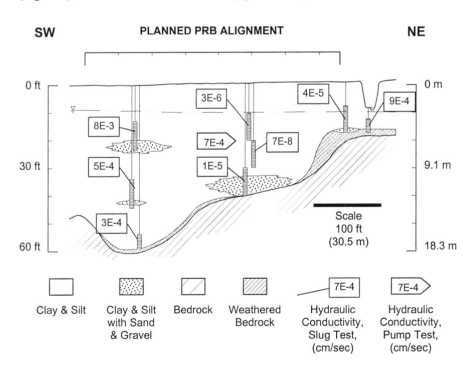

FIGURE 2. Hydrogeologic cross section at planned PRB alignment.

Characterization of the unconsolidated sediments in terms of hydraulic conductivity and heterogeneity was necessary to estimate the rate of groundwater flow through the planned PRB. Flow is a key design parameter influencing the required quantity of reactive iron. Both analytical and numerical modeling methods were used to obtain an estimated specific discharge through the PRB of up to 0.16 feet/day (0.49 m/day).

Groundwater analyses of samples collected in 1998 and 1999 from locations along the planned PRB alignment were compiled to obtain the summary of maximum VOC concentrations in groundwater (Table 1). All compounds exceeding remediation goals (from federal maximum contaminant levels under the Safe Drinking Water Act and Missouri groundwater quality standards) in one or more samples along the alignment are listed. Field analyses by direct-sampling ion-trap mass spectrometry (Davis et al., 1998) augmented laboratory analyses by U.S. EPA Method 8260.

TABLE 1. Maximum VOC concentrations at planned PRB alignment.

Compound	Maximum Concentration (ug/L)	Number of Analyses		Remediation Goal
		EPA 8260	Field	(ug/L)
Trichloroethene (TCE)	1,000	7	12	5.0
cis-1,2-Dichloroethene (cDCE)	740	7	12	70.0
1,1-Dicholoroethane (11DCA)	60	7	3	none
Tetrachloroethene (PCE)	53	7	12	5.0
1,1-Dicholoroethene (11DCE)	15	7	0	7.0
1,1,2-Trichloroethane (112TCA)	0.87	7	0	0.6
Vinyl chloride (VC)	0.28	7	0	2.0

TREATABILITY TESTING

The objective of treatability testing was to determine the residence time required to degrade the VOCs present (and their chlorinated breakdown products) in groundwater to concentrations below the remediation goals. Bench-scale column tests were designed to simulate specific conditions anticipated for potential construction alternatives suitable for installation of the PRB to depths of up to 60 feet (18 meters), including several methods involving the use of guar-based, biopolymer slurry (Hubble et al., 1997). The test was designed to gauge the potential effects of guar on the ability of iron to degrade the VOCs in site groundwater.

Three columns were prepared, using granular iron in two grain-size ranges from one supplier of commercial iron:

- Fe column--100% iron, -8 to +50 mesh (U.S. standard sieves, 2.36 to 0.3 mm), with a specific surface area of 1.1 m^2/g);
- Fe+Guar column--100% iron, -8 to +50 mesh, mixed with guar slurry;

- Fe/Sed+Guar column--50% by weight iron (-30 to +76 mesh, 0.6 to 0.2 mm), with a specific surface area of 1.9 m^2/g., mixed with unconsolidated sediment from the saturated zone of the site and guar slurry.

The Fe column simulated construction techniques that do not use guar, such as placement within a hollow mandrel or a dry treatment cell. The Fe+Guar column simulated placement within a trench supported by biopolymer slurry. The Fe/Sed+guar simulated placement by high-pressure jetting, which results in a mixture of iron and subsurface sediments. Prior to packing the columns, the slurry in the Fe+Guar and Fe/Sed+Guar mixtures was broken down using an enzyme breaker. Groundwater from the site containing chlorinated VOCs was run through each column at a controlled rate of flow.

Sharp declines in concentration with distance through the column were evident in the steady-state concentration profiles for PCE (Figure 3a) and TCE (Figure 3b). Similar declines were observed for cDCE in the Fe and Fe+Guar columns, but the decline was much more gradual in the Fe/Sed+Guar column (Figure 3c). The observed concentration declines indicate the rate of degradation.

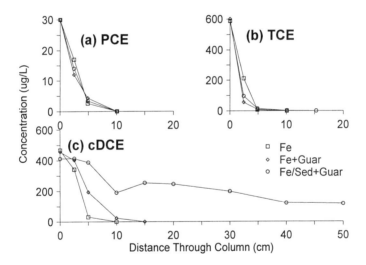

FIGURE 3. Steady-state concentration profiles.
(a) PCE. (b) TCE. (c) cDCE.

Decay constants were computed from the slopes of the concentration profiles in accordance with a first-order kinetic model. Half-lives for the major VOCs found in site groundwater, and corresponding correlation coefficients (r^2), are provided in Table 2. Half-lives of 11DCE were estimated based on two data points due to low influent concentrations and rapid disappearance within the columns.

The half-lives determined from the column tests were used in a mathematical model representing sequential VOC degradation assuming influent with the observed maximum concentrations shown in Table 1. The model is an expression of the chemistry that is observed in the solution phase, wherein specific parent compounds are converted to less chlorinated products according to various degradation pathways (Eykholt, 1998; Arnold and Roberts, 1999) and molar conversion percentages. Laboratory-based half-lives were doubled to account for field effects including temperature. Half-lives of compounds present at very low concentrations in the site water that were not determined by the site-specific column tests were estimated based on experience with those compounds in other laboratory studies and the observed effects of guar and sediment on the other VOCs in this study.

Table 2. Half-lives of VOCs from steady-state concentration profiles.

VOCs	Fe			Fe+Guar			Fe/Sed+Guar		
	Influent Conc. (ug/L)	Half -Life (hr)	r^2	Influent Conc. (ug/L)	Half -Life (hr)	r^2	Influent Conc. (ug/L)	Half -Life (hr)	r^2
PCE	30	0.71	0.913	30	0.91	0.999	30	0.53	0.971
TCE	588	0.39	0.911	601	0.43	0.993	588	0.40	0.978
cDCE	468	0.64	0.835	457	1.1	0.934	413	9.0	0.852
11DCE	27	0.88	1.000	11	1.2	1.000	10	1.6	1.000

The model allows calculation of the residence time required for iron to degrade specific concentrations of VOCs to remediation goals. The residence time determined using results from the Fe column, simulating placement without slurry, was 8 hours. A slightly longer residence time of 12 hours was determined in the Fe+Guar column, simulating placement in a biopolymer slurry trench. The residence time in the Fe/Sed+Guar column, simulating a PRB constructed by high-pressure jetting of guar slurry containing fine-mesh iron was much greater. A residence time of 87 hours within the mixture of iron and unconsolidated sediments was required.

BARRIER DESIGN
        Design parameters include the assumed influent concentrations and remediation goals of Table 1, and the required residence times discussed above. Additional parameters are summarized in Table 3.

Table 3. Design parameters.

Parameter	Value	Basis
PRB dimensions & area	380 feet (116 m) long 20 to 60 feet (6 to 18 m) deep 15,150 feet^2 (1407 m^2)	Full lateral extent of plume exceeding remediation goals, with PRB keyed in bedrock
Specific discharge	Up to 0.16 feet/day (0.49 m/day)	Numerical flow model
Average linear velocity (within continuous PRB)	Up to 0.40 feet/day (12 m/day)	Specific discharge, porosity of 0.4
Required iron quantity per surface area of PRB	21 lb/ft^2 (10 g/cm^2) 32 lb/ft^2 (16 g/cm^2) 99 lb/ft^2 (48 g/cm^2)	Fe column Fe+Guar column Fe/Sed+Guar column

Several potential construction methods were considered for installation of the PRB. The required iron quantities shown in Table 3 were tripled to provide a factor of safety when determining iron placement requirements for the various alternatives. Installation by biopolymer slurry trenching was selected as the preferred alternative. A long-stick hydraulic excavator will be used to dig the trench. The viscous slurry will shore the excavation until the treatment material has been placed, at which time a solution containing enzymes will be circulated through the PRB to break down the long-chain polymers. The excavated trench will be two to three feet (0.6 to 0.9 m) wide. The granular iron will be mixed with sand to make up the difference between the iron and trench volumes.

The slurry trenching approach requires a moderate quantity of iron compared to other alternatives, allows simple verification of iron emplacement, and minimizes potential changes to existing flow paths within the heterogeneous system. Biopolymer slurry trenching has recently been used to install pilot-scale and full-scale PRBs to depths greater than 30 feet (9 m). Although construction of this PRB in the summer of 2000 will extend the application of biopolymer slurry trenching for a PRB installation to a depth of 60 feet (18 m), permeable collection trenches have previously been installed to even greater depths.

REFERENCES

Arnold, W. A., and A. L. Roberts. 1999. "Pathways and Kinetics of Chlorinated Ethylene and Chlorinated Acetylene Reactions with Fe(0). *American Chemical Society, New Orleans, LA, Preprint Extended Abstracts, Division of Environmental Chemistry.* 39(2): 158-160.

Davis, W. M., M. B. Wise, J. S. Furey, and C. V. Thompson. 1998. "Rapid Detection of Volatile Organic Compounds in Groundwater by In Situ Purge and Direct-Sampling Ion-Trap Mass Spectrometry," *Field Analytical Chemistry and Technology.* 2(2): 89-96.

Eykholt, G. R. 1998. "Analytical Solution for Networks of Irreversible First-Order Reactions." *The Journal of the International Association of Water Quality.* 33(3): 814-826.

Hubble, D. W., R. W. Gillham, and J. A. Cherry. 1997. "Emplacement of Zero-valent Metal for Remediation of Deep Contaminant Plumes." *Proceedings: International Containment Technology Conference*, pp. 872-878. St. Petersburg, Florida.

DUAL PERMEABLE REACTIVE BARRIER WALLS REMEDIATE CHLORINATED HYDROCARBON CONTAMINATION

Kenneth J. Goldstein, CGWP, Malcolm Pirnie, Inc., White Plains, NY
Stephanie O'Hannesin, M.Sc., Envirometal Technologies, Inc., Waterloo, Ontario, Canada
Shane McDonald, CPG, and Chris Gaule, Malcolm Pirnie, Inc., Mahwah, NJ/Albany, NY
Grant A. Anderson, CPG and Russel Marsh, P.E., U.S. Army Corps of Engineers, Baltimore, Maryland
Maira Senick, Watervliet Arsenal, Watervliet, NY

ABSTRACT: Dual reactive barrier walls were installed at the Watervliet Arsenal, Watervliet, NY in December 1998 to remediate chlorinated hydrocarbon contamination (CHC) in overburden and weathered bedrock groundwater. A reactive barrier wall is an innovative in-situ remedial technology based on the use of commercially available metallic iron filings. As groundwater flows through the wall, the *zero-valent iron* reductively dehalogenates CHCs through the corrosion process into non-toxic chloride ions, ethenes and ethanes. Two reactive walls totaling approximately 285 ft. in length were installed in parallel through overburden and weathered bedrock using conventional excavation methods and trench boxes. A mixture of iron and sand at a ratio of 1:1 by weight was placed into each trench. Trenches are approximately 12 feet in depth and 30 inches wide and are keyed into competent bedrock. Analytical samples obtained one year after installation indicate CHC concentrations below detection limits within the reactive walls.

BACKGROUND

The Watervliet Arsenal (WVA) is a 140-acre government-owned installation located in the City of Watervliet, New York, which is west of the Hudson River, and five miles north of the City of Albany. A large, swampy, 14-acre area located to the west of the Main Manufacturing Area of the WVA, known as the Siberia Area, was purchased by WVA in the early 1940's and immediately filled in with debris consisting of slag, cinders, wood, brick and other debris of unknown origin. Once filled in, two areas were used for burning combustible material (i.e., scrap lumber and other solid waste) and liquids until 1967. During the period of 1994 through 1998 an RCRA Facility Investigation was conducted to assess the nature and extent of contamination in the area.

The site consists of three unconsolidated deposits overlying shale bedrock. The unconsolidated deposits consist of an upper fill unit, approximately four feet thick, the second unit is a clayey silt, approximately two to six feet thick which extents to weathered bedrock. The third unconsolidated unit is a fluvial sand and gravel, and is found primarily in the Northwest Quadrant of the site. The majority of the unconsolidated deposits at the site are saturated and are hydraulically connected with the weathered bedrock. During the majority of the year the water table is present in the overburden deposits, but during seasonal low water table conditions the water declines into the weathered bedrock.

The results of the RFI indicated that the majority of the groundwater contamination detected in the Siberia Area was confined to the Northeast Quadrant and was primarily CHCs. Figure 1 presents the limits and geometry of the CHC plume. The contamination in this area has migrated along the shallow groundwater flow paths in the overburden and weathered bedrock towards the site sewer line, which bisects the Siberia Area in a north-south direction. The maximum concentrations of volatile organic compounds detected in these units are as follows:

- Vinyl chloride 1,700 µg/l
- cis-1,2-Dichloroethene 4,200 µg/l
- 2-Butanone 11 µg/l
- Trichloroethene 1,500 µg/l
- Tetrachloroethene 1,100 µg/l
- Xylene 43 µg/l

- 1,1-Dichloroethene 7 µg/l
- trans-1,2-Dichloroethene 11 µg/l
- Benzene 20 µg/l
- Toluene 10 µg/l
- Ethylbenzene 7 µg/l

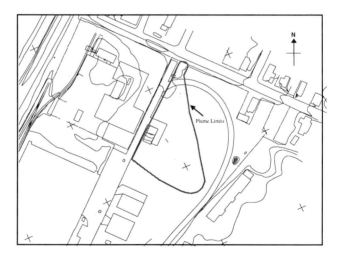

FIGURE 1. Extent of CHC Plume, Siberia Area, Watervliet Arsenal

Due to the location of the groundwater contamination, in WVA's primary shipping and receiving area, and the need to have un-restricted access to the Northeast Quadrant for storage, a remedial technology was required which would not compete for the valuable space. In order to meet these requirements a permeable reactive wall was considered to be the appropriate remedial technology. In order to determine the applicability of this technology to the Siberia Area groundwater contamination several, phases of study were completed. The following presents a summary of the methods utilized and there results.

BENCH SCALE TREATABILITY TESTS

In order to determine that the permeable reactive wall concept was suitable for treating the contaminated groundwater in the Siberia Area and to develop design parameters for such a system, bench scale treatability testing was conducted. The testing was performed by EnviroMetal Technologies, Inc., using contaminated groundwater collected from a monitoring well in the Siberia Area and zero-valent iron particles as the reactive agent. The specific goals of the bench scale treatabilty testing were to provide design parameters such as the volume of zero-valent iron required to achieve residence times, the degradation rates of the CHCs, and the effects of inorganic precipitation on the walls performance.

The test column used during the study consisted of 50-cm tube with several sampling ports along its length to collect samples throughout the testing period. The column was packed with 100% zero-valent iron. The site groundwater was then pumped through the reactive column at a flow rate of 1.0 ft/day, which was significantly higher than the site groundwater flow rates. This flow rate was used to conduct the testing in the required time frame of the project. Based on the analytical data collected from the sampling ports, and the input flow velocity, degradation rates were calculated using the first-order kinetic model. These degradation rates were later recalculated using the site specific groundwater flow velocity of 0.15 ft/day.

The longest residence time required to degrade the VOCs to the New York State MCL was associated with vinyl chloride and was estimated at 66 hours. At a flow velocity of 0.15 feet per day and a residence time of 66 hours (~2.5 days), a reactive wall, consisting of 100% reactive iron, would have to be 0.41 feet in thickness to degrade the vinyl chloride from an initial concentration of 1,700 µg/L to the MCL's of 2 µg/L. Figure 2 presents the required residence times to meet NYS MCLs. This wall thickness calculated during the testing did not account for a safety factor. Therefore, a safety factor of 2 was applied to the results of the treatability testing for the full-scale application, which resulted in a design thickness of 0.82 feet of pure reactive iron. The results of the treatability testing also indicated that

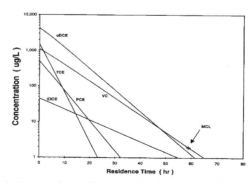

FIGURE 2. Bench Scale Treatability Residence Time Results

precipitation of carbonates and other inorganic precipitates would be minimal.

GROUNDWATER MODELING

Groundwater modeling was conducted to optimize the location and orientation of the reactive walls. The groundwater model simulated the groundwater flow directions (vertical and horizontal) under various reactive wall configurations. The simulations included the use of a funnel and gate scenario versus a continuous reactive wall. A total of 11 configurations of the reactive wall (continuous wall and funnel and gate) system were simulated. Particle tracking used during the modeling evaluated the walls ability to intercept the contaminated groundwater without flow escaping treatment.

The results of the modeling indicated that the use of funnel and gate system would not be an appropriate solution. It was found that the impermeable portion of the funnel increased the hydraulic gradient at the upgradient face of the wall and forced the contaminated groundwater to pass under the wall and into the underlying bedrock, escaping treatment. As a result of the model, a continuous reactive wall was selected. In order to achieve maximum treatment of the groundwater plume it was determined that two reactive walls would be required, one at the downgradient extent of the plume and the other closer to the source area. Figure 3 presents the location of the reactive walls and particle tracking results generated during the modeling.

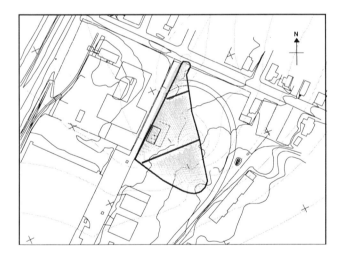

FIGURE 3. Reactive Wall Orientations and Particle Tracking Results

CONSTRUCTION

The designed thickness of pure granular iron in the reactive walls was approximately one foot (0.82 feet required to meet the residence time for degradation). The minimum trench width using conventional excavation and shoring equipment was determined to be 30-inches. Therefore, granular iron and clean washed sand was mixed together to make up the additional volume within the trench in the proportions of 1.5 volumes of washed clean sand to one volume of granular iron. The actual mixing of the materials was done on a weight basis rather than volume basis. The unit weight of the granular iron was approximately 150 pounds per cubic foot and the sand weight was approximately 100 pounds per cubic foot. A mix containing 150 pounds of sand (1.5 cubic feet) to 150 pounds of granular iron (1.0 cubic feet), or a 1:1 ratio by weight would be required in the 30-inch wide trench.

The weigh ratio was controlled by mixing the iron and sand at a cement mixing batch plant. To ensure the quality control on the iron:sand mix the cement batch plant provided the weigh tickets for each truck of material delivered to the site. Granular iron and washed sand was mixed in a rotary drum of a transit mix truck and delivered to the site (Figure 4). In addition to these controls Malcolm Pirnie also

FIGURE 4. Granular iron and sand delivered from transit mixer

conducted magnetic segregation of grab samples through out the construction process to confirm the mix ratio.

A conventional track-mounted excavator with a 24-inch bucket was used to excavate both reactive wall segments. Elevations of the bottom of the trench were determined with a level at a minimum of 20-foot intervals and tied to known elevation datum on site. The target trench width was 30 inches. To accomplish this, the trenches were excavated to a width of approximately 32 to 36 inches and one-inch thick steel sheeting plates 10-feet high and 20 feet long were placed vertically on each side of the trench to maintain the width. The opposing shoring plates were held in position using pairs of expandable hydraulic shoring posts oriented horizontally. Fine sand was placed between the shoring plates and the walls of the trench, as necessary, to fill larger voids in the trench walls. Figure 5 presents the

FIGURE 5. Partial wall with opposing sheet piles and shoring posts

opposing shoring plates and expandable shoring posts used to construct the walls. The final reactive walls consisted of two walls with a total length of 285 feet; each wall was keyed into the top of competent bedrock, at a depth of approximately 12 feet below the ground surface. Monitoring wells were installed inside the reactive walls, at three locations per wall, to monitor the reactions occurring inside the walls and to monitor for the production of precipitates during the operational life of the walls. During the construction of the trenches a total of 326,360 pounds of sand and 330,940 pounds of reactive iron were used. Figure 6 presents the constructed profile along the length of one the walls installed. The cost of construction for this project was approximately $278,000.

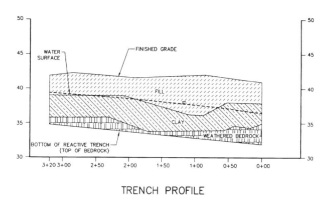

TRENCH PROFILE

FIGURE 6. Reactive Wall Profile

GROUNDWATER MONITORING

The results of groundwater samples collected approximately one year after construction show that concentrations of the CHCs entering the upgradient side of the source area wall range from low parts per billion (ppb) to greater than 2 parts per million (ppm). Concentrations of CHCs in the reactive wall are below the regulatory limits. The system continues to be monitored, both hydraulically and analytically, on a semi-annual basis.

This project was recently awarded the Design Excellence Award by the Consulting Engineers Council of New York State.

FIELD DEMONSTRATION OF PERMEABLE REACTIVE BARRIERS
TO CONTROL URANIUM CONTAMINATION IN GROUND WATER

David L. Naftz (USGS, Salt Lake City, Utah)
Christopher C. Fuller, James A. Davis, and Michael J. Piana
(USGS, Menlo Park, California)
Stan J. Morrison ([a]Environmental Sciences Laboratory, Grand Junction, Colorado)
Geoffrey W. Freethey and Ryan C. Rowland (USGS, Salt Lake City, Utah)
[a]The Environmental Sciences Laboratory is operated by MacTec-ERS for the Department of
Energy, Grand Junction Office

ABSTRACT: Three permeable reactive barriers (PRBs) were installed near Fry
Canyon, Utah, in August 1997. The overall objective of this project is to
demonstrate the use of PRBs to control the migration of uranium (U) in ground
water. A funnel and gate design was used to construct the three PRBs which
consist of (1) bone-char phosphate (PO_4), (2) zero-valent iron (ZVI) pellets, and
(3) amorphous ferric oxyhydroxide (AFO). During the first 28 months of PRB
operation (September 1997 through December 1999), the ZVI PRB was the most
effective at lowering U concentrations in the contaminated ground water. The
median U removal in the ZVI PRB was higher than 99.5 percent. Geochemical
modeling techniques were used to define and quantify the amount and type of
mineral precipitates that may be forming in the ZVI PRB. Modeling results
indicate that most of the mineral precipitation occurs within the first 1.0 foot of
barrier material. On the basis of water-chemistry data collected during May 1999,
the downgradient two-thirds of the ZVI PRB has not been affected by mineral
precipitation. Geochemical modeling results were consistent with solid-phase
analyses of reactive material collected from the ZVI PRB.

INTRODUCTION

The use of permeable reactive barriers (PRBs) for the remediation of
ground water contaminated with chlorinated organics has received considerable
attention in recent years (Gu et al., 1999). In contrast, the application of PRBs to
remove uranium (U) and other radionuclides from ground water has been limited.
As of 1999, 46 field projects utilized PRBs to treat contaminated ground water;
however, only 7 of these field projects are treating U in water.

One of these seven field projects is located near Fry Canyon, Utah (Figure
1). Three PRBs were installed at this site in August 1997 to determine if long-
term treatment of U-contaminated ground water would occur. Funding for this
PRB demonstration project was provided by the U.S. Environmental Protection
Agency (USEPA)/Office of Emergency Response and Remediation, and Office of
Radiation and Indoor Air.

Objectives. The overall objective of this project is to demonstrate the use of
three different PRBs to control the migration of U in ground water. Specific
objectives of this paper are to (1) present the initial results of PRB performance
for each of the three PRB materials during the first 28 months of field operation

and (2) utilize geochemical modeling techniques to determine chemical reactions that may potentially decrease the long-term effectiveness of the zero-valent iron (ZVI) PRB. Results contained in this report will be useful to personnel involved with the clean up of ground water contaminated with uranium and (or) the plugging and passivation of ZVI PRBs.

Site Description. A shallow colluvial aquifer contaminated by previous U-upgrading operations near Fry Canyon, Utah (Figure 1), was selected for the long-term field demonstration of selected PRBs. This site is located on land managed by the U.S. Bureau of Land Management (BLM).

The shallow colluvial aquifer consists of silt to gravel-size particles derived from nearby sandstone and shale formations. Maximum saturated thickness of the colluvial aquifer ranges from about 2 to 5 feet (ft). Based on laboratory and field tests, hydraulic-conductivity values range spatially from 5 to 85 feet per day (ft/d) and transmissivity values range spatially from 10 to 200 square

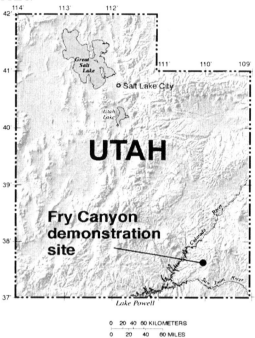

FIGURE 1. Location of the Fry Canyon demonstration site in southeastern Utah.

feet per day (ft^2/d). The in-situ effective porosity for the aquifer is estimated to be 20 to 25 percent (Freethey, Spangler, and Monheiser, 1994). Underlying the colluvial aquifer is the Permian Cedar Mesa Sandstone, which is virtually impermeable compared to the colluvial aquifer. Concentrations of U in the contaminated part of the colluvial aquifer exceeded 16,000 micrograms per liter (µg/L). These concentrations were substantially higher than background U concentrations, which ranged from 60 to 80 µg/L.

METHODOLOGY

PRB Installation. A funnel and gate design was chosen to demonstrate the three PRBs. This design consists of three "permeable windows" or gates in which each of the reactive materials is placed. Each gate is separated by an impermeable wall, and impermeable wing walls are installed on each end of the multigate structure to channel the ground water into the PRBs. Dimensions of each gate structure are 7 ft long by 3 ft wide by about 4 ft deep. The three PRBs and no-flow walls were placed into the upper parts of the bedrock (Cedar Mesa Sandstone) underlying the colluvial aquifer. A 1.5-ft-wide layer of pea gravel was placed on the upgradient side of the PRBs to facilitate uniform flow of contaminated ground water into each gate structure. The three gates contained (1) bone-char phosphate (PO_4); (2) ZVI pellets; and (3) amorphous ferric oxyhydroxide (AFO).

The mechanism of U removal in each of the PRBs is a function of the type of barrier material. The PO_4 barrier material consists of pelletized bone charcoal that facilitates surface complexation of U (Fuller et al., 1999). The ZVI barrier material consists of pelletized iron designed to remove U by reduction of U (VI) to the less soluble U (IV). The AFO barrier material consists of pea gravel coated with amorphous ferric oxyhydroxide that removes U by adsorption. Materials were pelletized or used as a coating on gravel to increase the permeability of the gate structure relative to the permeability of the native aquifer material.

Water Sampling and Analysis. An extensive monitoring network was installed in each PRB consisting of 16, 0.25-inch- (in.) diameter poly vinyl chloride

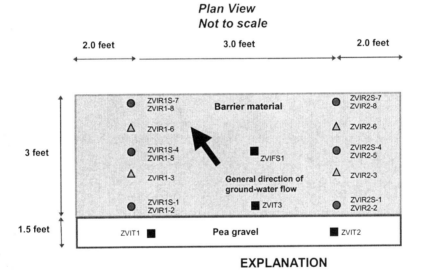

Plan View
Not to scale

FIGURE 2. Schematic diagram showing monitoring well placement in the zero-valent iron PRB. The monitoring design in the phosphate and amorphous ferric oxyhydroxide PRBs is the same.

(PVC) wells located along 2 parallel flow paths and 4, 2-in-diameter PVC wells (Figure 2) for sample collection and monitoring of water levels and selected water-quality parameters. Because of the proximity of wells to one another, limited purge volumes were extracted prior to sample collection. One gallon (gal) of water was removed from the 2-in-diameter monitoring wells and 0.26 gal of water was removed from the 0.25-in-diameter monitoring wells.

After purging, water samples were filtered on site using a 0.45-micrometer (μm) capsule filter and collected in field-rinsed polyethylene bottles. Samples for analysis of U, calcium (Ca), iron (Fe), potassium (K), magnesium (Mg), and sodium (Na) were acidified on site with ultra-pure concentrated nitric acid. Each monitoring point contains a dedicated sampling tube to minimize cross contamination.

Water analyses were conducted at the U.S. Geological Survey (USGS) Research Laboratories in Menlo Park, California. Dissolved U was measured by kinetic phosphorescence analysis (KPA). Ca, Fe, Mg, and Na concentrations were measured by inductively coupled plasma optical emission spectrometry (ICP/OES) using a Thermo Jarrel Ash ICAP 61 (Standard Methods, 1992). The K concentration was mesured by direct air-acetylene flame atomic absorption spectrometry (AA) using a Perkin Elmer AA 603. Sulfate (SO_4) concentrations were measured by ion chromotography (IC) using a Dionex Chromatograph CHB (Standard Methods, 1992).

The pH, Eh, and temperature of each water sample were measured in a flow-through chamber by using a Yellow Springs Instrument 600XL minimonitor that was calibrated daily with respect to pH and weekly with respect to Eh. Total alkalinity (as $CaCO_3$) of filtered (0.45 μm) water samples was mesured on site by using a HACH digital titrator and 1.6 normal sulfuric acid.

RESULTS AND DISCUSSION

U Removal. The percentage of input U removed by each of the PRBs was determined for water samples collected during the first 28 months of PRB operation (September 1997 through December 1999) (Figure 3). The ZVI PRB was the most efficient of the three PRBs for reducing input U concentrations. The median percentage of U removal was always higher than 99.5 percent during the 11 monitoring periods (Figure 3).

Although the median values for percentage of U removal were less in the PO_4 and AFO PRBs, substantial decreases in U concentration still occurred during the first 28 months of operation (Figure 3). The median values for percentage of U removal in the PO_4 PRB decreased to less than 70 percent in September 1998; however, there has been a steady increase in the percentage of U removed from December 1998 to December 1999. The AFO PRB has had a steady decrease in the median percentage of U removal since installation (Figure 3). The median percentage of U removal after 28 months of operation is 70 percent.

Plugging Reactions in Zero-Valent Iron PRB. ZVI PRB is removing the highest percentage of input U; however, core analysis from this barrier indicates that mineral precipitation is occurring. Minerals qualitatively identified in cores

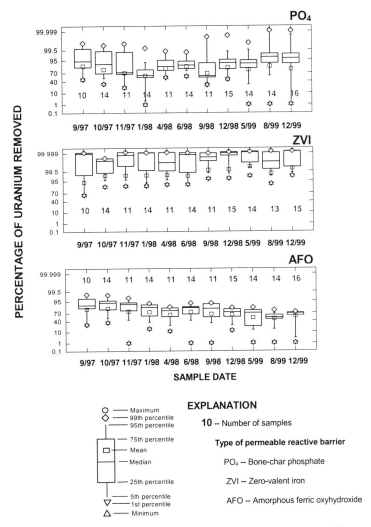

FIGURE 3. Box plots showing the percentage of input uranium removed for each permeable reactive barrier during the first 28 months of operation at Fry Canyon, Utah.

collected from the ZVI PRB during May 1999 included iron sulfide, calcite, and metallic iron (ZVI material). Mineral precipitation can decrease the permeability and reactivity of the ZVI barrier material and result in a decrease in contaminant removal efficiencies (Gu et al., 1999). In a series of laboratory column experiments, Gu et al. (1999) identified a number of mineral precipitates forming in ZVI material. Mineral precipitates included iron hydroxides, carbonates, and sulfides.

Mass-balance modeling techniques were used to define and quantify the amount and type of mineral precipitates that may be forming in the ZVI PRB at Fry Canyon. The USGS geochemical model NETPATH (Plummer, Prestemon, and Parkhurst, 1994) was used to determine the quantity of each phase (mineral or gas) that would likely form or dissolve in ground water flowing between monitoring wells in the ZVI PRB. The general chemical reaction for this modeling is in the form of:

Initial solution composition + *"Reactant Phases"* → *Final solution* + *"Product Phases"*   (1)

where the terms "Reactant Phases" and "Product Phases" refer to phases entering or leaving the water during the course of a chemical reaction. The possible "Reactant" and "Product" phases were based on the qualitative mineralogical analysis of the ZVI material, speciation calculations (Table 1), geochemical inferences made from the anaerobic and aerobic corrosion of Fe, and qualitative observations made during field visits.

Water samples collected during May 1999 from wells along three different hydrologic flow paths in the ZVI PRB were used in the NETPATH modeling. The wells from each flow path in upgradient to downgradient order are: (flow path 1) ZVIT1 → ZVIR1-2 → ZVIR1-3 → ZVIR1-6; (flow path 2) ZVIT2 → ZVIT3 → ZVIFS1; and (flow path 3) ZVIT2 → ZVIR2-2 → ZVIR2-5 (Figure 2).

Six chemical constraints and eight plausible phases were considered during the NETPATH modeling. The six chemical constraints included carbon (C), sulfur (S), Ca, Mg, and Fe concentrations and the term referred to as redox state (RS). The RS is a method of keeping track of electron transfer in the redox reactions considered in the NETPATH modeling. Conventions defining RS can be found in Plummer, Peterson, and Parkhurst (1994). The eight plausible phases considered during the NETPATH modeling are (1) calcite (precipitation (ppt.) only); (2) magnesite (ppt. only); (3) dissolved organic carbon (DOC) (dissolution (diss.) only); (4) siderite (ppt. only); (5) zero-valent iron (diss. only); (6) hydrogen gas (ppt. only); (7) iron sulfide; and (8) dissolved oxygen (diss. only).

On the basis of the chemical constraints and plausible phases that were selected, the small amount of dissolved oxygen (0.8 mg/L) in the input water was consumed by the aerobic corrosion of the ZVI material (Figure 4). Anerobic corrosion of the ZVI material caused the generation of hydrogen gas accompanied by the precipitation of carbonate minerals (calcite, magnesite, and siderite). Oxidation of DOC in conjunction with the observed decrease in total inorganic carbon accounted for the relatively large amounts of carbonate precipitation along each flow path. The observed decrease in SO_4 concentration along each flow path was accounted for by microbial mediated sulfate reduction and the formation of iron sulfide. Electron donors used during the SO_4 reduction were a combination of DOC and hydrogen gas.

This geochemical model is not the only set of plausible phases that can explain the observed changes in water quality along the hydrologic flow paths; however, other data also support this model. The generation of hydrogen gas has been indirectly confirmed by the observation of gas bubbles in water from wells completed in the ZVI PRB. The corrosion of Fe is supported by the 10 to 15 percent decrease in Fe concentration from cores collected in the ZVI PRB relative

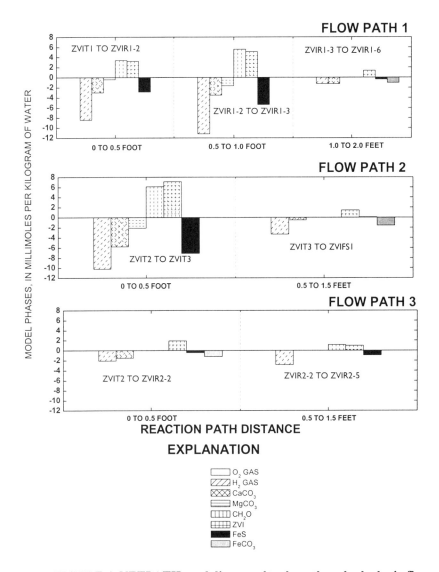

FIGURE 4. NETPATH modeling results along three hydrologic flow paths in the zero-valent iron permeable reactive barrier, Fry Canyon, Utah. A positive value indicates dissolution and a negative value indicates precipitation of a solid phase or generation of a gas phase.

to unreacted ZVI material. The precipitation of calcite in the ZVI PRB is supported by scanning electron microscope images in combination with energy dispersive analysis of core samples collected from the barrier. In addition, total inorganic carbon concentration increased from 718 milligrams per kilogram (mg/kg) in an unreacted sample of ZVI material to 7,780 mg/kg in a core sample collected from the ZVI PRB during May 1999. Selected samples from the ZVI material indicate up to a six-fold increase in Ca and two-fold increase in Mg concentration relative to the unreacted ZVI material. The DOC concentration in ZVI feed water has not been measured; however, leaching of the bone-char phosphate may produce high DOC concentrations. Tracer tests conducted in April 1999 indicate that water from the PO_4 PRB is moving along the gravel pack and entering the ZVI PRB. The total organic carbon concentration decreased from 19,600 mg/kg in unreacted bone-char barrier material to 3,660 mg/kg in a core sample collected from the PO_4 PRB during May 1999 (20-month time period). Sulfate reduction is supported by the detection of iron sulfide solid phases using SEM/EDA in core samples from the ZVI PRB.

Additional confirmation of the NETPATH modeling is provided by the computer program WATEQF (Plummer, Jones, and Truesdell, 1976). The program was used to calculate the ionic activities in water samples upgradient and within the ZVI PRB (Table 1). The degree of saturation, defined as the ratio of the ion-activity product divided by the equilibrium constant for the specific mineral, was determined for each mineral considered as a plausible source or sink for selected chemical constituents. The log of this ratio is referred to as the saturation index (SI). A positive SI indicates that thermodynamic conditions favor mineral precipitation from solution, whereas a negative SI indicates conditions that favor mineral dissolution. A SI near zero indicates that the solution is in equilibrium with respect to the mineral of interest.

TABLE 1. Saturation indicies of selected plausible phases in water samples upgradient and within the zero-valent iron permeable reactive barrier in Fry Canyon, Utah, during May 1999.

Well number	Calcite	Siderite	Gypsum
ZVIT1	0.495	0.240	-0.257
ZVI T2	0.468	0.350	-0.388
ZVIT3	1.143	1.687	-1.287
ZVIFS1	1.396	1.180	-1.472
ZVIR1-2	0.616	1.673	-0.470
ZVIR1-3	0.923	1.732	-0.997
ZVIR1-6	1.017	1.549	-1.299
ZVIR2-2	0.610	1.665	-0.479
ZVIR2-5	0.668	1.751	-0.513

The likely formation of calcite and siderite is supported by the larger positive SIs observed for these minerals in water samples from wells completed in the ZVI PRB (Table 1) relative to wells completed in the upgradient gravel pack (ZVIT1 and ZVIT2). Although magnesite is used in the NETPATH model to account for Mg loss along each of the flow paths, it is likely that magnesium is substituting in the calcite mineral phases that are observed. The SI for gypsum (Table 1) is negative, indicating that the observed decrease in sulfate concentration along each flow path is the result of sulfate reduction and not gypsum precipitation.

Results from the NETPATH model indicate that most of the mass transfer occurs along the first 1.0 ft in each hydrologic flow path (Figure 4). For example, 16.81 millimoles per kilogram of water (mmol/kg H_2O) of carbonates and sulfides are precipitated along the first 1.0 ft in flow path 1 compared with only 2.67 mmol/kg H_2O in the 1.0-to-2.0-ft section (Figure 4). This pattern of mineral precipitation indicates that approximately two-thirds of the ZVI PRB has not been affected by mineral precipitation. The amount of carbonate and sulfide precipitation indicated by the NETPATH modeling results can be used to estimate the rate of porosity reduction in the first 0.5 ft of flow path 1 in the ZVI PRB. By using a calculated ground-water input of 9.7 cubic feet per day (ft^3/d) into the ZVI PRB (Table 2), approximately 0.0017 cubic feet per day of ZVI PRB porosity is consumed by carbonate and sulfide mineral phases in the first 0.5 ft of reactive material. The total porosity in the first 0.5 ft of unreacted ZVI material along flow path 1 was 4.25 ft^3 (0.5 ft thick by 2.3 ft width by 3.7 ft deep). With the porosity reduction rate of 0.0017 ft^3/d, approximately 15 percent of the ZVI porosity will be lost each year. The yearly rate of porosity reduction will probably not be linear. As mineral precipitation continues, less of the ZVI material will be available for reaction causing the precipitation reactions to migrate farther into the PRB.

TABLE 2. Data used to calculate the volume of carbonate and sulfide mineral precipitates along flow path 1 in the zero-valent iron permeable reactive barrier, Fry Canyon, Utah. [ft^3/d, cubic feet per day; mmol/d, millimoles per day; g/cm^3, grams per cubic centimeter]

*Ground-water input (ft^3/d)	Carbonate precipitated (mmol/d)	FeS (troilite) precipitated (mmol/d)	Calcite density (g/cm^3)	FeS (troilite) density (g/cm^3)	Calcite volume (ft^3/d)	FeS (troilite) volume (ft^3/d)
9.7	933	785	2.71	4.61	0.0012	0.0005

*Based on tracer test conducted during April 1999.

REFERENCES

Freethey, G.W., L.E. Spangler, and W.J. Monheiser, 1994. *Determination of hydrologic properties needed to calculate average linear velocity and travel time of ground water in the principal aquifer underlying the southeastern part of Salt Lake Valley, Utah.* U.S. Geological Survey Water-Resources Investigations Report 92-4085. 30 p.

Fuller, C.C., J.R. Bargar, M.J. Piana, and J.A. Davis, 1999. Mechanisms of uranium uptake by apatite materials for use in permeable reactive barriers for the remediation of contaminated ground water (abstract). American Geophysical Union Fall Meeting, December 12-17, 1999, San Francisco, Calif.

Gu, B., T.J. Phelps, L. Liang, M.J. Dickey, Y. Roh, B.L. Kinsall, A.V. Palumbo, and G.K. Jacobs. 1999. Biogeochemical dynamics in zero-valent iron columns: Implications for permeable reactive barriers. *Environ. Sci. Tech.* 33(13): 2170-2177.

Plummer, L.N., B.F. Jones, and A.H. Truesdell. 1976. *WATEQF -- A FORTRAN IV Version of WATEQ, A computer program for calculating chemical equilibrium of natural waters.* U.S. Geological Survey Water-Resources Investigations Report 76-13. 63 p.

Plummer, L.N., E.C. Prestemon, and D.L. Parkhurst. 1994. *An Interactive Code (NETPATH) For Modeling NET Geochemical Reactions Along a Flow PATH Version 2.0.* U.S. Geological Survey Water-Resources Investigations Report 94-4169. 130 p.

Standard Methods. 1992. *Standard methods for the examination of water and wastewater.* 18th Edition. Greenberg, A.E., Clesceri, L.S., and Eaton, A.D., Editors.

DISCLAIMER. *Use of brand names in this article is for identification purposes only and does not constitute endorsement by the U.S. Geological Survey.*

HYDRAULIC FRACTURING TO IMPROVE
IN SITU REMEDIATION

William W. Slack (FRx, Inc., Cincinnati, Ohio)
Lawrence C. Murdoch (Clemson University, Clemson, South Carolina)
Ted Meiggs (Foremost Solutions, Golden, Colorado)

Abstract: Hydraulic fractures created from shallow boreholes can be filled with granular material to form layers that are tens of meters in maximum dimension. The material used to fill the fracture will control the properties of the resulting layer, and a variety of properties can be achieved that will improve in situ remediation. Fractures filled with well-sorted sand create permeable layers that improve the performance of wells used to inject treating solutions or recover contaminants. Chemically active compounds can also be injected to create layers that will either degrade or immobilize in situ contaminants. The orientation of the hydraulic fractures also plays a key role in the effectiveness of remedial applications. Hydraulic fractures are flat-lying to gently dipping in many shallow settings, and this orientation is ideal for creating reactive barriers that impede the migration of contaminants in the vadose zone. Vertical hydraulic fractures placed perpendicular to the direction of ground water flow are well-suited to creating reactive barriers to plumes that move horizontally.

INTRODUCTION

Hydraulic fracturing allows granular material to be injected to form sheet-like layers in the subsurface for a variety of remediation applications. The fracturing process uses fluid pressure to open a crack in the subsurface soil or rock and then granular material is suspending in a viscous slurry and injected into the growing crack. The techniques derive most directly from methods of stimulating oil and gas production, and related processes have been used to enhance potable water wells. Hydraulic fracturing alone does little to remediate contaminated soil or groundwater. Rather, it can be used to enhance common or innovative remedial processes. The contaminant and remedial process determine which granular material should be used in the hydraulic fracture.

The first and still most widely used environmental application is to inject sand into hydraulic fractures created in clay or rock. This forms highly permeable layers that enhance flow from the extreme edge of the fracture to the well. As a result of the permeability contrast, pressure gradients and flux within the subsurface assume a linear form proceeding into the face of the fracture, and this pattern is much more efficient than the radial form surrounding conventional wells. As a result, the flow rate of a well at a given drawdown from a hydraulically fractured well is typically by one to two orders of magnitude greater than from a conventional well in a low permeability formation. In addition, the effective radius of influence usually expands to beyond two times the extent of the fracture. These effects were first noted in SVE wells that were completed with hydraulic fractures as part of development projects sponsored by the USEPA in the late 1980's and early 1990's. (Wolf and Murdoch, 1993; Murdoch et. al, 1994).

Similar increases in flow rate and radius of influence can be realized for free-product recovery or other fluid-based technologies in tight formations where they might otherwise be infeasible. Murdoch et al. (1994) describe the use of sand-filled hydraulic fractures to recover free-phase hydrocarbon product from low permeability soils. During that test, additional fractures were used to manipulate the underlying water table, thereby enhancing the cone of drawdown for hydrocarbon product. Since then various fluid-based applications have been demonstrated across the country in low permeability clays deposited as glacial drift, lacustrine or overbank sediments, residuum on limestone, saprolite, and in shales and siltstones. Two examples of interest involved (1) injection of hot fluids to promote vapor transport of contaminants and (2) injection of oxidizers to destroy contaminants in situ.

In addition to enhancement of fluid delivery and recovery processes, hydraulic fracturing has also been used to inject reactive compounds, creating layers of chemically or biologically active material that will degrade chemicals in situ. Porous ceramics (Stavnes, 1999) and oxygen-producing peroxides (Davis-Hoover, et al, 1993) have been used to stimulate aerobic degradation, granules of potassium permanganate have been injected to create highly oxidizing conditions that destroy organic solvents and other compounds, and zero-valent iron has been used to reductively dechlorinate solvents (Seigrist et al., 1999). These processes have utilized horizontal fractures. The flat-lying orientation is significant because it allows reactive barriers to be created that will intercept downward-moving contaminants in the vadose zone, thereby extending the reactive barrier concept that is now widely used in the saturated zone. The projects discussed by these references were demonstration scale. Now, a full-scale project has been completed at a site in New Mexico.

Hybrid techniques that combine hydraulic fracturing with high-pressure jets are promising methods for creating vertical barriers filled with reactive materials in the saturated zone, particularly at depths where excavation is infeasible. A recent demonstration showed that such hybrid techniques were capable of creating a vertical reactive barrier filled with zero-valent iron at depths below 50 feet.

DELIVERY APPLICATIONS

In the summer of 1999 hydraulic fractures enabled cost-effective closure of an underground storage tank (UST) site in the Piedmont physiographic province in south central Pennsylvania. As with many such sites, the release of petroleum lubricants and fuels, principally gasoline, into subsurface soils and groundwater generated a plume of BETX. The site is underlain by thin (<2m) colluvium over clay-rich saprolite derived from underlying mica schist. The material is highly heterogeneous, so pockets of significant contamination can be found as much as 50 m from the source yet measurements among various locations on the site indicate limited hydraulic communication. Recovery processes traditionally have not performed well at this type of site. In situ chemical oxidation can be attractive if the oxidants can be delivered to the contaminated regions.

Seven sand-filled hydraulic fractures were created at various locations across the site. Table 1 lists the characteristics of the fractures. Two of these were created beneath the pit that formerly contained the UST. The two fractures underneath the pit targeted contaminated soils while the other five addressed groundwater on down-gradient adjoining properties.

Fenton-type oxidant solutions were injected into each fracture while observing dissolved oxygen, temperature, and volatile organic compound concentrations in offset monitoring wells. Injection terminated when appropriate changes in monitored values were observed. The total mass of injected oxidant was approximately 10 percent of the amount required to remediate the entire site, according to stoichiometric and mass balance calculations. As a result, we expect the major effects to occur within the vicinities of the injection wells, primarily near the former UST pit.

TABLE 1. Fracture sizes for enabling delivery of in situ oxidation chemicals.

Frac ID	Depth (m)	Mass Sand (kg)	Final Injection Pressure (KPa)
1a	4.3	770	75
1b	8.6	1500	120
2	7.2	1140	460
3	5.4	820	270
4	3.3	550	170
5	4.9	730	220
6	3.9	500	160

Effectiveness of the oxidation treatments that utilized hydraulic fractures was characterized by measuring the concentration of volatile organic compounds (VOC) in monitoring wells and soil samples. Soil samples taken beneath the UST location show the most striking results. Concentrations were initially the greatest in this region, but no BTEX compounds could be detected beneath the former UST pit after oxidants were injected. This result is particularly significant because the region beneath the former UST was suspected of acting as a source for the ground water plume. The results from six monitoring wells (Fig. 1) suggest that most of the VOCs initially present in groundwater were destroyed within about 8 m of the injection well, and roughly half were destroyed within 40 m of the injection well. These results are consistent with expected influence based on the mass of injected oxidant.

The duration of this project was 4 months, from the time the hydraulic fractures were created until the final soil sample was taken. Project completion was probably accelerated significantly by using hydraulic fractures to distribute the oxidant. Budgetary constraints prevented the use of a control well, so the effects of the fractures cannot be quantified at this site. Nevertheless, our experience at other sites underlain by clay-rich saprolite suggest that oxidant injection into conventional wells would have markedly delayed completion.

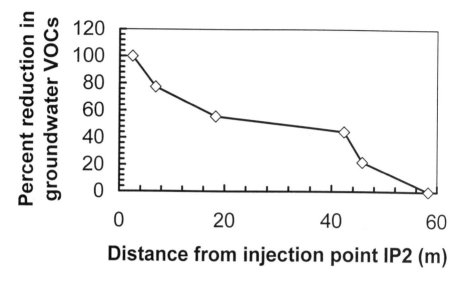

Figure 1. Ratio of initial and final concentration of BTEX compounds in water as a result of in situ chemical oxidation treatment.

REACTIVE BARRIERS

Jet assisted hydraulic fracturing has been used to construct a reactive barrier, or permeable treatment zone (PTZ) at depths of 50 – 70 ft (15 – 21 m) below ground surface at a site in central Alabama. In prior years, the site was an automotive salvage yard where old and damaged vehicles were stored and dismantled. Many of the surrounding properties housed similar industries. Presently all of the properties are vacant, and the land is a broad grassy field. Multiple industrial contaminants have been found in subsurface soils and groundwater of the site and appear to be migrating onto adjoining, pristine property.

The land within the site slopes gently to the north, which is the direction of flow in the surficial aquifer. Surface soils, which extend to 3 m below ground surface (bgs), are a mix of plastic and lean clays that contain a minor fraction of poorly graded particles ranging in size from fine sands to gravel. A 12-m-thick zone of poorly graded gravel underlies the surface soil on the southern half of the site and pinches out into silty sands and sandy silts by the northern property boundary. On the southern half of the property, 10 m of silty sand and sandy silt underlie the gravel, and this unit is continuous with the shallower sands and silts encountered at the northern edge. The silts, sands, and gravel in aggregate compose the surficial aquifer. The surficial aquifer rests on a thick (> 15 m) clay bed that serves a confining layer for the underlying deep regional aquifer. The water table occurs at 6 m bgs (Fig. 2).

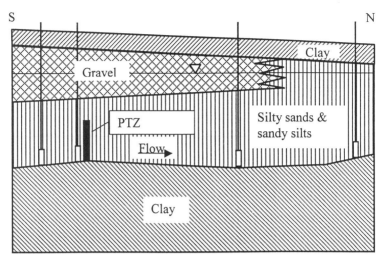

FIGURE 2. Schematic geological cross-section. The four monitoring wells indicated were completed in zones where significant TCE or PCE was detected; contaminants were found only near the bottom of the surficial aquifer, suggesting an up-gradient source.

Chlorinated solvents comprise the principal contaminants of concern, although minor amounts of other volatile organic compounds and several heavy metals have been detected. Five rounds of soil and groundwater sampling identified a plume of trichloroethylene (TCE) and tetrachloroethylene (PCE) at the top of the confining/semiconfining layer of the bottom of the surficial aquifer. The plume extends from the up-gradient edge of the property to the down-gradient edge. Interestingly, contamination appeared to be greatest in an area a few tens of meters wide along the southern property boundary. Furthermore, these contaminants were detected only along the bottom of the surficial aquifer, suggesting an off-site, up-gradient source. Nonetheless the property owner elected to construct an iron-filled PTZ across the path of the plume, and to locate the PTZ within the greatest concentration of TCE.

The contaminant distribution with depth rendered this site an ideal candidate for hydraulic fracturing construction techniques. In contrast to trenching methods, hydraulic fractures can be easily deployed at depths exceeding 15 m. Furthermore, the absence of contamination in the upper 15 m of the site established a zone that would be unnecessarily disturbed by trenching but easily skipped by hydraulic fracturing.

The primary objective was to evaluate the performance of hydraulic fracturing as a PTZ construction technique. Consequently, the PTZ did not need to be constructed across the entire plume. Instead, a demonstration scale PTZ 15 m in length was created. The specific location of the PTZ was selected by reference to a contour map of TCE concentration. The PTZ location was completely encircled by TCE concentration greater than 700 µg/L.

The PTZ was constructed by injecting iron-laden slurry into five boreholes. The boreholes were arranged along the length of the PTZ, so the joining of five individual panels, which ranged in length from 2 m to 5 m formed the wall. The panels were actually V- or Y-shaped in plan in order to promote the intersection of the tips of adjacent panels. High-energy jets that were directed into the wall of the borehole established the orientation of the V or Y. The jet action eroded an oriented cavity or kerf in surrounding soil, and subsequent pressurization within the borehole propagated a fracture in the same orientation. The location and orientation of the panels is shown in Figure 3.

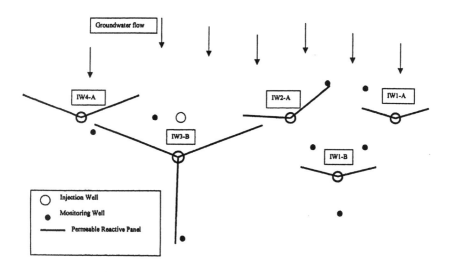

FIGURE 3. Plan view of monitoring and injection wells and estimated extend of permeable reactive panels that compose the PTZ.

Two techniques were used to install the boreholes and initiate the panels. Borehole IW1A was advanced by augering to 15 m and rotary drilling from 15 m to 24 m. A steel casing was pulled down along with the rotary bit. The casing was jacked out while injecting high-pressure (55 MPa) water through two nozzles that oriented to form a V at a 120-degreee angle between the slots. Soil cuttings were returned to the surface. Slots were cut for a height of 2 m. A packer was set to isolate the injection zone from the surface and iron-laden slurry was injected.

The second technique, which was used to construct the other four panels, was adapted because the first technique proved too time consuming. A nominal 8-inch (20 cm) hollow stem auger was used to drill all of the boreholes to depth and to set and grout a nominal 4-inch (10 cm) PVC liner in each hole. The high pressure (55 MPa) water jet was used to cut a 1-cm-wide slot through the PVC casing and surrounding grout. The jets were manipulated to cut slots 1.5 m high. The slots were oriented to form Vs with 120-degree angles or, as in the case of IW3B, to cut an Y, which had arms separated by 120 degrees. After cutting the slots, a packer was closed to isolate the injection interval from the surface and slurry was injected. A pad of neat guar gel (no iron), which was intended to dilate the fracture, was injected ahead of iron laden slurry.

Both techniques ultimately were successful in jet cutting and hydraulically fracturing sandy silt and silty sand soils at this site. Table 2 lists the quantities of material injected and the final injection pressure.

The linear extent and apparent thickness of the injected slurry panels 1A, 1B and 2A were measured using a radio-wave imaging method (RIM). The technique is attractive in that is gives a direct measure of panel thickness at preset ray-paths. However, it is somewhat limited by the fact that it can measure only along the direct line between observation wells. RIM demonstrated that the injected panels were approximately 5 to 7 cm thick in the middle and tapered to less than 2 cm above and below the injection slots. However, the panels were shown to extend a few meters away from the injection wells.

The combination of RIM measurements and volumetric data allowed the estimation of the extents of the panels as shown in Figure 3. Most impressive is the Y-shaped panel installed at IW3B. Assuming an average 7 cm thickness in each arm of the Y and a total slurry volume of 4.5 m^3, each arm of the Y was 6 m high and 3 m wide. The second most successful injection occurred at IW4A where approximately 1.5 m3 of slurry was injected to create a panel 7 cm thick, 6 m high and 3 m wide.

TABLE 2. Construction method and quantity of material injected into each panel. Construction method used either extracted steel casing (extracted) or fixed PVC casing (PVC)

Borehole	Construction Method	Volume of Slurry (m^3)	Final Injection Pressure (kPa)
IW1A	Extracted	0.2	7150
IW1B	PVC	0.43	6460
IW2A	PVC	0.70	2380
IW4A	PVC	1.5	2550
IW3B	PVC	4.5	4770

In summary, forty tons of Master Builder medium iron filings (-16+50 mesh) were injected into five panels, four of which were V shaped while the fifth was Y shaped. The five panels were arranged in a line to form a PTZ of suitable length. Radio wave imaging methods verified the orientation and size of individual panels. The geochemical performance of the PTZ has not yet been verified.

CONCLUSIONS

Hydraulic fractures have been used to improve the effectiveness of remediation for the almost 10 years, but recent projects have demonstrated some new applications. Hydraulic fractures filled with sand have been used to improve the delivery of Fenton-type oxidants, which rapidly degrade organic contaminants. The use of hydraulic fractures appears to have facilitated the complete degradation of BTEX contaminants in a source area beneath a former UST pit, and significant reduction of contaminant concentrations was observed several tens of m from the injection point. A control well was not used at this site, but experience at similar sites suggests that that distribution of oxidants would have been limited to a few ft from a conventional well.

Flat-lying hydraulic fractures filled with reactive material offer the possibility of creating in situ reactive barriers to intercept downward-moving contaminants in the vadose zone. A new hybrid technique that combines high energy jets and hydraulic fracturing can be used to create vertical layers of reactive material, which are applicable to the saturated zone. The hybrid technique was used at a site in Alabama to create a vertical reactive wall filled with zero-valent iron in an aquifer composed of granular sediments between 15 and 25 m bgs. The technique apparently created vertical layers of granular iron 2 to 7 cm thick that extended approximately 3 m on either side of the borehole, according to geophysical monitoring and mass balance calculations.

REFERENCES

Davis-Hoover, W. J., L. C. Murdoch, S. J. Vesper, H. R. Pahren, O. L. Sprockel, C. L. Chang, A. Hussain, and W. A. Ritschel. 1993. "Hydraulic Fracturing to Improve Nutrient and Oxygen Delivery for In Situ Bioreclamation." In R. E. Hinchee and R. F. Olfenbuttel (Eds), *In Situ Bioremediation: Application and Investigations for Hydrocarbon and Contaminated Site Remediation*, Butterworth-Heinemann, Stoneham, MA.

Murdoch, L. C., D. Wilson, K. V. Savage, W. W. Slack, and J. E. Uber. 1994. *Handbook of Alternative Methods for Delivery and Recovery*, US EPA EPA/625/R-94/003.

Siegrist, R. L., K. S. Lowe, L. C. Murdoch, T. L. Case, and D. A. Pickering. 1999. "In Situ Oxidation by Fracture Emplaced Reactive Solids." *Journal of Environmental Engineering*. 125(5): 429-440.

Stavnes, S. 1999. "Bioremdiation Barrier Emplaced Through Hydraulic Fracturing." In US EPA *Groundwater Currents*, Issue No. 31, March 1999.. http://www.epa.gov/swertio1/products/newsltrs/gwc/gwc0399.htm

Wolf, A. and L. C. Murdoch. 1993. "A Field Test of the Effect of sand Filled Hydraulic Fractures on Air Flow in Silty Clay Till." In *Proceedings of the 7th NGWA National Outdoor Action Conference*. National Ground Water Association, Dublin, OH.

FIELD-SCALE TEST OF AN INNOVATIVE PRB PASSIVE DRAIN

Thomas A Krug, Peter Dollar, Todd McAlary (Geosyntec Consultants, Guelph, Ontario, Canada), and Neil Davies (Geosyntec Consultants, Atlanta, GA)

ABSTRACT: A field-scale test of a Permeable Reactive Barrier (PRB) passive drain system (the "PRB system") was conducted at an aerospace manufacturing facility (the "Site") from August 1999 to January 2000. The PRB system was installed to treat trichloroethene (TCE) and its degradation products (collectively referred to as volatile organic compounds or VOCs) that were migrating in the shallow groundwater toward and into a storm sewer at the Site. The PRB system takes advantage of the hydraulic gradients induced by the existing storm sewer system to draw groundwater through a zone of zero-valent granular iron. The concentrations of VOCs in groundwater are reduced as groundwater moves through the granular iron and into a manufactured sump at the center of the PRB system. The field-scale test began in August of 1999 with the installation of the PRB system in an area of the Site with elevated concentrations of VOCs. The performance of this PRB system was monitored over a period of several months to evaluate the suitability of the design. The results of monitoring demonstrate the capability of the PRB system to significantly reduce the concentrations of VOCs in groundwater flowing into the storm sewer.

INTRODUCTION

This technical note describes a field-scale test (the "test") of an innovative permeable reactive barrier (PRB) passive drain system (the "PRB system") that was conducted at an aerospace manufacturing facility (the "Site"). The test involved the installation and monitoring of a PRB system in the vicinity of a former spill of trichloroethene (TCE). The PRB system takes advantage of the hydraulic gradients induced by the existing storm sewer system at the Site to draw groundwater containing dissolved phase volatile organic compounds (VOCs) through a zone of zero-valent granular iron so that VOC concentrations are reduced before entering the PRB system sump. Upon verification that the VOC reduction is sufficient for direct discharge to surface water, the treated groundwater would be allowed to flow directly into the storm drain. Installation of the PRB system involved the replacement of an existing storm sewer manhole with a pre-fabricated manhole structure surrounded by a circular ring of granular iron placed below the water table.

Objective. The overall objective of the test was to establish whether the PRB passive drain system could provide a cost effective remedy for treating groundwater migrating into storm sewers at the Site. Specific objectives were to: (i) design and install a PRB passive drain system; and (ii) collect sufficient monitoring data from the PRB system to characterize the groundwater residence time and VOC degradation rates to enable additional PRB systems to be designed and installed, if desired.

Site Characteristics. The Site is located on a river delta in an area that is slightly below sea level. In the 1960's a significant quantity of trichloroethene (TCE), a dense non-aqueous phase liquid (DNAPL), was reported to have been released to the subsurface. A groundwater extraction and treatment system was installed to remediate shallow groundwater. VOCs are, however, migrating in shallow groundwater toward and into the storm sewer that runs through the Site.

The regional geologic setting consists primarily of alluvial and deltaic deposits. The stratigraphy consists of a series of horizontal clay, silts and sand layers to a depth of greater than 300 meters below ground surface (m bgs). The shallow subsurface geology (approximately 0 to 13.5 m bgs) is characterized by surficial silty sands from 0 to 1.2 m bgs, a clay layer interbedded with silt, sand and peat lenses from 1.2 to 4.8 m bgs, and a silty clay grading to a more permeable silt and sand unit between 4.8 to 13.5 m bgs.

Groundwater and surface water at the Site drain into a canal which is below sea level that is pumped up to an adjacent canal above sea level. The water table is located approximately 0.9 to 1.2 m bgs in the vicinity of the PRB system. The storm sewers are buried about 1.8 to 2.4 m bgs and the elevation of water inside the storm sewers is maintained in the range of one meter lower than the ambient groundwater levels. Some sections of the storm sewers may be surrounded by backfill materials (sand or gravel) which may cause the groundwater to flow preferentially through this zone. It is not known how extensive the backfill may be or if different areas are hydraulically connected. The hydraulic conductivity of the native soil in the area of the PRB is approximately 1.6×10^{-4} centimeters per second (cm/s).

Groundwater containing VOCs flows into the storm sewers as a result of leaks in the sewer pipes and manholes. Due to the age of the storm sewer system, materials of construction, and site characteristics, this leakage is practically unavoidable. In an attempt to reduce inward leakage of VOCs, selected segments of the storm sewers were lined with an *InSituForm*™ liner system. However, because the seal between the liner and the old sewer pipe is imperfect, the influx of groundwater and VOCs into the sewer continues. In addition, significant leakage has been observed at the pipe/manhole connections. Assuming that groundwater will continue to flow toward the storm sewers, this test was designed to evaluate a PRB system capable of passively treating the groundwater to reduce VOC concentrations in the storm sewer to acceptable levels.

The subsurface investigations at the Site identified VOCs (primarily TCE and degradation products cis-1,2-dichloroethene [cDCE] and vinyl chloride [VC]) in soil and groundwater samples at the Site. A sample of groundwater leaking into a manhole at the same location as the PRB system was reported to contain concentrations of TCE, cDCE and VC of 22.5 milligram per liter (mg/L), 23.2 mg/L and 6.8 mg/L respectively. The significant concentrations of cDCE and VC demonstrate that natural attenuation processes are active in the subsurface at the Site.

PRB SYSTEM DESIGN AND INSTALLATION

The PRB system design included the replacement of an existing storm sewer manhole with a purpose-fabricated structure designed to act as a "groundwater sink" or collection point surrounded by a ring of granular iron. A cross section and a plan view of the general configuration of the PRB system are presented in Figures 1 and 2.

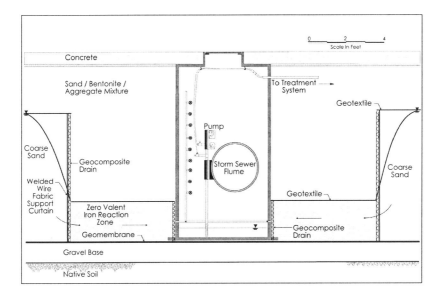

Figure 1. Cross Section of PRB Passive Drain System

The purpose-fabricated PRB system manhole at the center of the PRB system serves as a controlled collection point for groundwater containing VOCs by providing an inward hydraulic gradient toward the manhole. During the test, the infiltration rate of groundwater into the manhole was controlled by adjusting the flowrate of water pumped out of the PRB system manhole. The ring of permeable granular iron around the PRB system manhole creates geochemical conditions appropriate for the rapid reductive dechlorination of chlorinated VOCs in groundwater that flows through the granular iron and into the manhole. The PRB system also intercepts and collect groundwater that may be flowing along the preferential flow path created by any more permeable backfill material that could be present around the storm sewer pipes.

Storm sewer flow from upstream of the PRB system manhole is separated from groundwater flowing into the manhole to allow for monitoring of the flow of groundwater through the reactive granular iron separate from other storm sewer water. The separation of storm water from upstream of the PRB system manhole is achieved with an integral flume pipe in the PRB system manhole which allows storm sewer flow to pass through the PRB system manhole without mixing with groundwater infiltrating into the manhole.

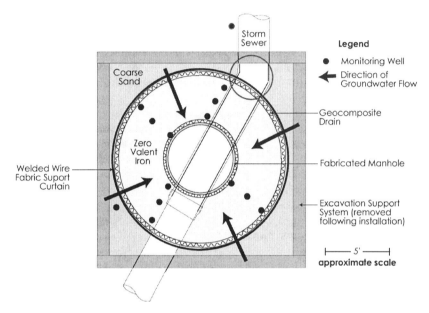

FIGURE 2. Plan View of PRB Passive Drain System

Three key elements of the design of the PRB system are: (i) the configuration (location, depth, etc.) of the PRB system; (ii) the characteristics of the granular iron in the PRB system; and (iii) the quantity of granular iron used in the PRB system. The configuration of thePRB system is determined by the desire to have groundwater flow through the granular iron and into the PRB system manhole. The configuration of the PRB system allowed the drawdown created by the PRB system manhole to be adjusted during the test if the groundwater flux into the manhole was different than anticipated by adjusting the rate that groundwater was pumped from the PRB system manhole. The characteristics of the granular iron in the PRB system determine the permeability, reactivity and other physical properties. The reactive granular iron consists of a zero-valent milled granular iron screened to be between –8 and +50 mesh (US Standard Sieve Mesh Numbers).

The amount of granular iron used in the PRB system design was calculated using the: (i) the estimated concentration of chlorinated VOCs in the groundwater flowing into the PRB system; (ii) the target treatment objectives for groundwater flowing into the manhole; (iii) the half lives for the target compounds (the reactivity of the granular iron); and (iv) the estimated volumetric flux of groundwater flowing through the granular iron into the manhole. The influent concentration of chlorinated VOCs used in the development of the design of the PRB system were based on the water concentrations observed in monitoring wells and manholes in the vicinity of the PRB system. The treatment objectives for the VOCs in groundwater flowing into the PCRW system manhole are based on the target concentrations for VOCs for the storm water collection system. The half lives for the degradation of chlorinated VOCs used in the development of the design of the PRB system were taken from a database of half lives for the degradation of chlorinated VOCs provided by EnviroMetal Technologies Inc. (ETI). The database of half lives was developed from an extensive database of bench scale studies on the degradation of chlorinated VOCs. The groundwater flux into the PRB system manhole was estimated based on data measuring the flux of groundwater into a manhole in the vicinity of the PRB system. The actual flux of groundwater entering the manhole structure was controlled by adjusting the rate that groundwater was pumped from the PRB system.

The residence time required for groundwater within the granular iron was calculated from the influent concentrations, the target concentrations, and the half lives to be 3.8 days. The quantity of granular iron to be used in the PRB system was estimated based on the residence time, the groundwater flux through the PRB system and the porosity of the granular iron using the following equation:

$$V = \frac{Qt}{n}$$

where: V = volume of granular iron   Q = flow rate
       t = residence time   n = porosity

Using the above equation, it was calculated that 10.2 cubic meters (m^3) of granular iron would be required around the manhole, assuming a flow rate ("Q") of 0.946 liters per minute (Lpm), residence time ("t") of 3.8 days and a porosity ("n") of 0.5.

A vertical thickness of 0.61 m was selected for the granular iron in the ring of granular iron around the manhole based on the expected water table elevations, sewer invert elevations and practical space constraints. The diameter of the PRB system manhole is 1.5 m. The outside diameter of the reactive granular iron ring around the manhole was calculated to be 4.8 m.

A total of twelve monitoring wells were installed within the iron filings material to allow for performance monitoring of the PRB system. The locations of the monitoring wells are shown in Figure 2.

The PRB system was installed in August of 1999. Steel bracing was used to stabilize the excavation while the manhole, granular iron and other system components were installed. Figure 3 shows a construction worker in the excavation for the PRB system getting ready to place one of the large bags of granular iron.

SCOPE OF OPERATION AND MONITORING

This Section describes the scope of the operation and monitoring activities that were conducted during the test to demonstrate the performance of the PRB system.

A tracer test was conducted to characterize the flow conditions in the PRB system, with particular emphasis on identifying preferential flowpaths for groundwater moving into the manhole at the center of the PRB system. A bromide tracer was added to water that was used to saturate the granular iron following the placement of the iron layer. After the placement and saturation of the iron the control valves were opened and the PRB system manhole sump pump was operated to initiate flow through the PRB system and to begin the tracer test. Groundwater samples were collected from monitoring points in the PRB system after the tracer test began. Bromide measurements were conducted in the field using an ion specific probe.

FIGURE 3. Installation of PRB Passive Drain System

During the tracer test and subsequent operation of the PRB system, the sump pump in the PRB system manhole was operated at a specified flowrate. The total flow pumped from the manhole was monitored to confirm the volumetric flow rate or flux of groundwater into the manhole. The water from the manhole was pumped to the feed tank of an existing air stripper for treatment until monitoring of the water quality determines that the water flowing through the PRB system will satisfy water quality requirements. The only operating equipment associated with the PRB system was the sump pump used to transfer water from the manhole to the air stripper and the flowmeter on this discharge line.

Groundwater monitoring activities conducted during the test included: (i) water level measurements and monitoring the flow rate of groundwater pumped from the PRB system manhole to evaluate the flow conditions in the PRB system; (ii) groundwater sampling and analysis for VOCs to assess degradation rates; and (iii) groundwater sampling and analysis for general groundwater quality parameters to evaluate changes in the geochemical conditions of the groundwater.

MONITORING RESULTS

The tracer test results and results of groundwater level measurements showed that there was considerable variability in the flow of groundwater through the PRB system along the 4 different transects. The data suggest that groundwater does not flow in a simple radial pattern from the outside to the inside of the granular iron in the PRB system.

The weekly average flowrate through the PRB system for weeks 1 to 9 ranged between 0.8 and 1.1 Lpm. The average flowrate for week 10 was 0.53 Lpm but following week 10, the flow was increased and the average flowrates for weeks 11, 12, 13, 14 and 16 ranged from 0.72 to 1.0 Lpm.

Figure 4a, 4b, and 4c show the concentrations of TCE, cDCE, and vinyl chloride in untreated and treated groundwater samples collected during each monitoring event during the test period. The "untreated" concentrations are the average of the concentrations measured in samples from each of the four "A" monitoring wells in the outside edge of the PRB system. The "treated" concentrations are the concentrations measured in groundwater samples collected from inside the manhole at the center of the PRB system and represent the combined flow of treated groundwater into the manhole. The results show increasing concentrations of TCE, cDCE, and VC in the untreated groundwater between weeks 4 and 8 and variable concentrations during the remainder of the test period. The results show consistent reductions in TCE, cDCE, and VC in samples of treated groundwater collected from the PRB system manhole. Based on the average of the untreated and treated concentrations of TCE, cDCE, and VC during the monitoring events, the PRB system provided a reduction in the concentration of TCE of 79%, of cDCE of 77%; and of VC of 89%.

DISCUSSION

The results of monitoring of the PRB passive drain system demonstrate that the system can provide for a significant reduction in the concentration of chlorinated VOCs leaking into the storm sewer at the Site. Additional analysis of the data from test in ongoing.

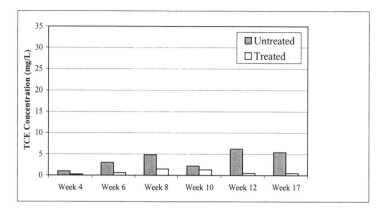

FIGURE 4a. Untreated and Treated Groundwater Trichloroethene (TCE) Concentrations

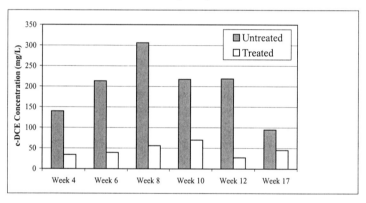

FIGURE 4b. Untreated and Treated Groundwater cis-1,2-Dichloroethene (c-DCE) Concentrations

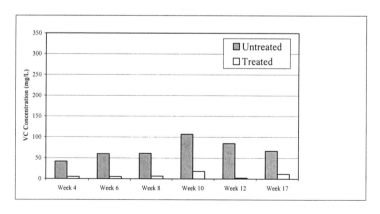

FIGURE 4c. Untreated and Treated Groundwater Vinyl Chloride (VC) Concentrations

DEEP REACTIVE BARRIERS FOR REMEDIATION
OF VOCs AND HEAVY METALS

Grant Hocking, Samuel L. Wells and Rafael I. Ospina
Golder Sierra LLC, Atlanta, GA, USA.

ABSTRACT: Azimuth controlled vertical hydraulic fracturing technology has constructed full scale in situ iron reactive permeable barriers at moderate to significant depth for remediation of groundwater contaminated with chlorinated hydrocarbons and metals. Zero valent iron reactive permeable barriers have been installed to remediate chlorinated solvent contaminated groundwater by abiotic degradation of the halogenated volatile organic compounds into harmless daughter products. Alternatively iron and other reactive materials have been used for precipitation, sorption or reduction of various groundwater metal contaminants. The azimuth controlled vertical hydraulic fracturing technology has constructed iron reactive barriers up to nine (9) inches in thickness over hundreds of feet in length and down to depths greater than 120 feet. The injection procedure and gel mixture chemistry ensures the barrier is constructed of near uniform thickness, of high permeability and porosity, with minimal impact on groundwater flow regimes, with minimal site disturbance and optimal iron degradation potential.

INTRODUCTION

Zero valent metals abiotically degrade certain compounds; such as, pesticides as described by Sweeny and Fisher (1972), and halogenated compounds as detailed in Gillham and O'Hannesin (1994). The abiotic reduction of trichloroethene (TCE), tetrachloroethene (PCE), vinyl chloride (VC) and isomers of dichlorethene (DCE) by zero valent iron metal is shown on Figure 1, with ethene and ethane being the final carbon containing daughter compounds (Sivavec and Horney, 1995; Orth and Gillham, 1996). The prime degradation pathway of TCE in the presence of iron is via chloroacetylene and acetylene to ethene and ethane, and only a small proportion < 5% (Orth and Gillham, 1996; Sivavec et al, 1997) to the less chlorinated hydrocarbons. The volatile organic compounds (VOCs) degraded by zero valent iron are listed in Table 1. The abiotic degradation of most of these compounds in the presence of iron can be approximated by a first order reduction process.

Certain metals, such as hexavalent chromium are reduced and thus precipitate in the presence of iron; whereas other metals are directly precipitated or absorbed by the iron and thus rendered immobile. Metals that can be removed from the groundwater flow regime in the presence of iron, include Al, Sb, As, Cd, Cu, Cr(VI), Pb, Mg, Hg, Ni, Se, Tc-99, U, V and Zn. A number of workers have constructed iron reactive barriers for the removal of metals, e.g. Gu et al (1998), Morrison (1998), Naftz (1998), Puls (1998), Su and Puls (1998), and in some cases a combination of metals and VOCs, Schlicker et al (1998) and Puls (1998).

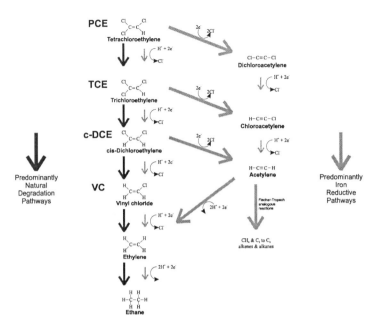

FIGURE 1. Degradation Pathways of Chloroethene Compounds.

TABLE 1. Chlorinated Compounds Abiotically Reduced by Iron.

Common Name	Common Abbreviation	Other Pseudonyms	CAS Number
Methanes			
Tetrachloromethane	CT, PCM	Carbon Tetrachloride	56-23-5
Trichloromethane	TCM	Chloroform	67-66-3
Tribromomethane	TBM	Bromoform	75-25-2
Ethanes			
Hexachloroethane	HCA	Carbon Hexachloride	67-72-1
1,1,1,2-Tetrachloroethane	1,1,1,2-TeCA		630-20-6
1,1,2,2-Tetrachloroethane	1,1,2,2-TeCA	Acetylene Tetrachloride	79-34-5
1,1,1-Trichloroethane	1,1,1-TCA	Methyl Chloroform	71-55-6
1,1,2-Trichloroethane	1,1,2-TCA	Vinyl Trichloride	79-00-5
1,1-Dichloroethane	1,1-DCA		75-34-3
Ethenes			
Tetrachloroethene	PCE	Perchloroethylene	127-18-4
Trichloroethene	TCE	Ethylene Trichloride	79-01-6
cis 1,2-Dichloroethene	cis 1,2-DCE	cis 1,2-Dichloroethylene	540-59-0
trans-1,2-Dichloroethene	trans 1,2-DCE		540-59-0
1,1-Dichloroethene	1,1-DCE	Vinylidene Chloride	75-35-4
Vinyl Chloride	VC	Chloroethene	75-01-4
Propanes			
1,2,3-Trichloropropane	1,2,3-TCP	Allyl Trichloride	96-18-4
1,2-Dichloropropane	1,2-DCP	Propylene Dichloride	78-87-5
Other Chlorinated			
N-Nitrosodimethylamine	NDMA	Dimethylnitrosamine	62-75-9
Dibromochloropropane	DBCP		96-12-8
Lindane		Benzene Hexachloride	58-89-9
1,1,2-Trichlorotrifluoroethane		Freon 113	76-13-1
Trichlorofluoromethane		Freon 11	75-69-4
1,2-Dibromoethane	1,2-EDB	Ethylene Dibromide	106-93-4

CONSTRUCTION METHOD

The azimuth controlled hydraulic fracturing technology can construct in unconsolidated sediments 1) vertical fractures at the required azimuth or bearing, 2) continuous coalesced fractures by injection in multiple well heads, and 3) thick fractures, by a process of tip screen out or multiple fracture initiations. The technology, Hocking et al (1998a & b), involves initiating the fracture at the correct orientation at depth and by controlled injection a continuous reactive barrier is created, see Figure 2. The hydraulic fracture reactive permeable barrier is constructed by injecting through multiple well heads spaced along the barrier alignment. A special down hole tool is inserted into each well and a controlled vertical fracture is initiated at the required azimuth orientation and depth. Upon initiation of the controlled fracture, multiple well heads are then injected with the iron gel mixture to form a continuous permeable iron reactive barrier.

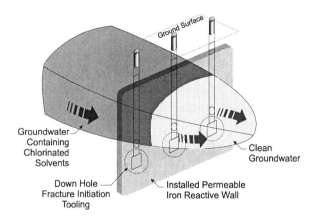

FIGURE 2. Hydraulic Fracture Iron Reactive Permeable Barrier.

The gel is injected into the formation and carries the iron filings to the extremes of the fracture. The gel is a water based cross link gel, hydroxypropylguar (HPG), which is a natural polymer used in the food industry as a thickener. HPG is used in the process because it has minimal impact on the iron's reactivity and upon degradation leaves an extremely low residue. The gel is water soluble in the uncross linked state, and water insoluble in the cross linked state. Cross linked, the gel can be extremely viscous, ensuring the iron filings remain suspended during the installation of the barrier. An enzyme breaker is added during the initial mixing to controllably degrade the viscous cross linked gel down to water and sugars, leaving a permeable iron reactive barrier in place.

With a single initiation and fracture injection, the typical thickness of the reactive barrier constructed by vertical hydraulic fracturing is 3" to 3.5" in dense sand and gravel formations. The ultimate thickness of the fracture is controlled by the formation stiffness and breakout pressure. In soft sediments, 8.5" thick fractures have been constructed in a single injection, see Figure 3. In dense compacted sediments, thick fractures can not be achieved in a single injection;

however, by multiple fracture initiations and injections, thicker fractures up to 9" thick can be constructed as shown on Figure 3.

 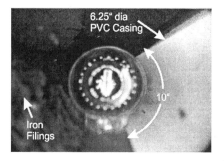

FIGURE 3. Thick Vertical Fractures in Soft and Stiff Sediments.

The iron reactive barrier installation is monitored in real time during injection to determine it's geometrical extent and to ensure fracture coalescence or overlap occurs. The quantities of iron reactive mixture injected are continuously monitored to ensure sufficient reactive iron is injected through the individual well heads. During injection, the iron gel mix is electrically energized with a low voltage 100 Hz signal. Down hole resistivity receivers are monitored to record the in phase induced voltage by the propagating fracture, see Figure 4, and utilizing an incremental inverse integral model, the fracture fluid geometry can be quantified during the installation process.

Active resistivity monitoring has the added benefit of determining when the individual fractures coalesce and thus become electrically connected. That is, by energizing each injected well head individually and in unison, the fracture electrical coalescence is clearly recorded. The imaging and inversion of the down hole resistivity data focuses on quantifying the continuity of the reactive barrier. Such monitoring enables construction procedures to be modified if necessary to ensure the barrier is installed as planned.

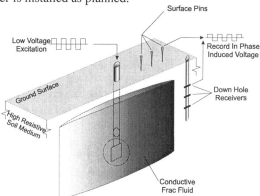

FIGURE 4. Fracture Mapping by Active Resistivity.

IRON REACTIVE BARRIER CONSTRUCTION

A former manufacturing facility in South-Central Iowa was contaminated with trichloroethene (TCE) in the soil and groundwater. Groundwater concentrations of TCE were detected up to levels of 14,000 μg/kg. The record of decision (ROD) was modified to an enhanced soil vapor extraction (SVE) system in the vadose zone and an in situ iron reactive permeable barrier for groundwater remediation. The iron reactive permeable barrier was installed by the azimuth controlled vertical fracturing technology. The remnant plume downgradient of the reactive barrier is expected to be in situ bio-remediated by the natural attenuation mechanisms at the site.

The site consists of medium to fine channel sands overlain by an over consolidated stiff to very stiff till. The iron reactive barrier system was constructed perpendicular to the groundwater flow direction and intercepted channel sands characterized as loose flowing sands with a permeability of approximately 1 Darcy. The iron reactive barrier is a source control barrier 240 feet long installed from a depth of 25 feet down to a total depth of 75 feet below ground surface with an average thickness of 3 inches. A plan view of the iron reactive barrier is shown on Figure 5, along with the hydraulic fracturing construction and instrumentation equipment.

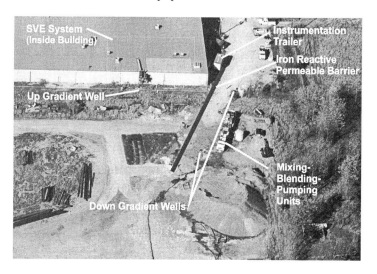

FIGURE 5. Plan View of Iron Reactive Permeable Barrier.

The geometry of the reactive barrier was quantified during injection of the barrier by the active resistivity method. Down hole resistivitiy receiver locations are shown as rectangles in cross-section on Figure 6 and in plan on Figure 6 as RR1 through RR7 and attached to the casings of monitoring wells GW-1 and GW-2. The resistivity receivers consist of copper collars attached to cables that are multiplexed to the instrumentation data acquisition system. The wells F1 through F16, shown on Figure 6 in both plan and section, are the fracture injection wells. An image of the injection of frac well F16 is shown on Figure 6, with the frac geometry determined by the measured induced voltages at the down hole receivers locations. All of the fracture injections at this site were recorded and their geometries delineated by active resistivity.

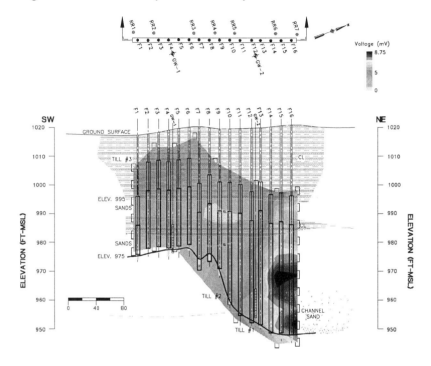

FIGURE 6. Induced Voltages from Propagating Fracture in Well F16.

The cross-section of the constructed iron reactive permeable barrier is shown on Figure 7. The barrier was keyed into the underlying till units #1 and #2 and extended to the lower surface of the upper till unit #3. The total cross-sectional area of the barrier was approximately 7,050 square feet. Due to the low groundwater flow velocity, the in situ reactive barrier has the capacity to degrade extremely high concentrations of TCE to below the MCL level. Of particular importance in selecting the remedy was that the reactive barrier system is complimentary and enhances the natural attenuation mechanisms active at the site. The reactive barrier was completed in October 1999 and recent groundwater down gradient monitoring data shows appreciable decline in TCE concentrations, with no daughter products detected, immediately down gradient of the barrier.

CONCLUSIONS
Permeable reactive barriers are suitable cost effective remedies for contaminated groundwater, both for plume remediation and as source control. Iron permeable reactive barriers have been most efficient in dehalogenating chlorinated

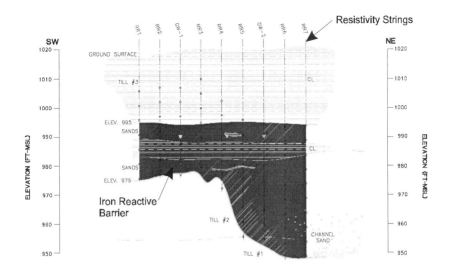

FIGURE 7. Cross-Section of Final Constructed Reactive Barrier.

solvents in groundwater and immobilizing metals and are a viable cost effective alternative to pump and treat. The iron reactive barrier system compliments and enhances natural attenuation mechanisms active at the site. The design, construction and performance monitoring of in situ reactive permeable barriers warrant special attention due to the functional design requirements of the systems and the low piezometric gradients across such systems. Particular attention needs to be paid to skin effects, and construction technique on iron reactivity and barrier permeability and porosity.

Azimuth controlled vertical hydraulic fracturing technology has placed permeable iron reactive barriers of moderate thickness up to 9" in highly permeable sands and gravel down to significant depths. The thicker reactive barriers are constructed by multiple initiations and re-injections, following the breaking of earlier injected gel. The real time monitoring of the injected geometry and materials provide the quality control and assurance required for construction of such systems.

The prime benefits of the fracturing installation method are cost savings over alternate installation techniques, flexibility to accommodate depth and thickness requirements, minimal site disturbance to overlying confining units and groundwater flow regimes, ability to be retrofitted if necessary, minimal waste volumes generated and deep application of the technology

REFERENCES

Gillham, R. W. and S. F. O'Hannesin. 1994. "Enhanced Degradation of Halogenated Aliphatics by Zero-Valent Iron", *Ground Water*, Vol. 32, No. 6, pp958-967.

Gu, B. D. Watson, W. Goldberg, M. A. Bogle and D. Allred. 1998. "Reactive Barriers for the Retention and Removal of Uranium, Technetium, and Nitrate in Groundwater", RTDF Meeting, Beaverton, OR, April 15-16.

Hocking, G., S. L. Wells, and R. I. Ospina. 1998a. "Field Performance of Vertical Hydraulic Fracture Placed Iron Reactive Permeable Barriers", Emerging Remediation Technologies for Soil and Groundwater Cleanup, Florida Remediation Conf., Orlando, FL, November 10-11.

Hocking, G., S. L. Wells, and R. I. Ospina. 1998b. "Performance of the Iron Reactive Permeable Barrier at Caldwell Superfund Site", RTDF Meeting, Oak Ridge, TN, November 17-18.

Morrison, S. J. 1998. "Status of a Permeable Reactive Barrier Project for Uranium Containment at Monticello, Utah", RTDF Meeting, Beaverton, OR, April 15-16.

Naftz, D. 1998. "Status and Preliminary Results of the Fry Canyon Reactive Chemical Wall Project", RTDF Meeting, Beaverton, OR, April 15-16.

Orth, S. and R. W. Gillham. 1996. "Dehalogenation of Trichlorethene in the Presence of Zero-Valent Iron", *Environ. Sci. Technol.*, Vol. 30, pp66-71.

Puls, R. W. 1998. "Elizabeth City, North Carolina Permeable Reactive Barrier Site Update", RTDF Meeting, Oak Ridge, TN, November 17-19.

Schlicker, O., M. Ebert, R. Kober, W. Wust and A. Dahmke. 1998. "The effect of competing chromate and nitrate reduction on the degradation of TCE with granular iron", RTDF Meeting, Oak Ridge, TN, November 17-19.

Sivavec, T. M. and D. P. Horney. 1995. "Reductive Dechlorination of Chlorinated Ethenes by Iron Metal", Proc. 209th. American Chemical Society National Meeting, Vol. 35, No. 1, pp695-698.

Sivavec, T. M., P. D. Mackenzie, D. P. Horney and S. S. Bagel. 1997. "Redox-active Media Selection for Permeable Reactive Barriers", Int. Containment Conf., St. Petersburg, FL, February 10-12.

Su, C. and R. W. Puls. 1998. "Retention of Arsenic by Elemental Iron and Iron Oxides", RTDF Meeting, Oak Ridge, TN, November 17-19.

Sweeny, K. H. and J. R. Fisher. 1972. "Reductive Degradation of Halogenated Pesticides", U.S. Patent No. 3,640,821.

FUNNEL-AND-GATE AT A FORMER MANUFACTURED GAS PLANT SITE IN KARLSRUHE, GERMANY: DESIGN AND CONSTRUCTION

Dr. Hermann Schad, IMES Gmbh, Amtzell, Germany
Dr.-Ing. Bertram Schulze, ARCADIS Trischler und Partner, Freiburg, Germany

ABSTRACT: A funnel-and-gate system will be installed at the former manufactured gas plant site in Karlsruhe, Germany for long-term remediation of a groundwater contamination by polycyclic aromatic hydrocarbons. The system will consist of a 240 m long and 17 m deep funnel and eight gates which will be constructed via large diameter borings. The technical design of the system is based on a specific site investigation program including hydraulic and geophysical site characterization methods, column tests at the site under quasi in situ conditions in order to simulate the decontamination process in the gates and numerical groundwater flow modeling. The funnel is planned to be installed as a sheet pile wall. In approximately equidistant gaps large diameter borings will be lowered into which the gate tubes will be introduced. After filling in pea gravel at the groundwater inflow and outflow areas and slurry towards the funnel the outer casing will be pulled out again and the gates will be connected to the funnel segments. Approximately 10 l/s contaminated groundwater will be decontaminated with about 150 tons of activated carbon, for which regenerations cycles between 5 and 15 years, depending on the concentration of the contaminants, are expected.

INTRODUCTION

Groundwater contamination by polycyclic aromatic hydrocarbons (PAHs) is typical for many former manufactured gas plant sites. Most of the PAHs are very persistent in the subsurface, i.e. they are still present in high concentrations many decades after the contamination occurred and cannot be removed from the subsurface within a reasonable period of time by pump-and-treat. This persistence above all is caused by slow dissolution kinetics of the compounds from non-aqueous phase liquids, slow diffusion of the contaminants from low permeability zones (in which the pollutants have accumulated over decades) or resistant adsorption of the contaminants by the aquifer material. Fast remediation of such contaminations is only possible by excavating the contaminated soil. Since protection of groundwater resources downgradient from the contaminated area is the remediation goal, in-situ treatment may be focused on the plume rather than on the source. This can be achieved using the concept of permeable reactive barriers. Within the reactive zone the pollutants can either be degraded, sorbed or precipitated through biotic or abiotic processes.

GENERAL SITE DESCRIPTION

The former manufactured gas plant site of the city of Karlsruhe is located in the Rhine valley and covers an area of approximately 100.000 m². The aquifer has a thickness of about 12 m and consists of mostly sandy gravel which is underlain by a clay layer at a depth of 16 m below surface. The contamination of the site is dominated by PAHs with Acenaphthene (up to 600 µg/l) being the highest concentrated compound in the plume extending about 400 m downgradient from the site (FIG. 1). Several infiltration hot spots of dense non-aqueous phase liquids (DNAPLs) were located within the saturated zone at the site. Within the saturated zone these separate phase liquids will be dissolved over a period of at least several decades of years. The groundwater flow rate from the contaminated and highly permeable site amounts to about 12 l/s under natural conditions.

Based on the site investigation and a technical and economical evaluation of several remediation technologies funnel-and-gate was recommended as the most favourable remediation technique for that site.

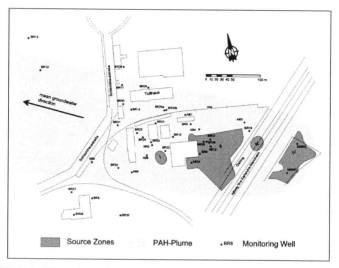

FIGURE 1. Plan view showing the site and the plume of contaminated groundwater emerging from the infiltration areas

REMEDIATION CONCEPT

The basic concept for long-term remediation of the site comprises passive in-situ groundwater treatment through a funnel-and-gate system down-gradient of the infiltration areas. The system will consist of a 240 m long funnel and a number of gates, where the contaminants will be removed from the groundwater. Due to high iron concentrations in the contaminated groundwater and in order to avoid microbial growth the anaerobic conditions prevailing in the aquifer shall be maintained in the gates. Regeneration cycles of the "reactive" sorptive material in the gates shall last at least several years. The hydraulic behaviour of the funnel-and-gate system must be such that the plume will be captured by the system for all relevant hydrological conditions.

In order to achieve these goals a comprehensive investigation program including hydraulic and geophysical field studies, numerical flow modelling, laboratory and field testing of different sorption materials was performed. Moreover, different construction techniques were evaluated with regard to their applicability at the site.

HYDRAULIC DELINEATION OF THE FUNNEL-AND-GATE SYSTEM

Based on a hydraulic and a geostatistical characterization of the aquifer a numerical flow model was used to design the funnel-and-gate system at the site. The hydrogeological aquifer characterization included (1) a number of pump tests in order to determine the aquifer transmissivity and its spatial variability, (2) flowmeter measurements to determine the variability of the hydraulic conductivity in vertical direction, (3) seismic tomography along two vertical profiles in order to determine heterogeneity structures of the subsurface, and (4) a statistical and geostatistical data evaluation.

A mean hydraulic conductivity (K) of 3.9×10^{-3} m/s was determined. Within the hot spot areas the hydraulic conductivity is lowered by about one order of magnitude. The variance of the log hydraulic conductivity values, determined from the flowmeter measurements amounts to 1.31. Aquifer heterogeneities could be characterized geostatistically with correlation lengths of 4 and 8 m in horizontal and 1.5 m in vertical direction. All this information was included in a flow model using a stochastic modeling approach. The generation of parameter fields was based on the assumption of "gaussian" properties of the log(K) field and was accomplished using the public domain simulation code SGSIM (Deutsch and Journel, 1992).

Model domain extensions of 2500 and 2300 m in horizontal directions and 20 m in vertical direction were applied. A finite difference discretization with grid spacings between 0.6 and 250 m was used. Whereas the central 450 x 500 m (see FIG. 2), including the site area, were modeled as a heterogeneous parameter field, an homogeneous K-value was used for the surrounding area. This was determined as the effective K-value of the heterogeneous model area. A satisfactory hydraulic equivalence of the heterogeneous and the homogeneous model areas was found for

$$K_{homogeneous} = 0.75 \times K_{arithmetic,\ heterogeneous} \quad (1)$$

Where $K_{arithmetic,\ heterogeneous}$ = arithmetic mean of the heterogeneous K-field
$K_{homogeneous}$ = effective mean hydraulic conductivity of the heterogeneous K-field.

A calibration of the heterogeneous parameter field was achieved by varying different kriging techniques and adapting the mean of the log (K) distribution in order to match the drawdown behaviour of experimental and numerical pump tests. Adapting the mean of the log(K) distribution was necessary since the point K-values derived from the flowmeter profiles were based on effective transmissivity values determined from pump test data.

Due to the lack of natural aquifer boundaries in the near vicinity of the site area, arbitrary boundary conditions had to be defined. Based on the determined K-value distribution and the measured hydraulic head values, groundwater flow rates into and out of the model domain were determined along all boundaries (s.

FIG. 2, left). These flow rates were applied for simulating the hydraulic effect of the funnel-and-gate system (s. FIG. 2, right). This procedure was applied for different hydrologic groundwater flow conditions (direction of flow and hydraulic gradient).

For each hydrologic scenario the length and the orientation of the funnel and the number of gates were varied. For each flow simulation the following parameters were determined from the model results:

- groundwater flow rate across the funnel area under natural conditions
- capture zone of the funnel-and-gate system,
- flow rates through all gates,
- variability of the flow rate along the vertical profile of the gates.

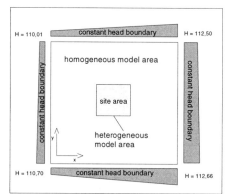 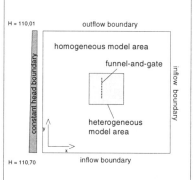

FIGURE 2. Model boundaries for different steps of the modeling task: determination of the flow rates into and out of the model domain under natural conditions (left) and application of these flow rates for the simulation of the funnel-and-gate system (right)

For a funnel-and-gate system (s. FIG. 3) consisting of 240 m funnel and eight gates, each of them 1,8 m wide and 12 m high, a total flow rate through all gates of approx. 10 l/s was determined. All contamination sources are located well within the capture zone of the system. Groundwater flow lines are almost parallel indicating that the effective conductivity of the entire funnel-and-gate system is close to that of the aquifer. In vertical direction for 2 m segments of the gates the flow rate varies between 15 and 30%. For a six-gate scenario the width of the capture zone reduces by about 45 m (20%) at the eastern part of the site, however all source areas would still be located within the capture zone. For only four gates this would be no more the case.

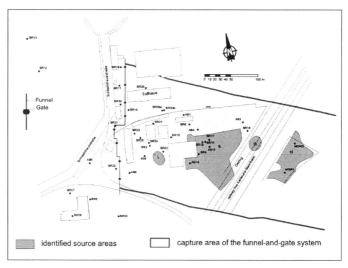

**FIGURE 3. Modeled capture area of funnel-and-gate
system with 240 m funnel and eight gates**

SORPTION OF THE CONTAMINANTS ON ACTIVATED CARBON

The adsorption of hydrophobic organic contaminants from the aqueous phase on hydrophobic surfaces generally increases with decreasing solubility of the compound (or increasing octanol/water partition coefficient, K_{OW}). Activated carbon (AC) therefore may be used for passive removal of PAHs which are the key problem at many former manufactured gas plant sites. For successful use in a funnel-and-gate system permeability and sorptive properties of the adsorptive wall material (adsorbent) must be optimized. Both the permeability and the sorption rates depend on the grain size of the adsorbent, the permeability increasing with increasing grain size, and the sorption rates decreasing with increasing grain size squared. Moreover no decrease in permeability or chemo-/biofouling of the adsorbent due to competitive adsorption of dissolved organic matter or the growth of a biofilm which may plug adsorbent pores should occur.

In order to identify the technologically and the economically relevant parameters for the selection of the adsorbent several experiments were performed including the determination of adsorption isotherms in the laboratory for different types of AC and long-term column tests at the site under quasi in situ conditions.

From the equilibrium adsorption isotherms it becomes evident that for areas with low PAH-concentrations a different type of AC will be most cost effective than for areas with high PAH-concentrations. This is illustrated in FIGURE 4 for five different types of AC.

The hydraulic conductivity of different types of AC was found to depend on the grain size distribution. The highest values were determined for the AC types D43/1 and TL830 (0.5 – 1 x 10^{-2} m/s). For both types there was no significant change in the hydraulic conductivity over a period of six months column testing at the site. During that time a 3100-fold exchange of the pore volume of the AC filled columns (1.6 m length and 0.2 to 0.4 m diameter) was achieved. From the concentration profiles along the columns it was concluded, that a specific filter capacity of 18 m³/kg can be expected for the central (highly contaminated) part of the plume.

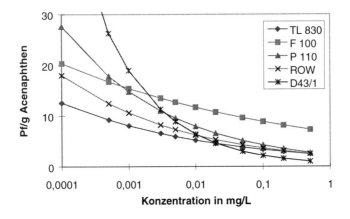

FIGURE 4. Specific costs (100 Pf = 1 DM) for different types of AC and different acenaphthene concentrations of the aqueous phase

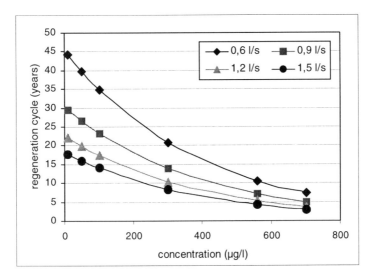

FIGURE 5. Calculated duration of regeneration cycles for the AC type TL830 (35 m³) for different flow rates at the gates and different inflow concentration of acenaphthene

Based on this result the length of regeneration cycles of the AC type TL830 was estimated using the Adsorption Simulation Software of the Michigan Technological University Houghton (PSDM – pore and surface diffusion model) for different Acenaphthene concentrations (s. FIG. 5).

A total of approximately 300 m³ AC will be filled into the eight gates. At the center of the plume with acenaphthene concentrations of 400 to 600 µg/l and benzene concentrations of up to 50 µg/l regeneration cycles of more than five years can be expected, whereas to both sides regeneration cycles could be in the range of 15 or more years.

Although no significant change of the dissolved inorganic constituents of the groundwater passing the test columns could be observed, over long periods iron precipitation could become a problem within the gates. Also long-term effects of chemofouling of the AC can not be excluded. During the six month column testing period no significant bacteria growth could be detected on the adsorbent. Sulfate and other redox susceptive parameters remained unchanged.

CONSTRUCTION OF THE FUNNEL-AND-GATE SYSTEM

The funnel-and-gate system will consist of an approximately 240 m long funnel and eight gates. The funnel will be constructed using sheet piles down to a maximum depth of 17 m. The aquifer material, consisting of mostly sandy gravel, was found to be densely bedded. It is underlain by a clay layer into which the funnel will key in. The sheet piles will be pressed rather than driven into the ground in order to prevent damages to buildings listed for preservation or gas supply pipelines in the near vicinity of the funnel. In a pilot test, this technique proved to be applicable at the site.

After installing a continuous funnel a total of eight "small" gates will be constructed. At each of the gate locations a few sheet piles have to be pulled out again. Within these gaps large diameter (d=2,5 m) borings will be excavated. The gate tubes consisting of perforated steel or HDPE (d=2,0 m) will then be introduced. After completing the gates (gravel pack, clay seal. etc.) the casing has to be removed in order to connect the funnel segments with the gates. This step marks one of the major challenges of the project. At each side of the gate one sheet pile will be driven into the ground connecting the funnel with the low permeable slurry seal. FIGURE 6 shows a schematic plan view of the gates.

The system will be constructed during summer 2000. The installation of the funnel will be finished within four to six weeks. For each gate a construction period of three weeks is planned. After installing the gates the adsorbent will be filled in.

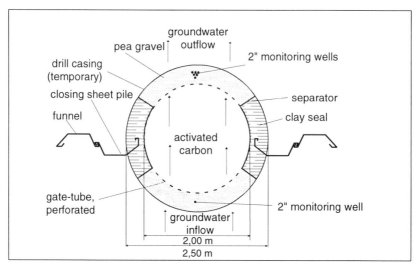

FIGURE 6. Plan view of the gates

MONITORING

The monitoring program will include hydraulic head measurements and groundwater sampling. In addition to the 2" monitoring wells within the pea gravel of the gates a 6" monitoring well will be installed five to ten meters downgradient of each gate. At this distance however, due to desorption processes in the aquifer it might last several months until the PAH-concentration will fall short of the maximum allowed concentration (0.15 µg/l). Sampling campaigns will be performed every month during the first year. For long-term monitoring sampling intervals of six months appear to be appropriate.

Acknowledgement: The project is financed by the Stadtwerke Karlsruhe GmbH. Funding was granted by the state of Baden-Württemberg.

REFERENCES

Deutsch, C.V. and A.G. Journel. 1992. *"GSLIB Geostatistical Software Library and User's Guide"*, Oxford University Press, New York.

BIOTIC ATTENUATION AND ZERO-VALENT IRON PERMEABLE BARRIER TECHNOLOGY

Hala A. Sfeir, A. Randall, D. Reinhart, M. Chopra, C. Clausen and C. Geiger
(University of Central Florida, Orlando, USA)

ABSTRACT: Glass columns (0.1 m I.D., 1 m in length) were constructed to provide direct comparison of zero-valent iron material with and without simultaneous biological reductive dechlorination such as might occur in an aquifer with significant biotic natural attenuation. The columns were filled with abiotic concrete sand (AS), biotic sand (BS), abiotic sand and 20 % by wt. peerless iron (AI), and biotic sand and 20 % by wt. peerless iron (BI). Approximately 600 g of biologically active soil from a TCE contaminated site that had significant reductive dechlorination of TCE to cis-DCE was added to each column. The influent was a deoxygenated groundwater with TCE added. Tracer studies were also done to confirm hydraulic retention times, porosity, and flow regimes. The half-life of TCE as a function of pore volume was calculated for BI, AI, and BS columns. The presence of microbial organisms and iron in the BI column seems to have had a positive effect on the destruction of TCE with an average half life of 0.22 days relative to both the abiotic iron column, with an average half life 0.43 days, and the biotic column containing no iron, with an average half life 0.68 days.

INTRODUCTION

Halogenated solvents are among the most commonly found groundwater contaminants in the world (Blowes *et al.*, 1995). Due to the limitations of pump-and-treat methods for groundwater remediation (NAS, 1994), considerable attention has now turned to the use of in situ permeable treatment walls (PTW). In this concept, a permeable "wall" containing the appropriate reactants is constructed across the path of a contaminant plume. As the contaminants passes through the reactive material under passive groundwater flow conditions, the halogenated organics are degraded, preventing contaminants from migrating further downstream. Among the advantages over pump-and-treat PTW offer reduced capital cost, low operating and maintenance costs, and conservation of water and energy.

Zero-valent zinc and iron significantly enhanced the reductive dehalogenation of aliphatic compounds with iron being particularly attractive due to its low cost and availability (Gillham et al., 1993). Batch tests in which aqueous solutions of a wide range of chlorinated methanes, ethanes, and ethenes were added to 100-mesh iron filings resulted in degradation rates that were three to seven orders of magnitude greater than natural abiotic rates reported in the literature (Gillham and Burris, 1992). Generally, the rates increased with the degree of chlorination and with increasing iron surface area to solution ratio. The chlorinated products of degradation subsequently degraded to non-chlorinated compounds. Similar results have been obtained by Vogan et al. (1995) who proposed that the corrosion of iron, while occurring independently of volatile organic compound degradation, likely provides the electron source needed for the reduction.

Biotic reductive dechlorination of TCE to *cis*-DCE frequently occurs in situ and has been reported since the early 1980s (Bouwer *et al.*, 1981; Bouwer and McCarty, 1983; Parson *et al.*, 1984; Vogel and McCarty, 1985). Its impact on permeable barriers has recently been observed in the UCF labs (Reinhart et al, 1998). As mentioned previously, reductive dechlorination in aquifer material upgradient of a permeable barrier can result in daughter products (*cis*-DCE and VC) with significantly longer half-lives than TCE. Half-lives of the chlorinated ethenes significantly increase as the number of chlorine substituents decreases. This effect has also been observed by numerous other investigators working with zero-valent iron, and *cis*-DCE and VC in particular have significantly longer half lives (Blowes *et al.*, 1995; Suthersan, 1996; Eykholt and Sivavic, 1995). Increased half-life is a potentially important phenomena with respect to the cost effectiveness of zero-valent iron permeable barriers. Biotic reactions could impact field kinetic rates, and thus retention time and mass of reactant, as well as construction costs. Thus the extent of intrinsic reductive dechlorination must be anticipated to correctly design the system.

This research examines the potential effects on performance, detrimental or beneficial, of biotic reductive dechlorination occuring within a zero-valent iron PTW.

Research Objectives. Research objectives are to (1) quantify TCE removal kinetics in the presence of zero-valent iron and active microbial consortia using laboratory columns, (2) evaluate the effect of biological activity on reactive iron kinetics, and (3) identify operative microbial consortia through measurement of byproducts under controlled and inhibited conditions.

MATERIALS AND METHODS

Column Studies. During Phase One, four parallel glass columns (0.1 m I.D., 1 m in length) were filled with abiotic concrete sand (AS), biotic sand (BS), abiotic sand and 20 % by wt. peerless iron (AI), and biotic sand and 20 % by wt. peerless iron (BI). Approximately 600 g of biologically active soils were added to each column. For the abiotic columns the soil was autoclaved at 120 °C for 25 minutes prior to adding to the column mixture.

Water preparation. Tap water (air stripped and chlorinated groundwater) was used to simulate groundwater. One hundred-L Tedlar bags placed in a121-L containers were filled with tap water. The water remained in bags for one week to permit dechlorination (confirmed through chlorine analysis) and then was purged with 99.9 % pure nitrogen for twelve hours to release all dissolved oxygen. After purging, pH increased from 7.5 to 8.3. Purging with 90% nitrogen and 10 % carbon dioxide for a few minutes reduced the pH to 7.5.

Stock Feed Preparation. A 4-L flask filled with deoxygenated and dechlorinated tap water was saturated with 99.9 % pure TCE (Fischer Scientific). A portion of the feed stock was pumped into a 1-L Tedlar bag, which was then emptied into the 100-L bag for a final concentration of ~ 5 mg/L.

Sampling Approach. Eight sampling ports were installed along the length of the column 10 cm apart. The column was fed in an up-flow fashion. Sampling events occur once per week using a 5-mL gas tight syringe at 5mL/min. The 5-mL sample is transferred to a vial for chlorinated hydrocarbon analysis in a gas chromatograph with a purge and trap and FID. Helium was bubbled through the sample for a period of eleven minutes to transfer the TCE onto a Vocarb 3000 trap. The desorb time from the trap was four minutes at 250 °C and the trap bake time seven minutes at 260 °C. A Hewlett-Packard gas chromatograph (Model 5890) equipped with a 0.25-mm id, 60-m long Vocol capillary column was programmed for a three-minute hold at 60 °C, and a 15 °C/min rise to 180 °C held for three minutes.

Tracer Study. Column tracer studies were conducted twice, once to evaluate initial hydrodynamic characteristics of the columns then a second time 10 months later to evaluate the impact of continued operation on the columns. Feed water, spiked with 5mg/L of $LiNO_3$, was introduced at 2.3 ml/min during the first study and 4.0 ml/min during the second. Sampling of column effluent occurred every hour for 28 hours.

Lithium concentration was measured by flame emission method using an Atomic Absorption Spectrometer (AA-475 series) at a wavelength of 670.8 nm. Triplicate check standards were prepared at 5mg/L, 10 mg/L, 50 mg/L and 100 mg/L as a measure of accuracy.

The retention time through the reactive zone (corrected for flow differences) and porosity for each column were calculated and results are presented in Table 1. These values suggest that flow persisted through all columns, although a reduction in porosity had occurred in the AI and BS columns and short-circuiting was possible.

TABLE 1. Retention Time and Porosity for four columns, Abiotic Iron, Biotic Iron, Abiotic Sand and Biotic Sand.

Column	Retention Time (hrs)		Porosity (%)	
	Study 1	Study 2	Study 1	Study 2
Abiotic Iron (AI)	28.5	13.1	48	22
Biotic Iron (BI)	25.0	26.2	42	44
Abiotic Sand (AS)	36.3	NA	61	NA
Biotic Sand (BS)	39.2	21.8	66	37

RESULTS AND DISCUSSION

The columns were operated from July 1998 to July 1999. A summary of the results is presented as follows.

♦ The AS column was shut down after more than 50 pore volumes passed through and no decline in the concentration of TCE was observed.

♦ The BS column initially had no significant microbial activity; the microbial consortium from the native soil associated with reductive dechlorination apparently did not survive. Subsequently the BS column was spiked with supernatant from a local wastewater treatment facility anaerobic digester resulting in stimulated methanogenic activity. This fact was confirmed by observed production of methane. A total of 114 pore volumes passed through the BS column before the column was shut down.

♦ Approximately 250 pore volumes passed through the AI column. Ultraviolet light applied to the influent was used as a control for microbial growth. As a further precaution, a known antimicrobial agent, sodium azide (150 mg/L), was fed continuously to the AI column after 121 pore volumes. At this point TCE degradation ceased. Sodium azide may have created conditions adverse to the zero-valent dehalogenation reaction. The AI column was shut down pending future investigation.

♦ The BI column was able to maintain microbial activity throughout the addition of 337 pore volumes. Methane production was identified in the BI column gas. At no time was hydrogen sulfide detected in any of the gas collection systems for the four columns probably due to low influent sulfate concentrations (below 2 mg/L).

After 150 pore volumes the average flow rate was increased from 2.27 mL/min to 4.0 mL/min for all three columns. The half-life of TCE as a function of pore volume is presented in Figure 1 for BI, AI, and BS columns. The presence of microbial organisms and iron in the BI column seems to have had a positive effect on the destruction of TCE relative to both the column containing no iron (BS) and the column containing no microbial life (AI) as seen in Table 2.

TABLE 2. Summary of TCE Disappearance Half Lives for Columns

Column	Average Half Life (days)	Half Life Range (days)
Abiotic Iron (AI)	0.43	0.37 – 0.48
Biotic Iron (BI)	0.22	0.09 – 0.44
Biotic Sand (BS)	0.68	0.22 – 0.69

On two occasions oxygen intrusion occurred to the influent of the BI column and once to the influent of the BS column. TCE degradation to cis DCE in the BI column appears to be more stable than the BS column in the face of oxygen intrusion. The presence of iron could have acted as a scavenger to oxygen protecting anaerobic bacteria or have increased the tolerance of the anaerobic bacteria to oxygen.

The primary degradation product in the BS column was cis-DCE. The BI column effluent had virtually no cis-DCE until over 220 pore volumes had passed. Eventually the BI effluent cis-DCE concentration increased from 0.009 to 0.89 mg/L. Both the BI and BS columns had similar TCE influent concentrations. Figure 2 presents the ratio of cis-DCE formation for the BI column.

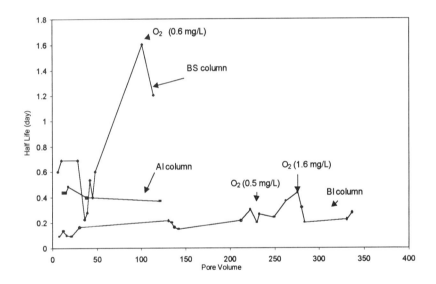

FIGURE 1. TCE Half Life in Abiotic Iron, Biotic Iron and Biotic Sand Column.

In the BI column ethylene formation as a by-product of the TCE degradation was decreasing gradually after 200 pore volumes. These observations suggested that TCE degradation was largely the result of microbial activity with a much smaller iron participation in the TCE degradation as time passed. This behavior was attributed to the depletion of the iron reactivity in the column.

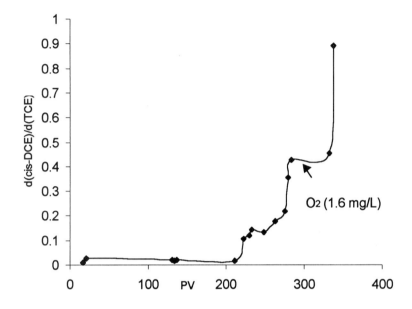

FIGURE 2. The Ratio of cis-DCE Formation to TCE Degradation in the Biotic Iron Column.

REFERENCES

Blowes, D.W., Ptacek, C.J., Cherry, J.A., Gillham, R.W., Robertson, W.D., "Passive Remediation of Groundwater Using *In Situ* Treatment Curtains", *Geoenvironment 2000*, Vol. 2., Geotechnical Special Publication No. 46, American Society of Civil Engineers, New York, New York, p. 1588-1607 (1995).

Bouwer, E. J., Rittmann, B. E., and McCarty, P. L., " Anaerobic Degradation of halogenated 1-, 2-carbon halogenated organic compounds", *Environ. Sci. Tech.*, 15, 596-599 (1981).

Bouwer, E. J., and McCarty, P. L., " Transformation of 1-, 2-carbon hologenated aliphatic organic compounds under methanogenic conditions", *Appl. Environ. Microbiol.* 45(4), 1286-1294 (1983).

Eykholt, G.R., and Sivavec, T.M., "Contaminant Transport Issues for Reactive-Permeable Barriers", *Geoenvironment 2000*, Vol. 2., Geotechnical Special Publication No. 46, American Society of Civil Engineers, New York, New York, p. 1608-1621 (1995).

Gillham, R.W., S.F. O'Hannesin, and W.S. Orth, "Metal Enhanced Abiotic Degradation of Halogenated Aliphatics: Laboratory Tests and Field Trials," presented at the Haz Mat Central Conference, Chicago, IL, March 9-11 (1993).

Gillham, R.W. and D.R. Burris, "Recent Developments in Permeable In Situ Treatment Walls for Remediation of Contaminated Groundwater," In *Proceedings of Subsurface Restoration Conference*, June 21-24, (1992).

Houser, T.J., Bernstein, R.B., Miekka, R.G., Angus, J.C., "Deuterium Exchange Between Trichloroethylene and Water. Infrared Spectral Data for Trichloroethylene," *J. Amer. Chem. Soc.*, Vol. 77, pp.6201-6203, (1955).

National Academy of Sciences, Alternatives for ground water cleanup. Report of the National Academy of Science Committee on Ground Water Cleanup Alternative. Washington, D.C., *National Academy Press*, (1994).

Parsons, F., Wood, P. R., and DeMarco, J., "Transformations of Tetrachloroethene and Trichloroethene in Microcosms and Groundwater, "*J. AWWA*. 76, 56-59 (1984).

Suthersan, S.S., Remediation Engineering *Design Concepts*, CRC/Lewis Publishers, Boca Raton, Fla. (1996).

Vogan, J.L., Gilham, R.W., O'Hannesin, S.F., Matulewicz, W.H., Rhodes, J.E., 'Site Specific Degradation of VOCs in Groundwater using Zero-Valent Iron', *Preprints of Extended Abstracts of American Chemical Society*, Vol. 35, pp. 800-804 (1995).

Vogel, T.M., and McCarty, P.L., "Biotransformation of Tetrachlorothylene to Trichloroethylene, Dichloroethylene, Vinyl Chloride, and Carbon Dioxide under Methanogenic Conditions," *Appl. Environ. Microbiol.*, 49(5), 1080 (1985).

DECHLORINATION OF CHLOROHYDROCARBONS IN GROUND-WATER USING NOVEL MEMBRANE-SUPPORTED Pd CATALYSTS

Katrin Mackenzie, Robert Koehler, Holger Weiss and Frank-Dieter Kopinke
(UFZ - Center for Environmental Research, Leipzig, Germany)

ABSTRACT Chlorinated organic compounds (COCs) such as PCE, TCE, and chloroaromatics are important groundwater contaminants. Most of the aliphatic COCs can be reductively dechlorinated in the aquifer by reaction with metallic iron ('rusty walls'). Unfortunately, this reaction fails completely for aromatic COCs. Their hydrodechlorination succeeds with Pd as a catalyst and hydrogen as the reductant. The potential of this technique for *in situ* groundwater treatment depends primarily on the long-term stability of the catalytic system. The application of non-protected Pd catalysts combined with electrochemical *in situ* generation of hydrogen failed in a long-term field test due to catalyst poisoning after several weeks of successful operation. A new approach will be presented, where a membrane-based Pd catalyst is used for the dechlorination of aliphatic and aromatic COCs. The polymer membrane protects the catalyst from deactivation and acts as a collector for hydrophobic pollutants. Laboratory-scale results and a field test show that our concept of membrane-supported catalysts is viable.

INTRODUCTON

The region Bitterfeld-Wolfen, located in the heart of Germany (Saxony-Anhalt, former East Germany), has been a center of especially chlorine-based chemical industry for more than 100 years. Particularly in the last few decades, huge amounts of COCs were released underground there. The groundwater is heavily contaminated with a mixture of organics, among them various COCs. The main contaminant at the test site is chlorobenzene with a concentration of up to 30 mg/l. A research project has been set up to examine and to further develop *in situ* groundwater decontamination techniques. Because of the extent of groundwater pollution in Bitterfeld, conventional remediation techniques, such as pump & treat, are not feasible. Passive *in situ* remediation techniques, such as permeable reaction walls are considered more viable.

RESULTS AND DISCUSSION

Abiotic reductive dechlorination. Since the beginning of the 90s, such permeable reaction walls are considered 'state of the art', implemented either as long packed walls filled with reactive granular material, or in a 'funnel & gate' configuration, where the reactive permeable wall is positioned in the center of an impermeable funnel-like bulkhead. The specific conditions at the test site require a technique for fast and complete dechlorination of a wide range of different COC classes. We chose the reductive dechlorination pathway in order to avoid unpredictable toxic intermediate products, which would have to be considered if an oxidative dechlorination method were chosen. In recent years, the conventional pump & treat technology received a strong competitor with the new concept of iron barriers. The so-called 'rusty walls', which were first developed by Gillham and co-workers in 1989 (Gillham & O'Hannesin, 1994) for clean-up of COC-contaminated sites, have proved to be a suitable technique. Metallic iron is able to reduce most aliphatic COCs to the chlorine-free substances (equation 1), although with very different rates.

$$R\text{-}Cl + Fe^0 + H_2O \rightarrow R\text{-}H + Fe^{2+} + OH^- + Cl^- \qquad (1)$$

For example, the half-lives for the dechlorination of PCE and TCE are markedly below 1 hour (normalized to 1 m^2 iron-surface per ml groundwater). For other compounds, such as cis-dichloro-ethylene or vinyl chloride, a much longer time period is needed for reduction. Unfortunately, iron fails completely as a reducing agent for compounds such as methylene chloride and the whole class of chlorinated aromatics.

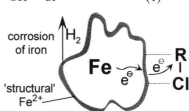

FIGURE 1. Direct electron transfer as reduction mechanism

Looking at the reduction mechanism using iron barriers, i. e. the direct electron transfer from the iron or the so-called structural iron(II)-cation on the iron surface (fig. 1), then it is not immediately apparent why chloroaromatics are not reduced. The reduction potentials for the two-electron-transfer reaction, resulting in chlorine-free hydrocarbons and chloride, are about + 500 mV (table 1). The standard redox potential of the iron electrode is about - 440 mV. This means that the free reaction enthalpy for the hydrodechlorination with iron is in the range of about 200 kJ/mol for each dechlorination step. This is actually a very high thermo-dynamic driving force, whereby the aromatic compounds are no exception.

TABLE 1. One-electron reduction potentials [in V] of selected COCs (vs. SCE) (Wiley et al. (1991) and Lide (1994))

Chloro-benzenes		Chloro-biphenyls		Chloro-naphthalenes		miscellaneous COCs	
n_{Cl}	- E$_{1/2}$	n_{Cl}	- E$_{1/2}$	n_{Cl}	- E$_{1/2}$		- E$_{1/2}$
0	-	0	2.76	0	2.55	1,1-DCE	2.5
1	**2.79**			1	2.33	TCE	2.25
2	2.55[1]	2	2.30[1]	2	2.15[1]	PCE	2.0
3	2.35[1]	3	2.20[1]	3	1.90[1]	Trichlorophenol	1.82
4	2.15[1]					Chloroform	1.68
5	1.92					Carbon tetrachloride	0.78
6	1.67	10	1.76	8	1.29		

[1] Mean for different isomers

If one assumes that the reduction is a sequence of two single-electron-transfers forming the radical anion in the first step, then the unexpected behavior of the aromatic hydrocarbons can be explained, especially that of chlorobenzene. The first step of the reduction during a direct electron transfer (e.g. from metallic iron) is therefore kinetically hindered, even though all thermodynamic prerequisites for a successful reduction are fulfilled. Since the thermodynamical conditions for the reduction of chlorobenzene are favorable in principle, a change of the reaction mechanism should lead to success.

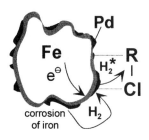

FIGURE 2. Catalytic hydrodechlorination as reduction mechanism

We are now using hydrogen as the reducing agent, which is for example always produced by iron corrosion as shown in the illustration (fig. 2). Hydrogen is collected and activated by a noble metal catalyst (preferentially Pd), and therefore usable for splitting-off the C-Cl bonds. The reaction is a hydrogenolysis resulting in a fast and complete dechlorination of a broad spectrum of COCs (equation 2).

$$R\text{-}Cl + H_2 \xrightarrow{\text{Pd catalyst}} R\text{-}H + HCl \qquad (2)$$

The end products of the reaction (the dechlorinated hydrocarbons) are left to be dealt with by the microbiology of the aquifer. The Pd catalyst is not dependent upon iron as a carrier: it only requires the availability of hydrogen. Hydrogen can in principle be made available by three methods: from an external source, from iron corrosion (as shown in fig. 2), or electrochemically. Most of the saturated COCs (e.g. chloroethanes) are not reduced with H_2/Pd.

The combination of in-situ generation of hydrogen and the use of Pd as catalyst was first described by McNab & Ruiz (1998). The authors used an electrolytic cell with graphite electrodes to generate hydrogen (and unavoidably also oxygen). The TCE-contaminated water is enriched with the gases and the dechlorination occurred at a commercially available Pd/Al_2O_3 catalyst. The system permits relatively high flow rates. The results of McNab and Ruiz show that catalytic dechlorination on a larger scale is successful in principle, but they also show that the system has a severely limited life. The catalytic activity is halved after only about 50 hours of operation. Although the catalyst may be regenerated by washing, nevertheless this makes the system unsuitable for *in situ* application. Very recently the authors reported about a field test using Pd/Al_2O_3 and externally supplied H_2 in a bore-hole reactor (McNab et al., 2000). To avoid rapid catalyst breakdown, frequent aeration of the reactor was necessary.

Electro-catalytic dechlorination in the field experiment. Under aquifer conditions, deactivation of noble metal catalysts occurs mainly due to mechanisms such as poisoning by heavy metal or sulfur compounds. However, suspended matter or biofilms also clog the catalytic surface. Similarly to McNab and Ruiz, we used a commercially available Pd catalyst (0.5% Pd on activated carbon, Degussa AG) and supplied H_2 by in situ water electrolysis, but in contrast to the authors, we used a separated electrolytic cell, where the anode and cathode compartments are passed subsequently. Figure 3 shows the construction scheme of the electro-catalytic reactor, which is a combination of electrolytic cell and fixed-bed catalytic dechlorination reactor. The reactor (described in detail by Koehler, 1999) has been designed as a 20-liter flow-through type electrolytic cell, where the groundwater passes anode and cathode compartments in succession. The compartments are divided by a porous polyethene diaphragm. At the cathode, hydrogen is produced and dechlorination of aliphatic COCs may already occur. The groundwater enriched with hydrogen then passes the catalyst bed, where all remaining COCs can be reduced.

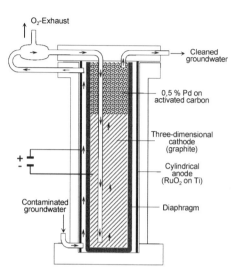

FIGURE 3. Scheme of the electro-catalytic reactor

The electro-catalytic reactor was put into operation with a groundwater flow of 1.3 L/h and a current of 1 A (cell voltage 5 V). This corresponds to a mean residence time in the catalyst bed of about 75 min. or a catalyst load of 0.6 v/vh. Up to the operation day 50, only insignificant amounts of chlorobenzene and benzene are detected in the water leaving the reactor (removal > 99.5%). This alone is no proof of chemical reactions, because both compounds are adsorbed by the catalyst carrier. Starting from operation day 59, benzene was analyzed in the outlet water as clear evidence of the reduction pathway. An activity test of the catalyst (sampled from various layers of the catalyst bed) at day 64 in the laboratory showed that the catalytic activity was undiminished.

After 6 months of operation the reactor was disassembled, because chlorobenzene started to break through. Activity tests of the catalyst showed the almost total loss of catalytic activity (fig. 4).

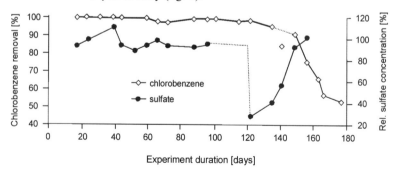

FIGURE 4. Removal rate for chlorobenzene compared to the relative sulfate concentration in the reactor efflux

Inspection of the reactor internals showed that the catalyst was covered with a brownish coating - washing the catalyst with diluted HCl evolved H₂S. The reason for the catalyst poisoning is depicted in figure 4 and can be found from the analysis of inorganic groundwater constituents: At day 121 a drastic decrease in the sulfate concentration from 800 mg/L to 220 mg/L was analyzed, whereas the concentrations of sulfate are equal under normal conditions at reactor in- and outlet. Sulfide is known to be a very effective catalyst poison and is presumably produced by microbial sulfate reduction. The reason for the spontaneous sulfate reduction is unclear. It may have been due to the temporary breakdown of the groundwater flow.

Because the water electrolysis (and therefore the gas production) continued during this operational disturbance, this led to an enrichment of hydrogen in the cathode compartment (total withdrawal of O_2 and hypochlorite, which effectively inhibit the microbial sulfate reduction).The field experiment showed that the electro-catalytic reactor was able to almost completely remove chlorobenzene from groundwater over a time period of several months. The experiment also showed that the catalyst can be dramatically influenced by catalyst poisons and microbiological interference. The potential of the catalytic dechlorination technique for *in situ* groundwater treatment depends mainly on the long-term stability of the catalytic system. Therefore, a way to protect the catalytic system is required!

Membrane-supported dechlorination catalysts. Our approach favors membrane-supported Pd catalysts. The basic hypothesis is simple and plausible: the catalyst is embedded in a hydrophobic polymer membrane, which protects it from deactivation (fig. 5). The membrane acts simultaneously as a shield against hydrophilic, ionic catalyst poisons and as an absorber ('concentrator') of hydrophobic COCs. The membrane materials preferred are silicon polymers (e.g. poly-(dimethylsiloxane) = PDMS), because of their high diffusivities. For example, the diffusion coefficient of benzene in PDMS is only 3 times lower than that in water. Using a silicon matrix, one is not dependent upon additional catalyst

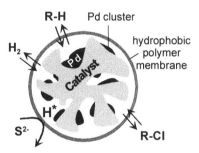

FIGURE 5. Operating mode of a catalyst particle coated by a hydrophobic membrane

carriers. Pd may also be embedded in a highly dispersed form within the PDMS matrix ($d_{cluster} \leq 5$ nm). In order to keep the transport hindrance of the silicone bulk phase as low as possible, we aimed to use thin polymer coatings. The membrane catalysts were designed in co-operation with the working group of Dr. D. Fritsch, (GKSS Geesthacht) as a silicone-coated version of commercially available supported catalysts, as foil (wall thickness about 750 µm), as 7 µm-coating on polyacrylonitrile fleece and as hollow fibers (wall thickness between 0.2 and 1 mm). We have achieved some very promising results on the laboratory scale with the systems outlined. The Pd/PDMS membranes (foil type) showed no reduction of the catalytic activity in the presence of sulfite ions, where conventional Pd catalysts were completely deactivated. In presence of sulfide ions, a reduction of catalytic activity was observed. The deactivating effect of sulfide depends on the pH of the solution. In neutral solutions, a significant proportion of sulfide is present as H_2S, which penetrates the silicone layer. Under such conditions, iron must be added as a sulfide scavenger. However, no complete destruction of the catalyst function occurred. An unprotected Pd catalyst loses its catalytic activity immediately after addition of sulfite and sulfide ions. In our experiments, the tube-like form proved to be an even more elegant method of applying membrane-supported Pd catalysts. The preparation of tube-like hollow silicone fibers containing 0.7 to 1.1% Pd led to catalysts which were very suitable for dechlorination applications (fig. 6).

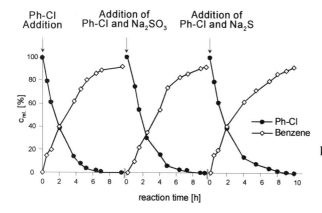

FIGURE 6. Chlorobenzene reduction using Pd/silicone (hollow-fiber type) with and without the addition of catalyst poisons (fiber: 4.2 x 1.0 x 500 mm, 0.7% Pd; Pd: S = 1 : 1, pH=10)

We have successfully tested this system in the laboratory on various scales, where it proved to be more efficient and robust than all other catalytic systems investigated. Figure 7 shows the operation mode of a tubular dechlorination module. The tubular form of the catalytic system has the advantage of bringing hydrogen directly to its reaction partner, so that both reactants are present in high concentrations at the reaction site.

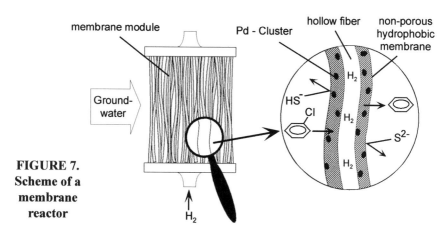

FIGURE 7. Scheme of a membrane reactor

The hydrogen can be supplied either from an external source or *in situ* by water electrolysis. A pilot scale catalytic reactor is now installed at a depth of 19 m as part of a modular construction. The reactor module contains a 60 m-Pd/fiber catalyst. The groundwater flows vertically from below through the reactor; pressure and temperature are the same as in the groundwater aquifer. The fiber catalyst is fed with externally generated hydrogen from the inside of the tube. Groundwater will be sampled frequently to inspect the performance of this new type of remediation method, which now must prove its worth under real groundwater conditions at the test site.

Laboratory experiments. Parallel to the field experiments, studies at the laboratory scale were carried out. We determined specific activities for the reduction of chlorobenzene with various palladium catalysts. It turned out that the measured reaction rates are transport-limited in many cases. This means that the intrinsic reaction rates at the Pd surface are sufficiently high, even for the polymer-embedded clusters. The potential for improvement of the catalyst performance is focussed on the reactor design (optimized flow conditions).

Furthermore, the activity data alone can not sufficiently predict the performance of the catalytic system as a whole. Figure 8 shows an example for the influence of the catalyst matrix on the selectivity of the system.

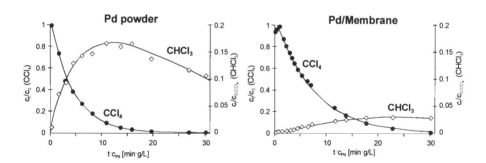

FIGURE 8. Influence of the catalyst matrix on the dechlorination selectivity

We used the hydrodechlorination of carbon tetrachloride with Pd catalysts as model reaction. In principle, carbon tetrachloride can react along two pathways: directly in one step without the release of intermediates, and over a sequential

$$CH_4 \xleftarrow{\text{fast}} CCl_4 \xrightarrow{\text{fast}} CHCl_3 \xrightarrow{\quad /\!/\quad} CH_2Cl_2 \xrightarrow{\quad /\!/\quad} \cdots \cdots \quad (3)$$

channel, where we observe intermediates (equation 3). Indeed, we find that both reaction channels occur simultaneously. Chloroform does not use both reaction pathways; there is no sequential reduction. This can easily be proved by the absence of methylene chloride, which would be stable under the reaction conditions. Unfortunately, the hydrodechlorination of chloroform is slower by one order of magnitude than the direct reduction of carbon tetrachloride to methane. Therefore, the formation and reaction of chloroform dictate the rate of the whole dechlorination process. In figure 8 the concentrations of carbon tetrachloride and chloroform during the reaction are depicted for two catalyst systems: left, using Pd powder and right, using our Pd-membrane. The specific activities of both catalysts are very similar for this reaction. However, the selectivity for the unwanted intermediate chloroform is very different. About 20% of carbon tetrachloride is transformed to chloroform using Pd without the hydrophobic matrix but only 2 % are converted using our new catalytic system. This is a very welcome experimental result, but there is no plausible explanation for it yet.

The comparison of simple activity patterns can smudge the whole picture of the performance of a catalytic system. Under field conditions, we believe the resistance against biofouling will determine the practical suitability of this new catalyst.

REFERENCES

Gillham, R. W. and S. F. O'Hannesin. 1994. "Enhanced degradation of halogenated aliphatics by zero-valent iron." *Ground Water* 32(6): 958-967

Köhler, R. 1999. "Elektrochemische und katalytische Dechlorierung von Chlor-kohlenwasserstoffen im Grundwasser." Dissertation, Fakultät für Verfahrens- und Systemtechnik, University Magdeburg, Germany

Lide, D. R. (Ed.). 1996. CRC Handbook of Chemistry and Physics, 76th Edition, CRC Press, New York

McNab, W. W. and R. Ruiz. 1998. "Palladium-catalyzed reductive dehalogenation of dissolved chlorinated aliphatics using electrolytically-generated hydrogen." *Chemosphere* 37(5): 925-936

McNab, W. W., Ruiz, R. and M. Reinhard. 2000. "In-situ destruction of chlorinated hydrocarbons using catalytic reductive dehalogenation in a reactive well: Testing and operational experience." *Environ. Sci. Technol.* 34(1): 149-153

Wiley, J. R., Chen, E. C. M., Chen, E. S. D., Richardson, P., Reed, W. R., and W. E. Wentworth. 1991. "The determination of absolute electron affinities of chlorobenzenes, chloronaphthalenes and chlorinated biphenyls from reduction potentials." *J. Electroanal. Chem.* 307: 169-182

RDX DEGRADATION IN COMBINED ZVI-MICROBIAL SYSTEMS

Mathew J. Wildman, Byung-Taek Oh, Andrew C. Hawkins, and *Pedro J. Alvarez*
(The University of Iowa, Civil & Environmental Engineering, Iowa City, IA)

ABSTRACT: RDX was rapidly removed from aquifer microcosms amended with zero-valent-iron (ZVI) and in flow-through columns packed with steel wool. This suggests that permeable reactive ZVI barriers might be a viable approach to intercept and degrade RDX plumes. The rate and extent of RDX degradation in the presence of ZVI was enhanced by adding anaerobic bacteria that feed on cathodic hydrogen (i.e., H_2 produced during anaerobic iron corrosion by water). It appears that some microorganisms utilized cathodic H_2 as an electron donor to reduce RDX. Microorganisms apparently participated also in the further degradation of heterocyclic intermediates produced by the reaction of RDX with ZVI. Reductive pretreatment of RDX with ZVI alone also reduced its toxicity to microorganisms and enhanced its subsequent biodegradability under either aerobic or anaerobic conditions. Therefore, a combined or sequential ZVI-biological treatment approach might improve treatment efficiency.

INTRODUCTION

RDX (Hexahydro-1,3,5-trinitro-1,3,5-triazine) is the British code name for Research Department Explosive (Testud *et al.*, 1996). RDX is a persistent and highly mobile groundwater contaminant that represents a major remediation challenge at numerous munitions manufacturing and load-assemblage-package facilities. Apparently, RDX-laden wastewater from washing production equipment was often discarded into unlined trenches or lagoons (Garg *et al.*, 1991). RDX is classified as a Class C (possible human) carcinogen, and can cause unconsciousness and epileptiform seizures. The Office of Drinking Water has set a limit for lifetime exposure to RDX at 0.1 mg l^{-1}. RDX is also used as a rodenticide, and the Surgeon General recommends a 24-h maximum RDX concentration of 0.3 mg l^{-1} to protect aquatic life (McLellan *et al.*, 1988).

Current practices to treat RDX-contaminated soil include incineration, composting, alkaline hydrolysis/oxidation, and aqueous thermal decomposition (Garg *et al.*, 1991). These *ex situ* approaches, however, are not cost-effective to treat large volumes of RDX-contaminated groundwater. In addition, complete destruction of RDX is not always achieved, giving rise to the possibility that products of equal or greater toxicity may accumulate. Thus, there is a need for a remediation strategy that is easy, cost-effective, less prone to accumulate toxic by-products, and addresses both chemical and microbiological advantages and constraints.

Considerable attention has been directed recently at the use of zero-valent iron (ZVI) to remove redox-sensitive priority pollutants from groundwater. Semipermeable ZVI barriers are particularly attractive for *in situ* remediation in that they conserve energy and water, and through long-term low operating and maintenance costs, have the potential to be considerably less costly than conventional cleanup methods (O'Hannesin and Gillham, 1998). Although iron barriers are mainly used to treat waste chlorinated solvents and hexavalent chromium plumes, it has been recently reported that ZVI can also chemically reduce RDX in contaminated soil (Singh *et al.*, 1998). Thus, ZVI barriers could also be useful to intercept and degrade RDX plumes.

Recent studies have also shown that microorganisms could enhance the treatment efficiency of ZVI barriers (Till *et al.*, 1998; Weathers *et al.*, 1997). Specifically, anaerobic ZVI corrosion by water produces cathodic hydrogen:

$$Fe^0 + 2H_2O \rightarrow Fe(OH)_2 + H_2 \qquad (1)$$

This hydrogen can serve as electron donor for the biotransformation of a wide variety of reducible contaminants. Thus, a hydrogenotrophic microbial consortium could be established around ZVI barriers to exploit cathodic depolarization and RDX degradation as metabolic niches. In addition, RDX reduction by ZVI might release NO_2^- (which can be used as N source by bacteria) and create potentially more biodegradable byproducts (Singh *et al.*, 1998). Therefore, RDX reduction by ZVI might enhance the participation of microorganisms in the cleanup process.

This study investigated the potential benefits of integrated microbial-ZVI systems to treat RDX-contaminated groundwater. Emphasis was placed on evaluating if bioaugmentation of ZVI can enhance RDX degradation kinetics and reduce the toxicity of degradation products. Experiments were also conducted to determine if treatment of RDX with ZVI alone would enhance its subsequent biodegradability under either aerobic or anaerobic conditions. In doing so, information was obtained to assess the potential for natural attenuation of any RDX byproducts that could escape an ZVI barrier, and to assess the feasibility of a sequential ZVI-biological treatment approach.

METHODS

RDX degradation assays. Batch reactors were prepared using 250-ml amber serum bottles capped with screw-cap Mininert valves. Bottles were filled with 100 ml of an acetate-enriched methanogenic culture (250 mg l^{-1} as volatile suspended solids, VSS) and fed 12 mg l^{-1} RDX. Five reactor sets were prepared in triplicate: sterilized cultures (poisoned with 350 mg l^{-1} $HgCl_2$), sterilized culture plus Fisher ZVI powder (10g, 2.0176 m^2 g^{-1}, 325 mesh), viable culture alone, viable anaerobic culture plus hydrogen gas (5 ml at 1 atm), and viable culture plus ZVI powder. Batch reactors fed H_2 gas were used as positive controls to determine if the cathodic H_2 (produced during anaerobic corrosion of iron by water) could serve as an electron donor for microbial reduction of RDX.

Biodegradability of ZVI-treated versus untreated RDX. Experiments were also conducted to determine if treatment of RDX with ZVI alone would enhance its subsequent biodegradability under either aerobic or anaerobic conditions. This information is relevant to assess the potential for natural attenuation of any RDX metabolites that could escape a ZVI barrier, and to assess the feasibility of a sequential chemical-biological treatment process. To do so, we first determined how the ratio of the biochemical oxygen demand (BOD) to the chemical oxygen demand (COD) changed after treatment with ZVI. An increase in this ratio is commonly considered as an increase in biodegradability.

The ultimate BOD of ZVI-treated and untreated RDX was determined using a HACH BODTrak™ instrument. This apparatus is equipped with six bottles. Two of these bottles were fed 420 ml of an RDX solution (30 mg l^{-1}), two contained ZVI-treated RDX (i.e., a 30 mg l^{-1} RDX solution that was filtered after reacting with 100 mg l^{-1} ZVI powder for 4 days), one contained ZVI-treated DI water to control for any oxygen demand exerted by ferrous iron, and the last one was a blank to control for the oxygen demand exerted by the seed. No RDX was present in the samples treated with ZVI. All bottles were amended with BOD nutrient pillows and seeded with 20 ml of primary effluent from a wastewater treatment facility. The bottles were then sealed and incubated on the BODTrak™ instrument, which automatically monitored the BOD continually over 11 days. The samples were continually stirred at 20 °C using magnetic stir bars. The initial COD of these samples was also measured using a HACH kit.

Toxicity of ZVI-treated versus untreated RDX. A Microtox™ assay was used to compare the toxicity of RDX versus ZVI-treated RDX. This assay uses the bioluminescent bacterium, *Photobacterium phosphoreum*, and reports the "effective concentration" of a sample decreases light output by 50% (i.e., EC_{50}). Higher EC_{50} values correspond to higher toxicity (Microbics Corporation, 1992).

Biochemical methane potential of ZVI-treated and untreated RDX was also measured to evaluate differences in anaerobic toxicity and the feasibility of anaerobic post-treatment. Methanogenic batch reactors were prepared in 120-ml serum bottles with 25 ml of the acetate-enriched culture. Reactors were fed acetate (2100 mg l^{-1}) and either ZVI-treated or untreated RDX at 0, 1, 5, 10, or 15 mg l^{-1}. Reactors were incubated in a Coy anaerobic box at 25°C for 50 days. Methane production was monitored regularly over this time.

Aquifer columns. Continuous flow columns (2.5-cm diameter × 26.5-cm long) were used to evaluate the potential for ZVI to remove RDX in a flow-through system mimicking a permeable reactive barrier or filter. One column was packed with 8 g of steel wool, as described elsewhere (Till *et al.*, 1998). The second (control) column was packed with glass wool. Both columns were fed in an upflow mode at a Darcy velocity of 1 ft day^{-1}, using Harvard syringe pumps.

Analytical methods. RDX and its degradation products were measured by HPLC with a Gilson 307 isocratic pump equipped with a Spectra 100 UV-Vis Detector (240 nm), and an Alltech Econosphere C18 5U column. RDX standards were obtained from Chem Service (West Chester, PA). Standards for RDX metabolites (i.e., 1,3,5-trinitroso-1,3,5-triazacyclohexane (TNX), 1,3-dinitroso-5-nitro-1,3,5-triazacyclohexane (DNX), and 1-nitroso-3,5-triazacyclohexane (MNX)) were

obtained from Ronald Spanggord, at SRI International, Menlo Park, CA.
CH_4 and H_2 gas concentrations were also monitored in the microcosms.
Headspace samples were collected using a 100-µL gas tight syringe and injected
into a gas chromatograph. CH_4 was analyzed with a HP 5890 Series II GC
equipped with FID detector and a DB-WAX capillary column (J&W Scientific).
H_2 was analyzed with a HP 5890 Series II GC equipped with TCD detector and a
Hayesep Q packed column (Alltech Associates).

RESULTS AND DISCUSSION

Bench-scale experiments suggest that that permeable reactive ZVI barriers
might be a viable approach to intercept and degrade RDX plumes, and that this
process can be enhanced by the participation of anaerobic microorganisms.

Figure 1 shows the residual RDX concentration in batch reactors amended
with ZVI, acetate-enriched anaerobic cultures (fed also 15 ml H_2 at the beginning
of the experiment), or both. RDX was degraded in all reactors prepared with either
the mixed culture or ZVI, but not in ZVI-free sterile controls. Only the reactors
amended with both ZVI plus the mixed culture No RDX was detected in reactors
removed all of the RDX within 1.5 days. In addition, heterocyclic metabolites
such as TNX, DNX, and MNX (up to about 2 mg l^{-1}) were found in reactors
prepared with viable bacteria or ZVI alone, even after 9 days. Yet, no metabolites
were detected in reactors amended with both bacteria plus ZVI at t = 1.5 days
(limit of detection 0.1 mg l^{-1}), suggesting that a more complete degradation of
RDX (beyond ring fission) occurred.

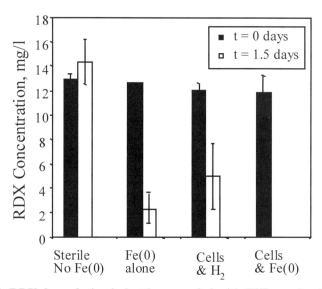

**Figure 1. RDX degradation in bottles amended with ZVI powder (100 mg l^{-1}),
anaerobic cultures (250 mg l^{-1} VSS), or both. Bars depict ± one standard
deviation from the mean of triplicate reactors.**

A similar experiment was conducted to determine if H_2 production by iron corrosion (equation 1) could enhance anaerobic RDX biodegradation. Three incubations were prepared with anaerobic cultures. ZVI was not added to preclude confounding effects associated with abiotic RDX reduction. RDX was degraded at similar rates in two reactors with live cells during the first 8 h, but not in the sterile control (Figure 2). At this time, H_2 (5 ml) was added to one of the two viable reactors. This stimulated RDX degradation. H_2, however, did not reduce RDX in sterile controls. This shows that H_2 produced by cathodic depolarization can be used as electron donor to support microbial reduction of RDX.

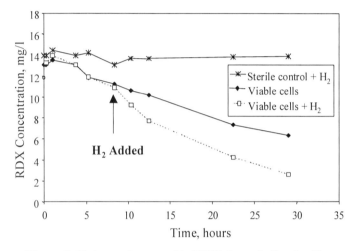

Figure 2. Enhanced anaerobic RDX degradation by H_2.

Microtox™ assay and comparison of BOD:COD ratios were used as indicators of the ability of ZVI treatment to reduce RDX toxicity and increase its subsequent biodegradability. Pretreatment of RDX with ZVI nearly doubled the COD, indicating that RDX was chemically reduced (Table 1). Untreated RDX did not exert any BOD, reflecting its recalcitrance to aerobic biodegradation. Treatment of RDX with ZVI increased the BOD:COD ratio from 0 to 0.31. The resulting ratio is relatively low. This indicates that while the reaction products are potential substrates for microorganisms, they are not readily biodegradable. Yet, they are more biodegradable than the parent compound, RDX.

Table 1. Comparison of the ultimate BOD and COD of various solutions. The increase in the BOD/COD ratio reflects increased aerobic biodegradability of RDX after treatment with ZVI.

Sample	Ultimate BOD, mg l^{-1} (corrected for seed)	COD, mg l^{-1}	BOD/COD ratio
ZVI-treated water (control)	0	-0.56 ± 1.8	0
Untreated RDX (30 mg/l)	0	11.1 ± 2.7	0
ZVI-treated RDX (30 mg/l)	6.6	21.3 ± 2.9	0.31

ZVI treatment also decreased RDX toxicity, as reflected by lower EC_{50} values (Table 2). No toxicity was detected in ZVI-treated RDX that was subject to subsequent aerobic biodegradation during the BOD assay. This illustrates a potential advantage of a sequential ZVI-biological treatment approach.

Table 2. Reduction of RDX toxicity after treatment with ZVI.

Sample	Microtox™ EC_{50} (mg l^{-1})
RDX	45.8 ± 5.2
ZVI-treated RDX	15.2 ± 4.3
ZVI-treated RDX after aerobic biodegradation	No Toxicity

Anaerobic toxicity assays were also conducted. These assays compared the biochemical methane potential of ZVI-treated versus untreated RDX. Methanogenic microcosms were fed acetate (2100 mg l^{-1}) and either ZVI-treated or untreated RDX, as described previously. Untreated RDX inhibited methanogenesis when fed at 10 mg l^{-1}, and no methane was produced when fed at 15 mg l^{-1} (Figure 3). In contrast, no inhibition was observed with 15 mg l^{-1} RDX that had been treated with ZVI. Thus, exposing RDX to ZVI increases the feasibility of anaerobic post-treatment. Interestingly, RDX degradation did occur in the reactors that did not produce methane. For example, the residual RDX concentration was 3 mg l^{-1} in the reactor fed 15 mg l^{-1}. Apparently, microbial reduction of RDX resulted in a buildup of metabolites that inhibited methanogens. This corroborates the finding of other researchers, who reported the inhibitory potential of RDX metabolites (McCormick *et al*, 1981).

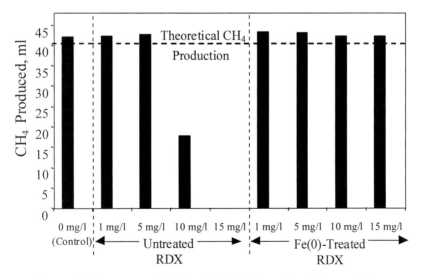

Figure 3. Treatment of RDX with ZVI also reduced its toxicity to a methanogenic consortium. The theoretical contribution to CH_4 production from 30 mg l^{-1} RDX is negligible compared to that from 2100 mg l^{-1} acetate.

Two continuous flow columns, one packed with steel wool and another with glass wool, were used to evaluate the potential for ZVI to remove RDX in a flow-through system mimicking a permeable reactive barrier or filter. The steel wool column (hydraulic retention time = 0.75 d) removed more that 98% of the influent RDX (Figure 4). In contrast, little RDX removal was observed in the glass wool control column. The high removal efficiency of the ZVI column was sustained over several months.

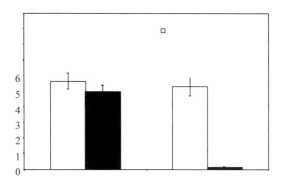

Figure 4. Influent and effluent RDX concentrations in flow through columns packed with steel wool. Bars represent ± one standard deviation of samples taken over a 15-day period.

CONCLUSIONS

Permeable reactive ZVI barriers might be a viable alternative to intercept and degrade RDX plumes. This treatment approach can be enhanced by the concurrent participation of anaerobic microorganisms that feed on cathodic hydrogen and increase the rate and extent of RDX biotransformation, or by iron reduces that depassivate the iron surface. In addition, reductive treatment of RDX with ZVI reduces its toxicity and enhances its subsequent biodegradability under either aerobic or anaerobic conditions. Consequently, a sequential scheme where a (thinner) pre-treatment ZVI barrier is followed by down-gradient natural attenuation of any products that break through might also be a viable alternative to treat RDX plumes.

ACKNOWLEDGMENTS

We thank the Center for Health Effects of Environmental Contamination at the University of Iowa and SERDP for their financial support. We are grateful to Craig Just for analytical methods development and to Gene Parkin, Michelle Scherer, and Jerry Schnoor for useful discussions.

REFERENCES

Garg R., D. Grasso, and G. Hoag (1991). Treatment of Explosives Contaminated Lagoon Sludge. *Hazardous Waste and Hazardous Materials*. 8:319-340.

Gorontzy T., O. Drzyzga, M.W. Kahl, D. Bruns-Nagel, J. Breitung, E. von Leow, and K.H. Blotevogel (1994). Microbial Degradation of Explosives and Related Compounds. *Critical Reviews in Microbiology*. 20:265-284.

McLellan W., W.R. Hartley, and M. Brower (1988). Health Advisory For Hexahydro-1,3,5-Trinitro-1,3,5-triazIne. Technical Report PB90-273533. Office of Drinking Water, U.S. EPA, Washington, DC.

McCormick, N.G., J.H. Cornell, and A.M. Kaplan (1981). Biodegradation of Hexahydro-1,3,5-Trinitro-1,3,5-Triazine. *Applied and Environmental Microbiology*. 42:817-823.

Microbics Corporation, (1992). Microtox® Manual.

O'Hannesin, S. F. and R. W. Gillham. (1998). Long-term Performance of an *In Situ* "Iron Wall" for Remediation of VOC's. *Groundwater*. 36: 164-170.

Singh J., S.D. Comfort, and P.J. Shea (1998). Remediating RDX-Contaminated Water and Soil Using Zero-Valent Iron. *Journal of Environmental Quality*. 27:1240-1245.

Testud F., J.M. Glanclaude, J. Descotes (1996). Acute Hexogen Poisoning After Occupational Exposure. Clinical Toxicology. 34:1:109-111.

Till B. A., L. J. Weathers, and P. J. J. Alvarez (1998). Fe(0)-Supported Autotrophic Denitrification. *Environmental Science and Technology*. 32:634-639.

Weathers L.J., G.F. Parkin, and P.J.J. Alvarez (1997). Utilization of cathodic hydrogen as electron donor for chloroform cometabolism by a mixed methanogenic culture. *Environmental Science and Technology*. 31: 880 -885.

BARRIER IMPLANTS FOR THE ACCELERATED BIO-ATTENUATION OF TCE

Jack K. Sheldon (Montgomery Watson, Des Moines, Iowa)
Kevin G. Armstrong (Montgomery Watson, Des Moines, Iowa)

ABSTRACT: A manufacturing facility located in the central United States, showing evidence of reductive dechlorination of TCE, was selected for a pilot study that was designed to evaluate a barrier implant technology. The concept of a barrier implant utilizes chemicals that can accelerate bio-attenuation and that have a long-term release profile. The compound used in this study was a polylactate ester (HRC™) that slowly releases fermentable lactic acid for reductive dechlorination. The semi-solid material was introduced in a series of four 4-foot (1.2-m) long perforated canister implants that were suspended inside a well. This constituted one unit of a potential barrier implant series. A previously existing monitoring well and a series of monitoring points were established in a 250-sq. foot (23.2-sq. m) study area. This area was located in a distal portion of the plume screened in a sand aquifer at a depth of approximately 30 to 45 feet (9.1 to 13.7 m) below ground surface.

During the study, a monitoring well, located approximately 5 feet (1.5 m) downgradient of the barrier implant well, showed the greatest change in TCE concentration of any of the locations sampled during the pilot study. Lactic acid was observed in that well six months after implant installation, indicating that the substrate had migrated into the aquifer from groundwater flow through the area. This coincided with a decrease in the sulfate concentration and an increase in sulfide. Dissolved oxygen (DO) concentration in the well decreased during the study to below 0.1 mg/L. Redox fell to -206 mV. A 70% reduction in TCE with a corresponding increase in daughter products was observed in that well, which showed that this method could produce the desired long-term enhancement of natural attenuation.

INTRODUCTION

HRC™ is a polylactate ester where the lactic acid moiety is hydrolyzed off the molecule and degraded through pyruvic acid to acetic acid. The hydrogen available at each step is then used by a specific population of microorganisms to drive the reductive dechlorination of trichloroethene (TCE) at an accelerated rate. The product is delivered to the subsurface through direct injection or through perforated canisters or barrier implants suspended inside a well. Delivery of HRC™, because it is a source of organic carbon, helps to consume existing oxygen and electron acceptors. A typical application for HRC™ would be a site showing evidence of natural attenuation due to biodegradation where a site owner would benefit from an increased rate of degradation.

HRC™ was selected for the pilot study because TCE breakdown products had been detected in the groundwater, field measurements showed low oxygen conditions (<0.5 mg/L) existed in the groundwater, and the groundwater flow velocity and site geology were favorable. The HRC™ product used at the site was a first generation version, a semi-solid material contained in perforated canisters. Based on laboratory studies by the manufacturer (Regenesis, San Clemente, CA), it was initially estimated that a minimum of six months would be required to see the effects of HRC™.

MATERIALS AND METHODS

The pilot test area consists of a single Delivery Well, a previously existing monitoring well, and a series of monitoring points. The 4-inch (10.2-cm) inner diameter (ID) Delivery Well and the previously installed 2-inch (5.1-cm) ID monitoring well, MW-14D, were installed using standard hollow-stem auger drilling techniques. The monitoring point network (MP-1 through MP-6) was established at a distal portion of the plume based on the groundwater flow system that was previously characterized. The 1-inch (2.5-cm) ID monitoring points were installed using direct push technology. All of the wells and monitoring points were constructed of polyvinyl chloride (PVC) risers and screens. The pilot test wells and monitoring points are screened in a sand aquifer at a depth of approximately 30 to 45 feet (9.1 to 13.7 m) below grade; plan and cross sectional views of the pilot test area are provided in Figure 1. Pilot study activities were implemented according to the schedule outlined in Table 1.

TABLE 1. Schedule of field activities for the barrier implant pilot study.

Activity	Date	Weeks Elapsed Since Placement of HRC™
Monitoring/Delivery Well Network Installation	February 9-11, 1998	Week -1
Background Groundwater Sampling (Round 1)	February 19, 1998	Week 0
Installation of HRC™ Canisters	February 19, 1998	Week 0
Round 2 Groundwater Sampling	March 3, 1998	Week 2
Round 3 Groundwater Sampling	March 19, 1998	Week 4
Round 4 Groundwater Sampling	April 14, 1998	Week 8
Round 5 Groundwater Sampling	May 8, 1998	Week 11
Round 6 Groundwater Sampling	August 4, 1998	Week 24
Round 7 Groundwater Sampling	October 21, 1998	Week 35

Groundwater purging and sampling were conducted using dedicated inertial pumps installed in each well. Purge water was analyzed using a multiparameter water quality meter with a flow-through-cell, monitoring temperature, pH, conductivity, DO, and redox. Samples were collected both pre- and post-purging for the February through May samples. Only post-purge

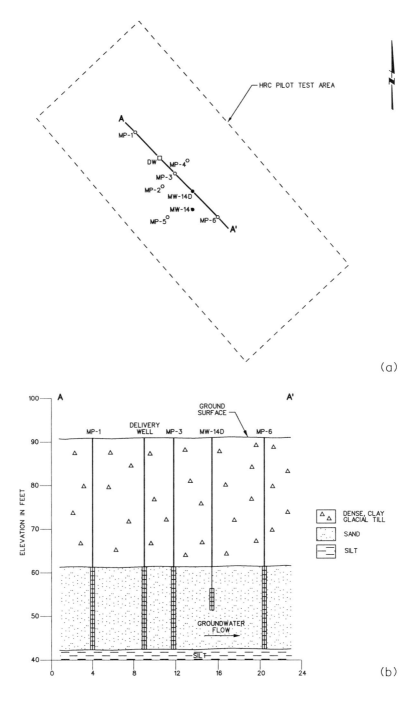

FIGURE 1. HRC pilot study design. (a) Plan view showing well placement and monitoring points. (b) Cross section showing well completion and monitoring points.

samples were collected for August through October sampling, since data from the pre-purge samples did not yield significantly different information. Post-purge samples were collected after stabilization of the field parameters to within approximately 10 percent over three consecutive readings. Samples were analyzed for the parameters outlined in Table 2.

Table 2. Monitoring parameters for the barrier implant study.

Parameter	Sampling Schedule
Chlorinated Volatile Organic Compounds	All Monitoring Events
Methane, Ethane, and Ethene Degradation Gases; Ammonia Nitrogen; Nitrate; Phosphorus; Sulfate; Sulfide; Iron (total and dissolved); Manganese; Chloride; Alkalinity; Total Organic Carbon (TOC); and Volatile Fatty Acids (lactic, propionic, pyruvic, and acetic)	All Monitoring Events Except Weeks 2 and 8
Aerobic and Anaerobic Microbial Counts	All Monitoring Events Except Weeks 2, 4, and 8

The HRC™ product was delivered to the subsurface by four 4-foot (1.2-m) long perforated polycarbonate canister implants. The perforations were approximately 1 mm in diameter and went completely around the canisters, spaced approximately 1 inch (2.5 cm) apart, and spanned the entire length of the canisters. The canister implants were linked together using carabiners and were lowered by a rope attached to the top canister implant so that the bottom canister implant was suspended approximately 3 feet (0.9 m) from the bottom of the well.

RESULTS

Following installation of the HRC™ canister implants, groundwater samples were collected from the monitoring array for selected parameters to monitor delivery of the HRC™ to the subsurface and ultimately the reductive dechlorination of the contaminants of concern. Groundwater sampling results from the pilot test are provided in Table 3.

Table 3. Summary of pilot test data (all units in mg/L).

Parameter	Week 0 Feb. 19, 1998	Week 2 Mar. 3, 1998	Week 4 Mar. 19, 1998	Week 8 Apr. 14, 1998	Week 11 May 8, 1998	Week 24 Aug. 4, 1998	Week 35 Oct. 21, 1998
Trichloroethylene	2.23	2.05	2.1	2.02	1.4	0.290	0.557
Cis-1,2-DCE	5.03	4.82	4.89	4.29	3.39	1.36	1.71
Trans-1,2-DCE	0.0674	0.0781	0.0786	0.0705	0.0403	0.0361	<0.001
Vinyl Chloride	0.134	0.147	0.141	0.127	0.0906	0.0337	<0.001
Lactic Acid	<200	-	<1	-	<1	2	<1
Pyruvic Acid	<2	-	<1	-	<1	<1	<1
Acetic Acid	<10	-	<1	-	<1	<1	<1
Sulfate	25.9	-	25.3	-	33	19.9	21.4
Sulfide	<0.1	-	<0.1	-	<0.1	0.14	0.11

Well MW-14D, located approximately 5 feet (1.5 m) downgradient of the delivery well, shows the greatest change in TCE concentration of any of the locations sampled during the pilot study. Lactic acid is observed during the sixth month of the study, indicating that HRC™ has migrated into the aquifer. This coincides with a decrease in the sulfate concentration and an increase in sulfide. Sulfate is an electron acceptor used in the reductive dechlorination of chlorinated compounds. An increase in sulfide is evidence of sulfate utilization and concentrations of sulfide in excess of 0.1 mg/L are considered significant indicators of biological activity. Also, the DO concentration in MW-14D decreased during the study to nearly 0.1 mg/L. Redox fell to -206 mV during the study. These measurements suggest the strongly reducing conditions necessary for degradation of TCE have been achieved at MW-14D.

A graph (Figure 2) of the Total VOCs and Percent TCE removed in the downgradient monitoring well, MW-14D, shows the most significant effects of HRC™ over the 8-month study. Cis-1,2-DCE was accumulating and was shown to degrade more than 75% during the study. Vinyl chloride began to accumulate early in the study, but its degradation also became apparent over the final half of the pilot study.

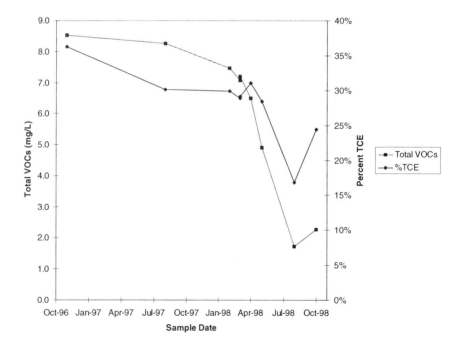

Figure 2. Changes in Total VOCs and Percent TCE during pilot study.

CONCLUSIONS

While clear evidence of biodegradation of TCE was observed at only one well during the pilot study, the pilot study did show that the HRC™ barrier implants could produce the desired enhancement of natural attenuation in the field. It was also determined from the study that the effects of HRC™ would last greater than six months, consistent with earlier laboratory tests.

Based on the analytical results, it appears that the HRC™ barrier implants had a significant effect in close proximity to the delivery well. The analytical results indicate that HRC™ may have traveled along a preferential flow path within the deep sand aquifer, resulting in increased reductive dechlorination and the detection of lactic acid at MW-14D, but not monitoring points closer to the delivery well.

ACKNOWLEDGEMENTS

This work was partially supported by Regenesis (San Clemente, CA) as a development activity for the new HRC™ bioremediation product.

REFERENCES

Koenigsberg, S.S. and C.A. Sandefur. "The Use of Hydrogen Release Compound for the Accelerated Bioremediation of Anaerobically Degradable Contaminants: The Advent of Time-Release Electron Donors." _Remediation._ 10(1): 31-53.

Sheldon, J.K., S.S. Koengisberg, K.J. Quinn, and C.A. Sandefur. "Field Application of a Lactic Acid Ester for PCE Bioremediation." In B.C. Alleman and A. Leeson (Eds.), _Engineered Approaches for In Situ Bioremediation of Chlorinated Solvent Contamination_, pp.61-66. Battelle Press, Columbus, OH.

KINETIC AND MECHANISTIC INVESTIGATION OF HALOCARBON REDUCTION AT IRON CATHODES

Tie Li (University of Arizona, Tucson, AZ)
James Farrell (University of Arizona, Tucson, AZ)

ABSTRACT: This research investigated the kinetics and mechanisms of reductive dechlorination of trichloroethylene (TCE) and carbon tetrachloride (CT) at iron cathodes. Reaction rates for TCE and CT were measured as a function of electrode potential, water chemistry, and influent reactant concentration in flow-through, porous iron electrode reactors, with and without the addition of palladium as an electrocatalyst. Rates of direct halocarbon reduction were also measured amperometrically using potential step chronoamperometry. Reaction rates for TCE and CT were first order in reactant concentration over the concentration range investigated. Typical reduction half-lives for TCE and CT in the iron reactor at neutral pH were 9.4 and 3.7 minutes, respectively. Palladium treatment at a level of 1 mg-Pd per square meter of electrode surface area increased reaction rates by a factor of three. When operated continuously, reaction rates in the palladized iron reactor were stable over a 9 month period of continuous operation. Amperometrically measured reaction rates for CT were the same as those measured analytically. This indicated that CT reduction occurred via direct electron transfer. In contrast, TCE reduction rates could not be measured amperometrically. However, the high current efficiencies in the electrode reactor indicated that the primary mechanism for TCE reduction was indirect, and occurred via reaction with atomic hydrogen produced from water reduction.

INTRODUCTION

Past use of chlorinated organic compounds has resulted in extensive groundwater contamination. Because most halogenated organic compounds are classified as carcinogens, mutagens and/or teratogens, remediation of contaminated groundwaters is of significant public health concern. Presently used treatment methods, such as adsorption to activated carbon and air stripping, simply transfer the contaminants from water to another medium, which then requires further treatment or disposal.

In recent years there has been considerable interest in developing destructive treatment methods for removing chlorinated organic compounds from contaminated waters. The use of zerovalent metals for reductive dechlorination has been a very active research area since Gillham and O'Hannesin proposed that metallic iron filings could be utilized in passive groundwater remediation schemes (Gillham and O'Hannesin, 1992; Gillham et al., 1994). In addition to *in-situ* treatment methods utilizing zerovalent iron, several investigations have focused on reductive dechlorination methods that can be employed in above ground canister treatment systems (Helvenston et al., 1997). Most of these methods use palladium as a reduction catalyst and hydrogen as the electron donor. Other investigators have attempted electrochemical reduction of chlorinated organic compounds using palladium supported on carbon and graphite cathodes (Munakata et al., 1997; Lyon, 1997). Although rapid dechlorination rates have been observed, the effectiveness of the electrocatalyst is short-lived, due to loss of palladium from the electrodes (Lyon, 1997; Helvenston et al., 1997).

There are both direct and indirect mechanisms for halocarbon reduction at metal electrodes. Direct reduction may occur by electron tunneling, or by formation of a chemisorption complex of the organic compound with surface metal atoms (Brewster, 1954). Electron tunneling may occur to halocarbons that are physically adsorbed at the electrode surface, or to halocarbons that are separated from the electrode by one or more layers of water (Bockris, 1970). Indirect reduction of organic compounds involves atomic hydrogen. Atomic hydrogen adsorbed at the metal surface may reduce organic compounds through formation of surface hydride complexes (Brewster, 1954). This mechanism is fast on metals with low hydrogen overpotentials, such as platinum and palladium, but is much slower on metals with high hydrogen overpotentials, such as nickel and iron.

A thorough understanding of the reaction mechanisms is needed for development of electrochemical methods for treating waters contaminated with chlorinated organic compounds. The overall objective of this research was to elucidate the reaction mechanisms involved in electrochemical reduction of chlorinated organic compounds by iron and palladized cathodes.

EXPERIMENTAL

The kinetics of trichloroethylene (TCE) and carbon tetrachloride (CT) reductive dechlorination were measured in flow-through reactors. A schematic diagram of the electrode reactors and experimental setup is depicted in Figure 1. The reactors employed cylindrical, porous iron cathodes (2 cm in diameter and 3 cm long) composed of iron filings and an aluminosilicate binder. In one of the reactors, palladium was added to cathode surface as an electrocatalyst. The anodes consisted of platinum wire screens or pieces of carbon cloth that were wrapped around the iron cathodes. Cation exchange membranes were used to separate the anodes and cathodes, and water passed through only the cathode compartment of each reactor. The reactors were contained within a glass tube fitted with a stainless steel pipe fitting at each end. Influent and effluent concentrations of TCE and CT were sampled at different flow rates.

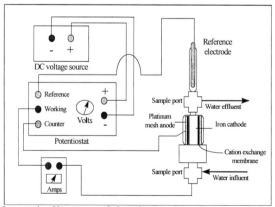

FIGURE 1. Schematic diagram of the electrode reactor and experimental setup.

Voltammetric analyses using chronoamperometry (CA) were performed in a custom three electrode cell using an EG&G (Oak Ridge, TN) Model 273A potentiostat and M270 software. All experiments utilized a Ag/AgCl reference electrode and a platinum wire counter electrode. An iron disk (Metal Samples Company, Munford, AL), and iron wires (Aesar, Ward Hill, MA) were used as working electrodes. Supporting electrolyte solutions were deoxygenated prior to use by purging with argon. An EG&G model 616 rotating disk electrode was used to amperometrically measure CT reduction rates and current efficiencies over a range of CT concentrations and electrode potentials. Rates of direct reduction were determined by comparing electrode currents in blank electrolyte solutions and those containing CT. Chloride balances in the electrochemical cells were used to determine the actual rates of TCE and CT dechlorination.

RESULTS AND DISCUSSION

The reactors were operated at high flow rates in order to approximate the differential reactor model (Levenspiel, 1972). Because of the small degree of reactant conversion (~30%), performance of the reactors could be modeled without accounting for changes in solution potential and reactant and byproduct concentrations along the length of the reactor. For a longer reactor, achieving a greater degree of reactant conversion, changes in solution potential and reactive species concentrations with distance into the reactor may necessitate a more mechanistic kinetic model (Trainham and Newman, 1977). However, if dechlorination rates are not limited by potential dependent factors (*i.e.*, do not follow Butler-Volmer kinetics), a simple first order reaction model may still apply for higher degrees of reactant conversion.

Both mass transfer and reaction rate limitations may affect the degree of reactant conversion in flow-through reactors. The steady state mass balance for a heterogeneous reaction in a plug flow reactor can be described by (Fogler, 1999):

$$u\frac{dC}{dz} + a_c\left(\frac{1}{1/k_{sa} + 1/k_f}\right)C = 0 \qquad (1)$$

Where C is the reactant concentration, z is the axial coordinate of position, u is the fluid velocity, a_c is the reactive surface area per unit pore volume of the reactor, k_{sa} is the first order reaction rate constant, and k_f is the first order film transfer coefficient. As defined in this manner, both k_{sa} and k_f are normalized for the surface area to solution volume of the reactor. The bracketed term in equation 1 is the observed rate constant, k_{obs}, which can be determined from measurements of the influent (C_o) and effluent (C_e) reactant concentrations, according to:

$$k_{obs} = \frac{\ln(C_o/C_e)}{a_c(L/u)} \qquad (2)$$

Where L is the length of the reactor. The effective half-life ($t_{1/2}$) for the reactant disappearance can then be calculated from:

$$t_{1/2} = \frac{\ln(2)}{a_c k_{obs}} \qquad (3)$$

Correlations in the chemical engineering literature indicate that k_f for a packed bed reactor increases with increasing fluid velocity. Under conditions where $k_f \gg k_{sa}$, k_{obs} will be independent of the flow rate, and $k_{obs} \approx k_{sa}$ (Fogler, 1999).

The reactors were operated at cathode potentials ranging from -755 to -1200 mV with respect to the standard hydrogen electrode (SHE). Although this potential range is below the stability domain of water, visible amounts of hydrogen gas were observed only at potentials below -900 mV (SHE). The low rate of hydrogen production can be attributed to the high overpotential of the hydrogen evolution reaction on iron and iron oxide surfaces (Bockris, 1970). At all potentials, the rate of water reduction was sufficiently low that there was no measurable pH increase between the influent and effluent water. Over the range of potentials investigated, the degree of cathodic protection was sufficient to maintain constant reaction rates in both the palladized and untreated iron reactors over a 9-month period of continuous operation.

Table 1 compares reaction rates for TCE and CT in terms of the disappearance half-life, and in terms of k_{sa} values. For the data in Table 1, the reactors were operated at flow rates where mass transfer limitations and dispersion effects on reactant conversion were negligible, and therefore, $k_{obs} \approx k_{sa}$. However, for reactors operated at higher palladium loadings and lower potentials, the k_{obs} values were always dependent on the flow rate, as shown in Figure 2.

TABLE 1. Reactant half-lives with 95% confidence intervals for CT and TCE reduction at a cathode potential of -755 mV (SHE). In the Pd-iron reactor, a palladium loading of 1 mg-Pd/m² of electrode surface area was employed.

Compound	Iron Reactor Half-life (min)	Iron k_{sa} (cm/min)	Pd-Iron Reactor Half-life (min)	Pd-Iron k_{sa} (cm/min)
CT	3.7 ± 0.2	9.3×10^{-6}	1.3 ± 0.1	2.7×10^{-5}
TCE	9.4 ± 0.4	3.7×10^{-6}	2.7 ± 0.2	1.3×10^{-5}

The increase in k_{obs} with increasing flow rate may be due to mass transfer or electrochemical effects. Since the solution potential and reactant and product concentrations vary along the length of the reactor, the overpotential for the TCE reduction reaction decreased with distance into the reactor. At higher flow rates, smaller amounts of reactant conversion contribute to smaller declines in reaction overpotential along the length of the reactor. We are currently developing a more mechanistic reactor model to discern between mass transfer and electrochemical reaction limitations.

Comparisons of amperometrically measured reaction rates with those measured analytically can be used to assess the relative rates of direct and indirect reduction. Chronoamperometry (CA) can be used to measure rates of direct halocarbon reduction, but cannot detect indirect reduction by atomic hydrogen. For example, by comparing the current in a blank electrolyte solution to that measured in a solution containing TCE or CT, the rate of direct halocarbon reduction can be determined. Figure 3 compares CA profiles for an iron wire electrode in a blank 10 mM $CaSO_4$ electrolyte solution with those in electrolyte solutions saturated with TCE (8 mM) or CT (5 mM).

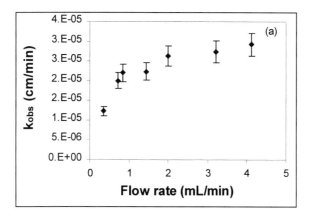

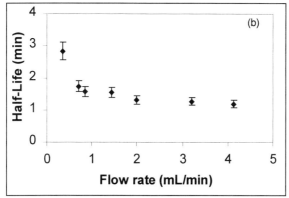

FIGURE 2. (a) Effect of flow rate on the observed rate constant for TCE reduction in an electrode reactor with 9 mg-Pd/m² of electrode surface area. (b) Effect of flow rate on the observed disappearance half-life for TCE. The reactor was operated at a cathode potential of –980 mV (SHE).

As shown in Figure 3, the presence of CT in the electrolyte solution increases the electrode current by almost one order of magnitude compared to the blank solution. This results in an amperometrically determined current efficiency (defined as the fraction of the total current going towards CT reduction) of 88%. This is similar to the analytically determined current efficiency measured in the electrode reactor, and indicates that CT reduction occurs primarily by direct electron transfer. The current efficiency for CT reduction at other concentrations is shown in Figure 4.

The CA profile for TCE in Figure 3 shows that the electrode current in the presence of TCE is slightly less than that in the blank the electrolyte solution. This indicates that TCE reduction cannot be measured amperometrically. However, as shown in Figure 4, the analytically determined current efficiency for TCE in the electrode reactor is actually greater than that for CT. This disparity between the analytically and amperometrically measured current efficiency indicates that TCE reduction occurs primarily via an indirect mechanism. This mechanism likely involves reaction with atomic hydrogen adsorbed to the surface of the electrode.

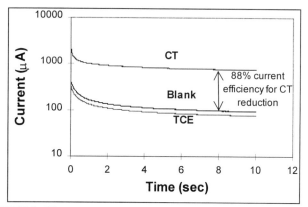

FIGURE 3. Chronoamperometry profiles at –755 mV for an iron wire electrode in a blank 3 mM CaSO₄ electrolyte solution and in 3 mM CaSO₄ solutions saturated with TCE (8mM) or CT (5 mM).

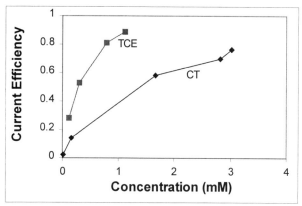

FIGURE 4. Current efficiencies for TCE and CT at an electrode potential of -755 mV (SHE).

REFERENCES

Bockris, J.O'M., and A. K. N. Reddy. 1970. *Modern Electrochemistry*. Vol. 2, Plenum Press, New York.

Brewster, J. H. 1954. "Mechanisms of Reductions at Metal Surfaces. I. A General Working Hypothesis." *J. Amer. Chem. Soc. 76*, 6361-6363.

Fogler, H. S. 1999. *Elements of Chemical Reaction Engineering*, 3rd ed., Prentice Hall: Upper Saddle River, NJ.

Gillham, R. W., and S. F. O'Hannesin. 1992. "Modern Trends in Hydrology." *International Association of Hydrologists Conference*, pp. 10-13. Hamilton, Ontario, Canada.

Gillham, R. W., and S. F. O'Hannesin. 1994. "Enhanced Degradation of halogenated aliphatics by zero-valent iron." *Ground Water 32*(6): 958-967.

Helvenston, M. C, R. W. Presley, B. Zhao. 1997. "Electro-reductive Dehalogenation on Palladized Graphite Electrodes." *Preprint of Extended Abstract, ACS Meeting*, pp. 294-297. San Francisco, CA.

Levenspiel, O. 1972. *Chemical Reaction Engineering*, 2nd ed.; John Wiley and Sons, New York.

Lyon, J. M. 1997. "Electrochemical Hydrodechlorination of Chlorinated Aromatic Compounds at a Palladium Modified Carbon Electrode." *Preprint of Extended Abstract, ACS Meeting*, pp. 143-144. San Francisco, CA.

Munakata N., W. McNab, W. Haag, P.V. Roberts, and M. Reinhard. 1997. "Effects of Water Matrix on Pd-Catalyzed Hydrodehalogenation." *Preprint of Extended Abstract, ACS Meeting*, pp. 168-170. San Francisco, CA.

Trainham, J. A., Newman, J. 1977. "Thermodynamic estimation of minimum concentration attainable in a flow-through porous electrode reactor." *J. Electrochem. Soc.* 7(4): 287-297.

TREATMENT OF HEAVY METALS USING AN ORGANIC SULFATE REDUCING PRB

Ralph Ludwig (U.S. EPA, Ada, Oklahoma, USA)
Keith Mountjoy and Rick Mcgregor (Conor Pacific Environmental
Technologies, Vancouver, B.C., Canada)
David Blowes (University of Waterloo, Waterloo, Canada)

ABSTRACT: A pilot-scale permeable reactive wall consisting of a leaf-rich compost-pea gravel mixture was installed at a site in the Vancouver area, Canada to evaluate its potential use for treatment of a large dissolved heavy metal plume. The compost based permeable reactive wall promotes microbially mediated sulfate reducing conditions such that dissolved metals are precipitated out as metal sulfides. The pilot-scale wall, measuring 10 m in length, 5.9 m in depth, and 2-2.5 m in width, has demonstrated good effectiveness in removing dissolved copper, cadmium, zinc, and nickel from ground water at the site over a 21-month period since installation. Performance has been particularly strong within the lower half of the wall where tidal influences are more limited and sulfate-reducing conditions are more easily maintained. Dissolved copper concentrations decrease from concentrations of over 4500 µg/L in the influent ground water to less than 10 µg/L within the lower half of the wall. Zinc, cadmium, and nickel concentrations decrease from average concentrations of over 2300 µg/L, 15 µg/L, and 115 µg/L, respectively to concentrations of less than 30 µg/L, 0.2 µg/L, and 10 µg/L, respectively within the lower half of the wall. The activity of sulfate reducing bacteria is evidenced by a significant increase in sulfide concentrations within the wall.

INTRODUCTION

As the number of successful permeable reactive barrier (PRB) installations at contaminated sites continues to increase, permeable reactive barrier technology is gradually being accepted as a viable alternative to conventional pump and treat. Much of the focus and field success to date has involved the use of zero valent iron-based permeable reactive barriers to treat chlorinated hydrocarbons such as the chlorinated ethenes. With the exception of chromium, limited work to date has focused on the use of permeable reactive barriers for treatment of heavy metals. This paper presents results from an organic-based sulfate reducing pilot-scale permeable reactive barrier installed at an industrial site in British Columbia to treat heavy metals associated with acid rock drainage.

The concept of using organic-based systems to treat acid rock drainage is not new. Engineered wetland systems have been used to treat acid rock drainage impacted surface water runoff at mining sites for many years. The use of organic-based permeable reactive barrier systems for treatment of acid rock drainage impacted ground water was first proposed in 1990 (Blowes, 1990). The first full-scale application of an organic-based sulfate-reducing permeable reactive barrier for treatment of acid rock drainage was at the Nickel Rim site near Sudbury, Ontario in 1995 (Benner et al., 1997). Organic-based systems rely primarily on the microbially mediated conversion of sulfates to sulfides by sulfate-reducing bacteria residing in the organic media. The simplified reaction involving reduction of sulfate and oxidation of a typical organic substrate such as lactate is given below.

$$3SO_4^{2-} + 2CH_3CHOHCOO^- + 2H^+ \rightarrow 3H_2S + 6HCO_3^- \qquad (1)$$

The reaction involves the production of both sulfide and bicarbonate. The bicarbonate produced plays an important role in regulating the pH environment of the sulfate-reducing bacteria. The sulfide produced is available to react with dissolved metals to form insoluble metal sulfides in accordance with the following reaction.

$$H_2S + Me^{2+} \rightarrow MeS_{(s)} + 2H^+ \qquad (2)$$

where Me^{2+} denotes a heavy metal such as Cd, Cu, Ni, Pb, Zn, etc. In order to ensure target metal removal from solution through the process of sulfate reduction, a sufficient quantity of sulfide must be produced to meet the demand of the heavy metal flux into the system. In a permeable reactive barrier application, under an ideal design scenario, the amount of sulfide produced would just equal the heavy metal flux into the wall. By avoiding excess production of sulfide, the organic media is not needlessly consumed and the lifetime of the wall is maximized.

Site Description. The test site is located in the Vancouver area, British Columbia and has been impacted by acid rock drainage as a result of historical ore concentrate handling and transfer practices occurring on site. The oxidation of sulfide minerals on site has resulted in the underlying ground water being extensively contaminated with heavy metals including dissolved cadmium (Cd), copper (Cu), nickel (Ni), and zinc (Zn). Copper in ground water at the site has been measured at some locations at concentrations exceeding 200,000 µg/L. Impacted ground water at the site discharges into a nearby marine inlet thus posing a potential threat to the shoreline ecosystem.

The geology at the test site is comprised primarily of deltaic deposits consisting of sands and gravel with some cobbles. The shallow aquifer, which is unconfined, begins at approximately 1 m below ground level (bgl) and extends to at least 20 mbgl. Hydraulic conductivities in the upper 15 m of the aquifer are in the 10^{-2} to 10^{-3} cm/sec range based on bail tests conducted (McGregor et al., 1999). The average hydraulic gradient has been calculated at 0.001 based on 71-hour water level averages. Metal contamination within the ground water is confined to the upper 15 m of the aquifer with the majority of the contamination being present in the upper 6 m.

MATERIALS AND METHODS

Reactive Mixture. Selected batch tests were conducted with leaf-rich compost (obtained from the City of Vancouver municipal composting facility) prior to wall installation to ensure the compost would support sulfate reduction. The final reactive mixture utilized in the wall consisted of 15% (by volume) leaf-rich compost, 84% pea gravel, and 1% limestone and was based on the results of previous laboratory and field studies (Benner et al., 1997; Waybrant et al., 1998). The large percentage of pea gravel was required to achieve a minimum desired hydraulic conductivity of 10^{-1} cm/sec within the wall. The limestone was added to ensure suitable initial pH conditions for the establishment of a sulfate reducing bacteria population within the wall. The compost, pea gravel, and limestone were thoroughly mixed by tossing and turning the materials in batches with a backhoe bucket. The mixing process for each batch was conducted until a visually-based homogeneous mixture of the components was obtained.

Pilot Wall Construction. The pilot wall was installed using cut and fill excavation methods approximately 50 m inland from the shoreline of the site to avoid ongoing construction activities along the inlet shoreline. As a result, the wall was installed in a location of known up-gradient and down-gradient soil and ground water contamination. The wall was constructed using a Komatsu Model 310 excavator to a depth of approximately 5.9 m and a length of 10 m. The width of the wall is approximately 2.5 m at surface, narrowing to 2 m width at the final depth. Excavation initially involved benching down approximately one meter to a depth just above the water table. A guar gum based slurry was used during trenching to prevent trench collapse and allow emplacement of the reactive media. The reactive media was placed into the trench using a Manotowc 4500 clam shell unit and Komatsu Model 310 excavator bucket.

A total of 17 multi-level wells were installed in and around the wall following construction as shown in Figure 1. Each multi-level well consisted of seven lengths of 1.27-cm internal diameter (ID) high density polyethylene tubing with nytex screen affixed to a 1.9 cm (ID) PVC Schedule 40 center stalk at seven discrete depths. This allowed for sampling of up to 119 sampling points at seven depths within, up-gradient, and down-gradient of the wall.

Wall Sampling. Six discrete sampling events occurred over an initial 21-month span following installation of the wall. The initial two sampling events covered all 17 multi-level wells. Sampling events thereafter were limited to a center-transect through the wall consisting of wells ML2, ML6, ML10, ML13 and ML16, as initial results indicated this transect was adequate to monitor wall performance. Sampling events consisted of ground water level measurements, and collection and analysis of ground water samples. Ground water samples were collected using a low-flow peristaltic pump with Teflon tubing and filtered through 0.45 μm cellulose acetate filters. Field measurements included pH, E_H (corrected to standard hydrogen electrode), temperature, conductivity, alkalinity, sulfide, and ferrous iron. Field measurement techniques and equipment including

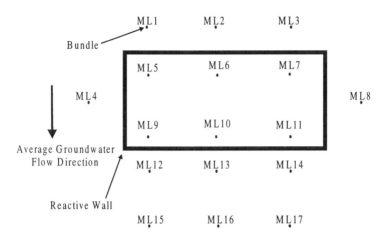

FIGURE 1. Location of monitoring bundles relative to the reactive wall (plan view).

QA/QC procedures employed are described in McGregor et al. (1999). Samples were analyzed for anions by ion chromatography and dissolved metals by ICP-OES and/or ICP-MS.

RESULTS AND DISCUSSION

Indicators of sulfate reduction within the pilot-scale wall 21 months after installation included an increase in dissolved sulfide concentrations, a decrease in the redox potential, and a decrease in metal concentrations relative to the influent ground water. Other indicators consistent with sulfate reduction included an increase in alkalinity and increase in pH although the dissolution of limestone within the reactive wall may have contributed significantly to these observed increases. Vertically averaged results for the center transect multi-level wells after a 21-month period are provided in Table 1. Metal concentration profiles through the center of the wall are shown in Figure 2. Figure 2 shows that treatment is generally greatest within the lower half of the wall where sulfate-reducing conditions are likely more easily maintained. The upper half of the wall shows poorer treatment presumably due to a greater susceptibility to influences from tidal fluctuations (i.e. wet/dry cycles and back flushing) and perhaps also oxygen intrusion from the surface. In addition, Figure 2 shows high concentrations of metals immediately down-gradient of the wall at shallow depths. This is attributed to the effects of recharge water from the surface that becomes laden with heavy metals as it infiltrates through the overlying sulfide impacted soils into the ground water on the down-gradient side of the wall.

As shown in Table 1, field measurements of pH and E_H at well ML2 indicate ground water entering the wall exhibits a relatively high redox potential (E_H of +430 mV), a pH of 6.36, and an alkalinity of 89 mg/L as $CaCO_3$. The ground water entering the pilot-scale wall is also characterized by high concentrations of copper, nickel, and zinc. As ground water passes through

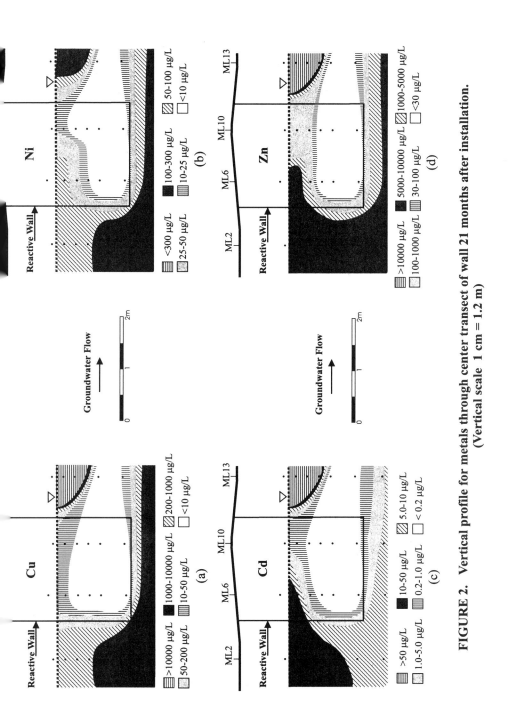

FIGURE 2. Vertical profile for metals through center transect of wall 21 months after installation. (Vertical scale 1 cm = 1.2 m)

TABLE 1.   Chemistry of the ground water entering, within, and exiting the reactive wall.

Sample Location*	pH	#E_H	Alk.	Sulfide	Cd	Cu	Ni	Pb	Zn
ML-2 Influent	6.36	430	89	<1.0	15.9	4510	118	3.8	2396
ML-6 Wall front	6.76	177	155	1704	2.7	4.4	23.4	4.3	567
ML-10Wall back	6.63	141	202	613	<0.1	10.5	5.4	1.9	82.2
ML-13 Effluent	6.57	175	180	130	<0.1	7.7	6.5	0.7	27.5

* Vertically averaged values for monitoring points. # E_H values corrected to standard hydrogen electrode. All units µg/L except pH, Eh (mV) and alkalinity (mg/L as $CaCO_3$).

the wall, alkalinity increases to an average of 155 mg/L (as $CaCO_3$) near the front end of the wall and an average of 202 mg/L near the back end of the wall. A slight increase in pH values is also noted, ranging from 6.76 near the front end of the wall to 6.63 near the back end of the wall. Dissolved sulfide concentrations within the wall increase to as high as 1704 µg/L and redox potential decreases to +141 mV.

As ground water flows through the pilot-scale wall, dissolved copper, nickel, cadmium, and zinc concentrations are significantly reduced. Copper is reduced from a vertically averaged concentration of 4510 µg/L in ground water entering the wall to averages of 4.4 µg/L and 10.5 µg/L, at the front and back ends of the wall, respectively. Figure 3 shows copper removal trends within the lower half of the wall for six sampling events spanning a 21-month period. As can

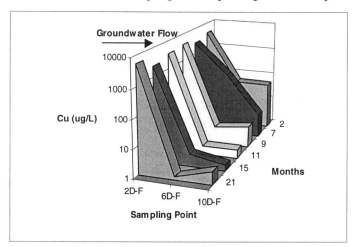

FIGURE 3.   Copper concentration trends with time within lower half of reactive wall.

be observed, a lower removal efficiency occurs over the first seven months followed by a significantly higher removal efficiency thereafter. This is presumably linked to a lag in the establishment of strong sulfate-reducing conditions within the wall.

Nickel concentrations are reduced from an average of 118 µg/L to averages of 23.4 µg/L and 5.4 µg/L at the front and back ends of the wall, respectively. Within the lower half of the wall, nickel concentrations are reduced to less than 10 µg/L (Figure 2b). Cadmium concentrations are reduced from an average of 15.9 µg/L to 2.7 µg/L and <1.0 µg/L, respectively at the front and back ends of the wall. Within the lower half of the wall, cadmium is reduced to concentrations of less than 0.2 µg/L (Figure 2c). Zinc concentrations are reduced from an average of 2396 µg/L to averages of 567 µg/L and 82.2 µg/L at the front and back ends of the wall. Within the lower half of the wall, zinc concentrations are reduced to less than 30 µg/L (Figure 2d).

Since the sulfate reduction process involves replacement of less dense organic substrate (specific gravity 1 to 2) with more dense metal sulfide precipitate (specific gravity 3 to 5), a decrease in permeability associated with metal sulfide precipitation within the pilot-scale wall would theoretically not be expected over time. Clearly, however, other precipitation reactions (e.g. hydroxides) may occur within the wall depending on the site-specific conditions in effect and these may ultimately impact the hydraulic conductivity of the wall.

The utility of an organic-based sulfate-reducing permeable reactive barrier system will depend on site-specific needs. For larger plumes where large barrier systems may be required, the low cost of using an organic substrate may be attractive. Organic-based sulfate-reducing permeable reactive barrier systems can also have the added ecological benefit of helping to restore down-gradient ecosystems by removing ferrous iron acidity from ground water and simultaneously generating a carbonate alkalinity plume. Removal of ferrous iron from the ground water prior to discharge into a surface water body prevents iron oxidation and the precipitation of ferric iron hydroxides, and production of acid that would otherwise occur in accordance with the following reaction.

$$4Fe^{3+} + O_2 + 10H_2O \rightarrow 4Fe(OH)_3 + 8H^+ \qquad (3)$$

The production of a carbonate alkalinity plume associated with the sulfate reduction process has been observed at the Nickel Rim site (Benner et al, 1997). There, influent ground water was converted from a net acid producing potential of 7.8 to 46 meq/L to a net acid consuming potential of 16 to 45 meq/L following passage through the organic substrate based reactive wall.

Wall longevity will be dependent on the reactive material maintaining its permeability and reactivity properties. Benner et al (1997) calculated that the organic-based reactive wall at the Nickel Rim Site consisting of 50% organic substrate (by volume) could be effective for a minimum of 15 years based on column study results. Metal sulfides precipitated out within the wall can be expected to remain stable provided they are not subjected to oxidizing conditions. As long as the metal sulfides remain below the water table, the oxidizing potential is likely to be limited.

Two additional monitoring events have occurred on the pilot-scale wall since the 21-month sampling event. Both of these sampling events continue to demonstrate sulfate reduction and metals removal within the wall. The chemistry from these sampling events is currently being validated and interpreted to determine recent wall performance. It is intended that the pilot-scale wall will continue to be monitored to determine the long-term performance of the wall and serve as an "early warning system" for "break-through" for full-scale installation.

CONCLUSIONS

The monitoring of geochemical ground water parameters of a compost-based pilot-scale wall over a 21-month period has indicated that the leaf-rich compost is providing a suitable organic carbon source for microbially mediated sulfate-reduction and that dissolved metals (Cd, Cu, Ni, and Zn) are being effectively attenuated by reactions within the pilot-scale wall. Continued monitoring is planned to evaluate the long-term performance of the wall.

ACKNOWLEDGEMENTS

The authors would like to thank Environment Canada for approval to write and submit this paper. The project management oversight of Eric Pringle and relentless efforts of Mike Choi including his dedication and continuity to quality field sampling and data analysis are also acknowledged.

DISCLAIMER

The views expressed in this paper are those of the individual authors and do not necessarily reflect the views and policies of Environment Canada or the U.S. Environmental Protection Agency.

REFERENCES

Benner, S.G., D.W. Blowes and C.J. Ptacek. 1997. "A Full-Scale Porous Reactive Wall for Prevention of Acid Mine Drainage." *Ground Water Monitoring and Remediation.* 17(4):99-107

Blowes, D.W. 1990. *The Geochemistry, Hydrology and Mineralogy of Decommissioned Sulfide Tailings: A Comparative Study.* Ph.D. Thesis, University of Waterloo, Waterloo, Ontario, Canada.

McGregor, R., D. Blowes, R. Ludwig, E. Pringle, and M. Pomeroy (1999) "Remediation Of A Heavy Metal Plume Using a Reactive Wall." In A. Leeson and B.C. Alleman (Eds.), *Bioremediation of Metals and Inorganic Compounds.* Batelle Press, Columbus, OH, 1999. 190pp.

Waybrant, K.R., D.W. Blowes and C.J. Ptacek. 1998. "Selection of Reactive Mixtures for Use in Permeable Reactive Walls for Treatment of Mine Drainage." *Environ. Sci. Technol.* 32:1972-1979.

INFLUENCE OF SEDIMENT REDUCTION ON TCE DEGRADATION

Jim E. Szecsody, Mark D. Williams, John S. Fruchter,
Vince R. Vermeul, John C. Evans, Deborah S. Sklarew
(Pacific Northwest National Laboratories, Richland, Washington)

ABSTRACT: A field-scale remediation technique for TCE and chromate is currently being implemented which uses a chemical treatment to reduce existing iron(III) in sediments. While reduction of some contaminants is well established, TCE data show that dechlorination is more complex, and the role of iron oxides to catalyze the reaction is not well understood. The purpose of this laboratory-scale study was to investigate the influence of temperature and partial sediment reduction on TCE dechlorination. Fully reduced sediments can degrade TCE at sufficiently fast rates (1.2 to 19 h) during static and transport experiments over $2^{\circ}C$ to $25^{\circ}C$ that a successful barrier could be made at the field scale. In contrast, partially reduced sediment resulted in up to a 3 order of magnitude decrease in the TCE dechlorination rate. While minimally reduced sediment had nearly no TCE reactivity, > 40% reduced sediment had considerably faster dechlorination rates. The second-order dependence of the TCE dechlorination rate on the fraction of reduced iron demonstrates the significant role of the iron oxide surface (as a catalyst or for surface coordination) for TCE dechlorination. Based on these results, the field-scale reduction was designed with specific reagent concentrations, temperature, and flow rates to efficiently create a reductive barrier.

INTRODUCTION

A field-scale TCE and chromate remediation technique is currently being implemented which uses a chemical treatment to reduce existing Fe^{III} in sediments. While chromate, nitrate, and U^{VI} species are readily reduced by Fe^{II} surface phases and immobilized, field and laboratory-scale TCE data show that dechlorination is more complex and indicate that the role of iron oxides to catalyze the dechlorination reaction is not well understood. The purpose of this laboratory-scale study was to investigate the influence of: a) temperature on sediment reduction, b) temperature on TCE reduction, and d) fraction of reduced sediment on the TCE degradation rate. The application of these experiments is to determine how to efficiently reduce sediment at the field scale ($11^{\circ}C$ to $21^{\circ}C$) for TCE degradation where high Fe-bearing sediments cannot be uniformly reduced. While single reactions generally have a well-defined relationship with temperature, sediment reduction (multiple dissolution, reduction, adsorption reactions) and TCE dechlorination (surface catalyzed reaction) may not. The relationship between the fraction of reduced iron and the TCE dechlorination rate will provide some evidence for the role of the Fe^{III} oxide surface to catalyze TCE dechlorination.

GEOCHEMICAL REACTIONS FOR REMEDIATION OF TCE

Iron Reduction Mechanism. The proposed technology utilizes an aqueous reductant to reduce existing iron in aquifer sediments (to Fe^{II} phases) to create a permeable reductive barrier (Fruchter et al, 2000; Chilakapati et al., 2000). Zero valent iron/mixed metal barriers also rely on the oxidation of ferrous iron (adsorbed or Fe^{II}, green rust; Genin et al., 1998) for dechlorination (Balko and Tratnyek, 1998; Johnson et al., 1998) or reduction of metals (Istok et al., 1999; Blowes et al., 1997), and not the oxidation of Fe^{O}. While aqueous Fe^{II} can reduce chromate (Eary and Rai, 1988), adsorbed Fe^{II} on an Fe^{III}-oxide, clay, or zerovalent iron surface is necessary for dechlorination reactions, although the role of the surface is not clearly understood.

The dithionite chemical treatment dissolves and reduces amorphous and some crystalline Fe^{III} oxides, producing adsorbed Fe^{II} [dominant phase], Fe^{II}-carbonate (siderite), and FeS. Although multiple Fe^{III} phases are present in natural sediments, a reaction that describes a single phase of iron that is reduced by sodium dithionite:

$$S_2O_4^{2-} + 2 \equiv Fe^{3+} + 2 H_2O <==> 2 \equiv Fe^{2+} + 2 SO_3^{2-} + 4 H^+ \quad (1)$$

has been used to describe most of sediment reduction. With some sediment a second parallel reaction is additionally used. Because aqueous Fe^{II} produced has a high affinity for surfaces, it is quickly adsorbed, so Fe^{II} mobility in mid- to high pH in low ionic strength systems is low. Disproportionation of dithionite that occurs in aqueous solution (~27 h half-life) or in contact with sediments:

$$2S_2O_4^{2-} + H_2O <==> S_2O_3^{2-} + 2 HSO_3^- \quad (2)$$

describes dithionite mass that is not available for iron reduction. The consequence of this reaction is to limit how slowly dithionite can be reacted with sediments; if dithionite is injected too slowly, a significant amount of the mass is lost to disproportionation.

Sediment Oxidation Mechanisms. The oxidation of the adsorbed and structural Fe^{II} in the sediments of the permeable redox barrier occurs naturally by the inflow of dissolved oxygen through the barrier, but can additionally be oxidized by contaminants that may be present such as chromate, TCE, nitrate, or uranium. In relatively uncontaminated aquifers, dissolved oxygen in water is the dominant oxidant of reduced iron species:

$$4 Fe^{2+} + O_2 + 4H^+ <=> 4 Fe^{3+} + H_2O \quad Eh = -1.85 \text{ v} \quad (3)$$

where 4 moles of Fe^{II} are oxidized per mole of O_2 consumed. At oxygen-saturated conditions (8.4 mg L^{-1} O_2, 1 atm, 25°C), 1.05 mmol L^{-1} Fe^{II} is consumed. Experimental evidence indicates that the oxygenation of Fe^{II} in solutions (pH >5) is generally found to be first order with respect to Fe^{II} and O_2 concentration and second-order with respect to OH-. The rate of oxidation of Fe^{II}(aq) by oxygen at pH 8 is a few minutes (Eary and Rai, 1988, Buerge and Hug, 1997), and in contrast, the oxidation of Fe^{II} phases in natural sediments was found to be 0.3 to 1.1 h. At sites with high oxidant contamination, barrier lifetime is a function of all of the constituents that can oxidize the Fe^{II} phases. For example, if chromate was present at a concentration of 41 mg L^{-1}, the equivalent mass of Fe^{II} would be oxidized relative to oxygen-saturated water.

TCE Degradation. The degradation pathways for most organic compounds including TCE are complex, involving multiple reaction steps. Of four possible abiotic degradation pathways for TCE, the two common pathways are reductive elimination and hydrogenolysis. Reductive elimination:

$$TCE + 2H^+ + 2e^- <=> chloroacetylene + 2Cl^- \quad Eh = 0.60v \quad (4)$$

describes the destruction of TCE to easily degraded (abiotically or biotically) chlorinated acetylene products. Abiotic degradation by hydrogenolysis:

$$chloroacetylene + H^+ + 2e^- <=> acetylene + Cl- \quad Eh = 0.50v \quad (5)$$
$$acetylene + 2H^+ + 2e^- <=> ethylene \quad Eh = 0.39v \quad (6)$$

proceeds rapidly as chlorinated acetylenes are unstable (Delavarenne and Viehe, 1969). The degradation of TCE to ethylene by reductive elimination involves the production of 5 moles of electrons, or 26 mg L^{-1} TCE is needed to oxidize the equivalent mass of Fe^{II} as oxygen-saturated water [1.05 mmol L^{-1} Fe^{II}]. Studies of TCE degradation pathways using zero-valent iron and various Fe^{II} minerals (Roberts and others, 1996; Sivavec and Horney, 1995; Orth and Gillham, 1996) indicate that reductive elimination is the major abiotic pathway, with some DCE isomers and vinyl chloride produced from the hydrogenolysis pathway.

EXPERIMENTAL METHODS

A series of batch and column experiments were conducted to determine the mass and rate of reduction of iron in sediment by the reduction solution. Batch experiments consisted of a single large septa-top glass bottle in which 14 to 200 g of sediment was mixed with the dithionite solution for 100s of hours in a temperature-controlled chamber ($2°C$ to $42°C$). At specific time intervals, a sample was withdrawn, filtered, and analyzed for dithionite remaining in solution. TCE degradation studies were conducted in batch systems. Batch TCE experiments consisted of reacting 1.0 mg L^{-1} TCE in water with reduced sediment in vials with no headspace for minutes to 240 h in a temperature-controlled chamber ($2°C$ to $25°C$). At specific time intervals, water was removed, filtered, and organic solutes analyzed by gas chromatograph mass spectrometry (GC-MS).

RESULTS

Sediment reduction and temperature. Reduction can be qualitatively observed from change in sediment color from tan to gray (within 10 h). Because dithionite is used for iron reduction (~5 h half-life) and disproportionation (~27 h half life), excess dithionite is needed to fully reduce sediment. An experiment at 0.1 mol/L dithionite (Fig. 1a) shows consumption from iron reduction with a shallow slope for the first 100 h, followed by consumption from disproportion with a steeper slope for >100 h. With a greater relative mass of reducible sediment (Fig. 1b), nearly all of the dithionite is consumed by reduction. Reduction and disproportionation rates were determined from simulations of time-dependent dithionite consumption from reduction (rxn 1) and disproportionation (rxn 2). In general, simulations with a single third-order reduction and first-order disproproportionation match the observed data, so the natural sediment could be modeled roughly as a single reductive dissolution reaction even though multiple iron phases were being reduced. A simulation without disproportionation (dashed line, Fig. 1a) shows the fraction of dithionite used for iron reduction only. For one sediment (<4mm fraction of silty gravel), batch reductions averaged 3.3±0.7 h at 25°C, and five column reductions averaged 8.0 ± 2.5 h. The cause of the difference in the observed reduction rate in batch and column systems may be due to mass transfer limitations in the packed porous media. Twelve batch data sets at over 2°C to 42°C showed that iron reduction averaged 2.27x slower for each 10°C decrease, and disproportionation was 3.04x slower for each 10°C decrease.

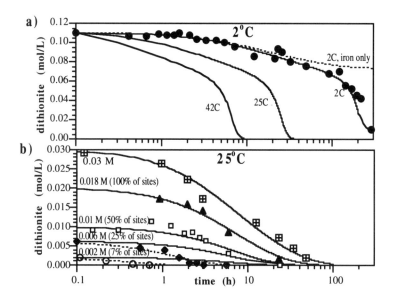

FIGURE 1. Batch reduction of sediment at: a) high dithionite/sediment ratio and 2°C, and b) variable dithionite/sediment ratio and 25°C.
There was evidence of multiple Fe^{II} phases created by the dithionite, and that these Fe^{II} phases remained immobile during reduction and subsequent oxidation. Iron extractions showed a decrease in Fe^{III} oxides and an increase in Fe^{II} phases (80-100% adsorbed Fe^{II}, 20% siderite) during reduction. This reducible fraction (0.4% Fe/g) was a small fraction of the total iron oxides (3.0%Fe/g). Modeling of experiments in which a small fraction of the surface sites were reduced had a faster rate (dashed lines, Fig. 1b), indicating the most reactive sites (likely amorphous Fe^{III} oxides) are more quickly dissolved/reduced. Reactive transport modeling of sediment reduction and oxidation during flow (not shown) showed similar results of two reducible sites (with differing rates) could better match breakthrough data. Extractions also demonstrated that there was little iron mobility, where <3% of iron was mobilized after 9 pore volumes of

reductant injection and <10% of iron was mobilized after 600 pore volumes of O_2-saturated water injection in a column. Aqueous Fe^{II} has a high affinity for Fe- and Al-oxides at this pH (7.7-8.3), which accounts for the low mobility.

TCE reduction and temperature. The major pathway for TCE degradation by dithionite-reduced sediments was shown to be reductive elimination (Fig 2a), where 99.5% of TCE mass is accounted for in chloroacetylene and acetylene. The degradation rate (3.5 hours) could be modeled with two reactions (4 and 5). Evidence for the lack of importance of the hydrogenolysis pathway for TCE degradation is shown by the very small increase in vinyl chloride (0.30 µg/L) and no change in DCE concentration (not shown). Similar experiments conducted from 2°C to 25°C (Fig. 2b; only acetylene data) showed a regular decrease in TCE degradation rate with lower temperature. Therefore, the regular 2.67x decrease in rate per 10°C indicates that TCE dechlorination can occur at aquifer temperatures at a predictable rate, even though TCE degradation is surface catalyzed.

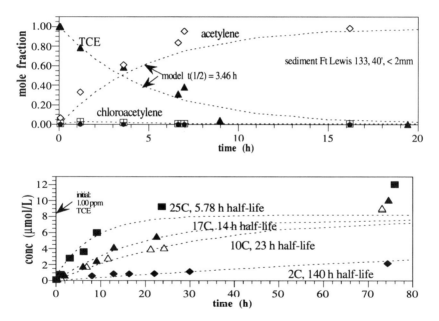

FIGURE 2. Degradation of 1.0 mg/L TCE at: a) 25°C, with TCE and degradation products shown, and b) 2°C to 25°C, with acetylene shown.

TCE reduction by partially reduced sediment. Because TCE degradation requires both an electron donor (adsorbed Fe^{II}) and a surface (iron oxide or clay), the rate of dechlorination is a simple function of the mass of reduced iron. This fact is significant at the field scale because sediments cannot be uniformly reduced. The TCE degradation rate is highly dependent on the fraction of reduced iron in sediment and varied from > 1000 h for 11% reduced to 1.2 h for 100% reduced sediment (Fig. 3), with a significant increase in rate between 30% and 45% reduced sediment. While the TCE degradation rate was found to be a first-order function of the Fe^{II} surface concentration (Fig. 2) for fully reduced sediments, the observed TCE degradation rates in partially reduced sediment (Fig. 3) was a second-order function of the Fe^{II} surface concentration. This may indicate the importance of both Fe^{II} available as an electron donor and the Fe^{III} oxide surface. The surface is necessary for the electron transfer reaction as laboratory experiments have shown that TCE and carbon tetrachloride are not dechlorinated in the presence of only $Fe^{II}(aq)$. The role of the surface is not well understood, although may act as a catalyst, a semiconductor, or provide the necessary surface coordination for the electron transfer reactions (Scherer et al., 1999; Wehrli, 1992). These batch results are consistent with a long term (4-month) column study in which fully reduced sediment quickly degraded TCE, but when the sediment became partially oxidized (<50% reduced), the TCE degradation rate decreased significantly (Thornton et al., 1998).

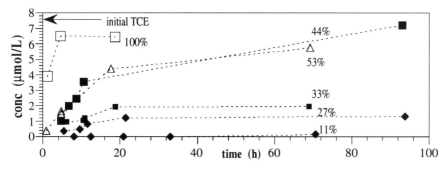

FIGURE 3. TCE dechlorination rate in partially reduced sediment, as shown by acetylene production rate with differing fraction reduced iron.

Application to field-scale TCE dechlorination. A field-scale demonstration of TCE remediation by reduced sediments was initiated in four dithionite injections designed to incrementally reduce sediment. The incremental injections provided partial iron reduction versus TCE dechlorination rate data to compare to laboratory data (Fig. 3), but with additional field scale complexities of physical/ chemical heterogeneities and fluctuation in ground water velocity and direction. Based on temperature studies (Fig. 1), injection water was heated to decrease the sediment/dithionite contact time needed for reduction. Because density sinking occurs with dithionite injection at high concentration, the injection strategy used was a heated, lower dithionite concentration (and larger volume), which resulted in more uniform and higher percentage iron reduction.

Three separate dithionite injections over the course of 13 months resulted in a ~5 m radial by 4.6 m vertical zone of >35% reduced sediment (in the 9th month; dashed line, Fig. 4), and a larger zone that was less reduced. In these field tests, the TCE degradation increased with increased percent reduction, as shown in one well (of 14) located 5.5 m down gradient of the injection well. The ~130 µg/L TCE present initial was degraded to 100 µg/L (0.76 µmol/L) TCE after the first reduction, 0.4 µmol/L after the second reduction (sediment was 22% reduced), and 0.1 µmol/L after the third reduction (sediment was 35% reduced).

The main degradation product, acetylene (see Fig. 3), showed a corresponding increase to 82% of the initial TCE mass when aquifer sediments were 35% reduced. TCE degradation was greater at downgradient wells within the reduced zone and less for upgradient wells, due to greater TCE residence time within the reduced zone. Sediment coring after one additional scheduled injection will provide additional evidence of the spatial distribution of reduced sediment and correlation with TCE degradation.

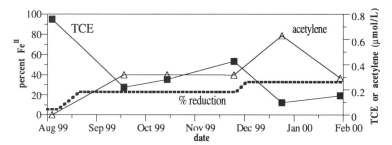

FIGURE 4. Field-scale demonstration of incremental reduction (dashed line) versus TCE degradation into acetylene at a monitoring well located 5.5 m down gradient of the dithionite injection well.

CONCLUSIONS

The purpose of this laboratory-scale study was to investigate the influence of temperature and partial sediment reduction on TCE dechlorination by reduced sediments. The reduction reaction, being third-order, reduces significant iron at short times, and is slower at later times. Efficient reduction of iron in sediment is achieved with limited sediment/dithionite contact times (< 32 h at 25°C, < 100 h at 12°C), because dithionite mass is removed mainly by disproportionation at long contact times. Dithionite-reduced sediments were shown to degrade TCE in ground water at sufficiently fast rates (1.2 to 19 h) during static and transport experiments that a successful barrier could be made at the field scale. The resulting TCE dechlorination mechanism is reductive elimination, based on the measurement of these primary pathway species:

TCE --> chloroacetylene --> acetylene --> etylene

A second dechlorination pathway (hydrogenolysis) appears to be minor, with < 0.5% of the TCE mass being converted. The rate of TCE dechlorination decreased regularly with temperature, which indicated that the iron oxide surface acted as an effective catalyst over 2° to 25°C. In contrast, the relationship between partially reduced sediment and TCE reactivity was highly nonlinear. While minimally reduced sediment had nearly no TCE reactivity, > 40% reduced sediment resulted in TCE reduction rates that were viable at the field-scale (< 65 h half-life). Based on these results, the field-scale reduction experiment was designed with specific reagent concentrations, temperature, and flow rates to efficiently create a permeable reductive barrier. A field-scale demonstration in a confined aquifer contaminated with TCE currently shows that the reduced zone (10 m diameter by 4 m vertical) is removing up to 80% of the TCE.

REFERENCES

Balko, B., and Tratnyek, P., 1998, Photo effects on the Reduction of Carbon Tetrachloride by Zero-Valent Iron, *J. Physical Chemistry*, 102(8): 1459-1465.

Buerge, I. J. and S. J. Hug, 1997. Kinetics and pH Dependence of Chromium(VI) Reduction by Iron(II), *Environ. Sci. Technol.*, 31(5):1426-1432.

Blowes, D.,Ptacek, C., and Jambor, J., 1997, In Situ Remediation of Cr(VI) Contaminated Groundwater Using Permeable Reactive Walls: Laboratory Studies, *Env. Sci. Technol.*, 31(12): 3348-3357.

Chilakapati, A., M. Williams, S. Yabusaki, C. Cole, 2000, Optimal design of an in situ Fe^{II} barrier: transport limited reoxidation, *Environ. Sci. Technol.,* submitted.

Delavarenne, S., and Viehe, H., 1969, *Chemistry of Acetylenes,* Marcel Dekker, New York, p 651-750.

Eary, L., and Rai, D., 1988, Chromate Removal from Aqueous Wastes by Reduction with Ferrous Ion, *Environ. Sci. Technol.,* 22: 972-977.

Fruchter, J., C. Cole, M. Williams, V. Vermeul, J. Amonette, J. Szecsody, J. Istok, and M. Humphrey, 2000, Creation of a subsurface treatment zone for aqueous chromate contamination using in situ redox manipulation. *Ground Water* Monit. Rev., in press.

Genin, J., Bourrie, G., Trolard, F., Amdelmoula, M., Jaffrezic, A., Refait, P., Maitre, V., Humbert, B., and Herbillon, A., 1998, Thermodynamic Equilibria in Aqueous Suspensions of Synthetic/Natural Fe^{II}-Fe^{III} Green Rusts: Occurrences of the Mineral in Hydromorphic Soils, *Environ. Sci. Technol.*, 32, 1058-1068.

Istok JD, Amonette, J.E., Cole CR, Fruchter JS, Humphrey MD, Szecsody JE, Teel SS, Vermeul VR, Williams MD and Yabusaki SB. 1999. In Situ Redox Manipulation by Dithionite Injection: Intermediate-Scale Experiments. *Ground Water*. 37:884-889.

Johnson, T., Fish, W., Gorby, Y., and Tratnyek, P., 1998, Degradation of Carbon Tetrachloride: Complexation Effects on the Oxide Surface, *J. Cont. Hyd.*, 29, 379-398.

Orth, W., and Gillham, R., 1996. Dechlorination of trichloroethene in aqueous solution using Fe^{o}, *Environ. Sci. Technol.,* 30(1):66-71.

Roberts, A., Totten, L. Arnold, W., Burris, D., and Campbell, T., 1996, Reductive Elimination of Chlorinated Ethylenes by Zero-Valent Metals, *Environ. Sci. Technol.,* 30(8): 2654-2659.

Scherer, M., Balko, B., and Tratnyek, P., 1999, The Role of Oxides in Reduction Reactions at the Metal-Water Interface, in *Kinetics and Mechanisms of Reactions at the Mineral/Water Interface*, eds D. Sparks, ACS Symposium Series #715, p 301-322.

Sivavec, T., and Horney, D., 1995, Reductive Dechlorination of Chlorinated Ethenes by Iron Metal and Iron Sulfide Minerals. in *Emerging Technologies in Hazardous Waste management VII*, p. 42-45. Atlanta, Ga., American Chemical Society.

Thornton, E., Szecsody, J., Cantrell, K., Thompson, C., Evans, J., Fruchter, J., and Mitroshkov, A., 1998, Reductive dechlorination of TCE by dithionite, in *Physical, Chemical, and Thermal Technologies for Remediation of Chlorinated and Recalcitrant Compounds*, eds G. Wickromanayake and R. Hinchee, p 335-340.

Wehrli, B., 1992, Redox Reactions of Metal Ions at Mineral Surfaces, in *Aquatic Chemical Kinetics*, ed. W. Stumm, Wiley Interscience, New York.

FIELD RESULTS FROM THE USE OF A PERMEABLE REACTIVE WALL

Ole Kiilerich, Jan Wodschow Larsen, and Charlotte Nielsen
(HOH Water Technology, Geminivej 24, DK-2670 Greve, Denmark)
Lars Deigaard (ScanRail Consult, Pilestræde 58, DK-1112 Copenhagen, Denmark)

ABSTRACT: A permeable reactive wall consisting of 75 tons of granular iron was constructed to reduce the concentrations of chlorinated aliphatics in a groundwater pollution plume. The performance of the wall has been closely monitored for 15 months, including chemical analysis of groundwater samples from monitoring wells in the wall and upgradient and downgradient of the wall. The results show that the half-lives of the chlorinated aliphatics vary from 18 to 75 hours. The pH rises from 7.7 outside the wall to 9.4 in the middle of the wall, causing precipitation of various inorganic minerals inside the wall. It is estimated that 1,000 kg of $Fe(OH)$, 200 kg of $CaCO_3$, 200 kg of $FeCO_3$, and 60 kg of FeS precipitate in the wall per year. Slug tests indicate that the average permeability inside the wall has been reduced from $6 \cdot 10^{-5}$ to $0.8 \cdot 10^{-5}$ m/s during the monitoring period, presumably due to the precipitation of inorganic minerals.

INTRODUCTION

The permeable reactive wall based on zero-valent iron was introduced as a groundwater remediation technique in the beginning of the 1990's (O'Hannesin and Gilham, 1992), and it is now being used at an increasing number of sites where the groundwater is contaminated with chlorinated aliphatics and other substances (Vogan et al., 1998). The method is patented by the Canadian company Envirometal Technologies Ltd.

A contaminated site at the Copenhagen Freight Yard in Copenhagen, Denmark was among the first sites in Europe where the permeable reactive wall was applied. The wall was constructed in July 1998 and this paper presents the results from the first 15 months of monitoring of the wall, with special emphasis on the problems arising from the precipitation of minerals and their effect on the permeability and dehalogenation capacity of the wall. The results are evaluated and compared to results from other studies described in the literature.

MATERIALS AND METHODS

Site Description. The site belongs to the Danish Railway Agency and the Danish State Railways. It contains a workshop for locomotive engines and ferry parts where trichloroethylene and possibly also perchloroethylene have been used. A map of the site including the location of the reactive wall and the monitoring wells is shown on figure 1.

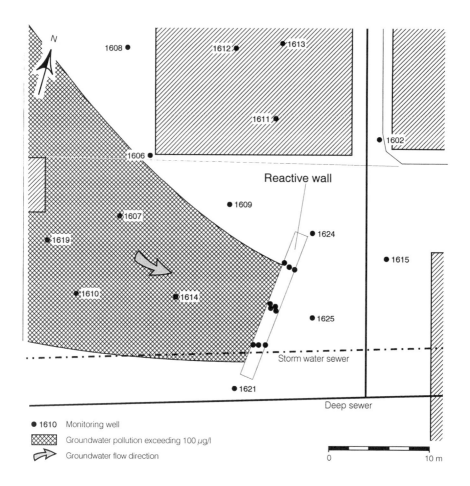

FIGURE 1. Site map showing location of monitoring wells and reactive wall.

The local geology consists of a top layer of 2 m sandy fill, underlaid by 1-2 m clayey fill mixed with peat, followed by 2-3 m of marine coastal sand, which acts as a local upper groundwater aquifer with a piezometric head about 2.5 m below ground level. The sand is underlaid by 1-2 m of clayey till, followed by 6 m of silty, fluvial sand in hydraulic contact with Danien limestone, which acts as a regional groundwater aquifer. The permeability of the upper aquifer was tested to be in the order of $6 \cdot 10^{-5}$ m/s. The groundwater flow gradient is about 0.007 and the average linear velocity is estimated to be about 40 m/year.

The upper groundwater aquifer is polluted by chlorinated aliphatics in concentrations of up to 4,000 µg/l. Dichloroethylene is the dominant compound but also perchloroethylene, trichloroethylene, and vinylchloride are found. The groundwater is anaerobic with elevated concentrations of dissolved iron, ammonium and methane.

Construction of the Reactive Wall. The reactive wall was designed to cut off that part of the pollution plume where the total concentration of chlorinated aliphatics exceeds 100 µg/l. The wall was constructed 15 m wide and 6 m high, containing 75 ton of granular iron in the deepest 2.5 m of the section and clayey fill on top. A wall thickness of 0.9 m was chosen to ensure a sufficient residence time. Figure 2 shows the principal layout of the wall.

Horisontal section.

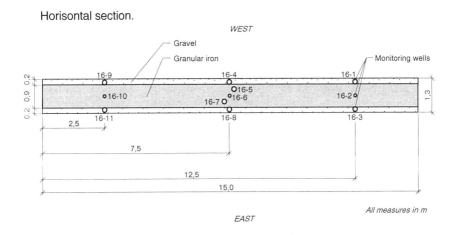

Vertical section.

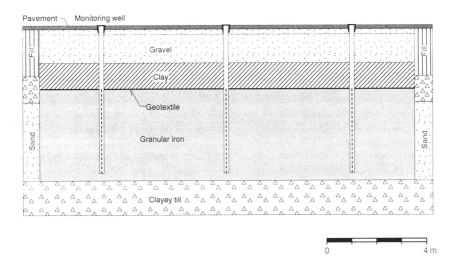

FIGURE 2. Horisontal and vertical section of the reactive wall.

Monitoring. In order to monitor the physical and chemical behavior of the wall, 3 sets of 3-5 monitoring wells each were installed in the wall to facilitate groundwater sampling in the middle, upgradient, and downgradient of the wall. Additional monitoring wells were installed around the wall for monitoring of the pollution plume outside the wall.

An extensive monitoring program was carried out from the construction of the wall in July 1998 until October 1999. The monitoring program consisted of quarterly measurements of groundwater levels and groundwater sampling with chemical analysis from 15 monitoring wells inside and outside the wall. Additionally, permeability tests were performed in 11 monitoring wells.

The chemical analysis of the water samples comprised chlorinated aliphatics and inorganic parameters. The chlorinated aliphatics included perchlororethylene (PCE), trichloroethylene (TCE), trichloroethane (TCA), 1,1-dichloroethane (1,1-DCA), 1,1-dichloroethylene (1,1-DCE), trans-1,2-dichloroethylene (trans-1,2-DCE), cis-1,2-dicholoethylene (cis-1,2-DCE), and vinylchloride (VC). The chemical analysis of chlorinated aliphatics were performed by extraction with xylene followed by gaschromatographic analysis with electron capture detector (PCE, TCE, and TCA) or mass spectrometry (DCA, DCE, and VC). The inorganic parameters included pH, conductivity, COD, oxygen, bicarbonate, ammonium, nitrate, calcium, chloride, iron, sulphate, sulphide, methan, manganese, fluoride, magnesium, sodium, and potassium.

The aquifer and wall permeabilities were investigated by means of pumping tests and slug tests. Pumping tests were used in July 1998, immediately after the construction of the wall. In later testing rounds only slug tests were used. The results from the slug tests were interpreted using the Bouwer and Rice method (Bouwer, 1989).

RESULTS AND DISCUSSION

Table 1 shows a summary of the chemical analysis of selected contaminants and inorganic species in samples taken from the middle section of the wall in March 1999 (ScanRail Consult, 2000). In general, similar results are found with the other samples taken in the monitoring period.

Inorganic Species. Table 1 shows that the concentrations of some of the inorganic species decrease in the wall, especially iron, sulphate, calcium and bicarbonate. This is probably due to the rise in pH from 7.7 to 9.4, which is caused by redox-reactions related to the granular iron. From the concentrations in table 1 the precipitation in the wall are calculated to be 200 kg of $CaCO_3$, 200 kg of $FeCO_3$, and 60 kg of FeS per year. These calculations are based on a wall thickness of 0.9 m, a porosity of the granular iron of 0.4, and an average linear groundwater flow velocity of 40 m/year, which yields a residence time in the wall of about 200 hours and a total yearly amount of 600 m³ of groundwater flowing through the wall.

Apart from the precipitation reactions that can be calculated on the basis of the dissolved inorganic species in the groundwater, a precipitation of Fe(OH) will probably take place due to the corrosion of the iron in the wall. In a study where a different type of iron was used, the corrosion caused the production of 0.6 mmol hydrogen per kg iron per day (Reardon, 1995). Assuming, in spite of differences in the type of iron, that the same scaling factor for hydrogen production is valid at the present site, a rough estimate can be made of the release of Fe^{2+} from the wall. With a total of 75 tons of iron in the wall this corresponds to a release of 2.5 kg of Fe^{2+} per day, or about 1,000 kg of Fe(OH) per year if all the Fe^{2+} precipitates as Fe(OH).

Table 1. Summary of analytical results for selected groundwater constituents.

Parameter		Upgradient of the wall (Well 16-4)	Middle of the wall (Well 16-6)	Downgradient of the wall (Well 16-8)
TCE	(μg/l)	20	0.2	0.01
trans-1,2-DCE	(μg/l)	58	3	<0.1
cis-1,2-DCE	(μg/l)	180	77	28
VC	(μg/l)	82	46	9
pH	-	7.7	9.4	9.6
NVOC	(mg/l)	8.8	12	8.1
Oxygen	(mg/l)	0.2	0.6	0.3
Nitrate	(mg/l)	0.2	0.2	<0.1
Ammonium	(mg/l)	9.8	10	9.2
Iron	(mg/l)	4.3	0.2	0.2
Manganese	(mg/l)	0.3	0.03	0.008
Sulphate	(mg/l)	110	9.8	8.2
Sulphide	(mg/l)	0.08	0.01	0.1
Methane	(mg/l)	3.2	11	11
Magnesium	(mg/l)	29	22	8.6
Sodium	(mg/l)	160	170	180
Chloride	(mg/l)	180	180	190
Calcium	(mg/l)	130	3.6	3.7
Bicarbonate	(mg/l)	610	330	290

The precipitation of minerals apparently take place between the upgradient well and the middle well, as there are only small variations in the concentrations of inorganic species between the middle well and the downgradient well. Fe(OH), however, may precipitate throughout the wall, as it does not depend upon the dissolved inorganic species conveyed by the groundwater. Experiences from other studies show that the main part of the precipitation of minerals in a reactive wall takes place in the first few cm of the wall (Vogel et al., 1998).

From table 1 it also appears that the methane concentration is rising from 3.2 to 11 mg/l in the wall. The rise in methane concentrations might be due to biological activity under the reduced redox conditions in the wall. In a similar study, increased methane concentrations were found immediate downgradient of the wall (Warner et al., 1998).

Chlorinated Aliphatics. Generally, table 1 shows that about 90% of the total content of chlorinated aliphatics are removed in the wall. In table 2 the estimated half-lives of single compounds are shown. The half-lives are calculated on the basis of the upgradient and downgradient concentrations in table 1 and a residence time of the groundwater in the wall of 200 hours. For comparison the estimated half-lives from two other studies are also shown.

Table 2. Calculated half-lives for single compounds.

Compound	$t_{1/2}(a)$ (hours)	$t_{1/2}(b)$ (hours)	$t_{1/2}(c)$ (hours)
TCE	18	14	9.5
trans-1,2-DCE	<22	55	22
cis-1,2-DCE	75	432	18
VC	63	374	18

a) This study
b) Gilham and O'Hannesin, 1994
c) Rafalowich et al., 1998

The estimated half-lives from this study are seen to be in the same order of magnitude as in the other two studies, except for cis-1,2-DCE and VC where the two other studies show higher and lower half-lives, respectively.

When estimating the half-lives on the basis of the other data from the monitoring period, some variations are found in the half-live values. However, the half-lives do not show any correlation with the time elapsed since installing the wall, neither do they depend on whether they are calculated on the basis of the concentration difference from the upgradient well to the middle well, or from the middle well to the downgradient well. This leads to the conclusion that the precipitation of inorganic species does not yet have any significant effect on the dehalogenation capacity of the wall.

Aquifer Permeability. Figure 3 shows the results of the aquifer permeability tests. The vertical crosses show the standard deviation for results that involve several slug tests or pumping tests of the same well.

The slug tests performed inside the wall (wells 16-2, 16-6, 16-7, and 16-10) show that the average permeability falls from $6 \cdot 10^{-5}$ m/s initially to $0.8 \cdot 10^{-5}$ m/s after 15 months of monitoring. The wells 16-2 and 16-10, for which there are results from the whole monitoring period, both show a continuous decrease in permeability during the first 12 months, after which the permeability seems to be more stable. The decreasing permeability in the wall might be due to the precipitation of inorganic minerals. In another study, porosity losses of 6 to 10% in the first 0.3 m of the wall were measured after 18-24 months (Vogel et al., 1998).

With regard to the test results in the aquifer outside the wall (wells 1602, 1609, 1614, 1615, 1621, and 1625), the average permeability has fallen from initially $9 \cdot 10^{-5}$ m/s initially to $6 \cdot 10^{-5}$ m/s at the end of the monitoring period.

However, this reduction in permeability is insignificant compared to the general uncertainty of the testing method. Furthermore, the permeability measurements do not show any consistent pattern between the wells, e.g. the permeability in well 1609 decreases whereas the permeability in well 1614 and 1615 increases. Hence, the results do not indicate any significant change in aquifer permeability.

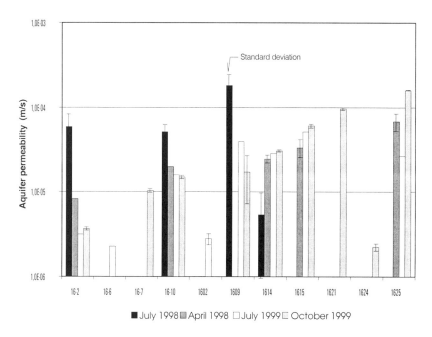

■ July 1998 ▨ April 1998 ☐ July 1999 ▨ October 1999

FIGURE 3. Results of the aquifer permeability tests.

Measurements of the groundwater level around the wall support the assumption of a decrease in the wall permeability, as they show a slight rising of the groundwater table immediate upgradient of the wall. Consequently, a minor part of the pollution plume appears to deviate from its former path and travel North of the wall instead of through the wall.

ACKNOWLEDGEMENTS

The site remediation discussed in this paper is part of a development project on environmental/economic evaluation and optimizing of contaminated sites remediation, carried out by the National Danish Railway Agency and the Danish State Railways (ScanRail, 2000). The project includes the development of a method for involving environmental assessment in the decision-making, including guidelines for an all-inclusive environmental accounting. The method was demonstrated at six sites in Denmark. The project started in March 1997 and finished in February 2000, and it was supported by the EU LIFE-program (no. 96ENV/DK/0016) and the Danish Environmental Protection Agency.

REFERENCES

Bouwer, H. 1989. The Bouwer and Rice Slug Test - An Update. *Ground Water* *27*(3): 304-309.

Gilham, R.W. and S.F. O'Hannesin. 1994. "Enhanced degradation of halogenated aliphatics by zero-valent iron. *Ground Water 32*(6): 958-967.

O'Hannesin, S.F. and R.W. Gilham. 1992. A permeable reaction wall for in situ de-gradation of halogenated organic compounds. Paper presented at *The 45th Canadian Geotechnical Society Conference* in Toronto, Ontario, October 25-28.

Rafalovich, A., S. O'Hannesin, G. Haling, C. Gaudette, and W. Clark. 1998. Reactive wall column testing to support remediation decision making. In G.B. Wickramanayake and R.E. Hinchee (Eds.), *Remediation of Chlorinated and Recalcitrant Compounds: Designing and Applying Treatment Technologies*, pp. 59-64. Battelle Press, Columbus, Ohio.

Reardon, E.J. 1995. Anaerobic corrosion of granular iron: Measurements and interpretation of hydrogen evolution rates. *Environ. Sci. Technol.* 29(12): 2936-2945.

Scanrail Consult. 2000. *Environmental/Economic Evaluation of Contaminated Sites Remediation. Evaluation of Demonstration Projects.* Report prepared for the Danish Railway Agency and the Danish State Railway, Copenhagen.

Vogan, J.L., B.J. Butler, M.S. Odziemkowski, G. Friday, and R.W. Gilham. 1998. Inorganic and biological evaluation of cores from permeable iron reactive barriers. In G.B. Wickramanayake and R.E. Hinchee (Eds.), *Remediation of Chlorinated and Recalcitrant Compounds: Designing and Applying Treatment Technologies*, pp. 163-168. Battelle Press, Columbus, Ohio.

Warner, S.D., C.L. Yamane, N.T. Bice, F.S. Szerdy, J. Vogan, D.W. Major, and D.A. Hankins. 1998. Technical update: The first commercial subsurface permeable reactive treatment zone composed of granular zero-valent iron. In G.B. Wickramanayake and R.E. Hinchee (Eds.), *Remediation of Chlorinated and Recalcitrant Compounds: Designing and Applying Treatment Technologies*, pp. 145-150. Battelle Press, Columbus, Ohio.

ULTRASONIC REGENERATION OF PERMEABLE TREATMENT WALLS: LABORATORY/FIELD STUDIES

Christian A. Clausen, Cherie L. Geiger, Debra R. Reinhart, Nancy Ruiz, Kristi Farrell, Patrick Toy, Nancy Lau Chan, Mark Cannata, Stephen Burwinkle (University of Central Florida, Orlando FL, USA), Jacqueline Quinn (NASA, Kennedy Space Center, FL, USA)

ABSTRACT: In-situ permeable treatment barriers containing iron as the reactive agent have gained popularity in the past decade as a near-passive, in situ remediation technology for halogenated solvents. Although iron has been shown to be effective for this purpose, a continuing problem is the loss of reactivity over time. This is due, at least in part, to a build up of corrosion products or other precipitates on the iron surface. The lifetime of the barrier could be significantly extended with a technology that can remove the materials blocking the iron surface.

The purpose of this research was to investigate the application of ultrasonic energy to rejuvenate an iron surface with the goal of enhancing/restoring the rate of trichloroethylene (TCE) degradation. Extensive studies (laboratory and field) were conducted to examine the impact of ultrasound on iron under various conditions. Results indicate that a sonication period as brief as 30 minutes has a significant positive impact on the first order rate constant for TCE degradation.

INTRODUCTION

Permeable treatment walls (PTW) containing iron as the reactive agent have been used over the past decade in an attempt to remediate groundwater contaminated with chlorinated organics in an in situ fashion (Gillham and O'Hannesin, 1994). PTWs installed in the subsurface path of contaminated groundwater have shown promise as a low maintenance, long term remediation method although research indicates that fouling can cause decreased effectiveness.

The primary responsible reaction for the corrosion of Fe^0 in an aerobic aqueous system is given in equation 1. In an environment with sufficient oxygen availability, Fe^{2+} may be further oxidized to yield Fe^{3+} by equation 2. In anaerobic conditions, Fe^0 oxidizes slowly by reaction with water as shown in equation 3 (Vogan et al, 1994).

$$2Fe^0 + O_2 + 2H_2O \leftrightarrow 2Fe^{2+} + 4OH^- \quad (1)$$
$$4Fe^{2+} + O_2 + 2H^+ \leftrightarrow 4Fe^{3+} + 2OH^- \quad (2)$$
$$Fe^0 + 2H_2O \leftrightarrow Fe^{2+} + H_2 + 2OH^- \quad (3)$$

The corrosion process proceeds rapidly in the presence of strong oxidants and acidic pH (pH 2-4). As the corrosion process continues, the reaction rate begins to decrease due to the build-up of corrosion products on the iron surface. This process, which is known as passivation, serves to protect the iron below the oxide layer. Under aerobic conditions, the oxide layer consists of precipitates like ferric oxides.

Halogenated aliphatic hydrocarbons, RX, may contribute to the corrosion of Fe^0. When electron transfer occurs, dehalogenation of the hydrocarbon and oxidative dissolution of Fe^0 are the net effects (Vogan et al, 1994). The overall reaction is given in equation (4).

$$Fe^0 + RX + H^+ \leftrightarrow Fe^{2+} + RH + X^- \qquad (4)$$

As can be seen from equation 4, the degradation of chlorinated hydrocarbons by iron is a corrosion reaction; therefore, as chlorinated hydrocarbons are degraded, passivation will occur as the iron develops a corrosion layer on the surface of the metal (Johnson and Tratyek, 1995). Many researchers agree that this is a surface area dependant reaction (Johnson et al., 1996). Therefore, as the surface is occluded, the rate of degradation of chlorinated compounds will decline. Adding to this problem are the constituents often contained in groundwater (Agrawal et al., 1995). Water with high alkalinity or moderate sulfide concentration will act as precipitating water and increase the formation of precipitates.

An important consideration when designing a PTW is that complete dechlorination must occur within the wall or intermediate chlorinated byproducts will be released into the environment (Matheson and Tratyek, 1994). For a given groundwater flow rate, an occluded surface may cause the de-chlorination rates to be reduced enough to allow the contaminants (or chlorinated by-products) to penetrate the wall or flow around it thus defeating the remediation process (Mackenzie et al., 1997). In order to extend the useful lifetime of the reactive iron, some treatment walls will require the use of an additional technology to remove corrosion products and precipitates and therefore, expose a 'fresh', unoccluded iron surface.

The in situ application of ultrasound to the wall is a possible solution to the prevention of passivity. Ultrasound has a variety of effects on substances based on energy used. High frequency or diagnostic ultrasound (2-10 MHz) is often associated with medical diagnostic equipment, materials testing for flaw detection and underwater ranging. Low frequency or power ultrasound (20-100 kHz) is used as an engineering tool for drilling, grinding, cutting and welding. Industrial uses include various cleaning applications as well as for dispersion purposes. Power ultrasound has also been used to enhance chemical reactions (Orzechowska et al., 1995).

MATERIALS

Two types of elemental iron were used in this work. Iron powder (100-mesh) was obtained through Mallinekrodt Chemical and recycled iron metal filings were obtained from Peerless, Inc. The chemicals used were trichloroethylene,   cis-dichloroethylene,   trans-dichloroethylene,   1,1-dichloroethylene, vinyl chloride, ethylene (all 99+% and stabilized, Acros

Organics) and HPLC grade methanol (Fisher Scientific). One-liter Tedlar® bags with single polypropylene fittings were obtained from SKC Inc.

METHODOLOGY

Iron was used in either an unwashed (as received state), acid-washed, or air oxidized condition. The procedure for acid washing the iron is as follows: approximately 50 mL of 5% sulfuric acid solution were added to five grams of dry, 100-mesh or Peerless iron. The mixture was allowed to stand for 10 minutes under periodic shaking and the acid solution was decanted. The resulting "wet" iron was washed repeatedly with de-ionized water to ensure that no acid remained. In the experiments using unwashed iron, the iron was simply used as received from the manufacturer. Oxidized iron was exposed to air for one week before use. The iron was transferred to a 1.0 L Tedlar® bag and sealed. Five hundred mL of natural ground water was added to the bag through the septum/port. The bag was then purged for 30 minutes with nitrogen gas. All remaining headspace in the bag was evacuated.

Bags that were exposed to ultrasound were then placed in a 600-W, 20-khz Branson ultrasonic bath with an ultrasonic intensity of 0.32 W/cm^2. Ultrasound was applied for periods of 30 minutes to 3 hours, depending on the experiment. To minimize any temperature effects, water flowed continuously through the bath during use.

TCE/methanol solution was injected via syringe, through the septum, to obtain approximately 10mg/L solution of TCE in water. The bag was placed on a shaker table for 15 minutes and a 13-mL sample was removed using a 20.0-mL gas-tight syringe. Subsequent 13-mL samples were taken at random time intervals for the next 14 days in an attempt to establish a TCE disappearance rate. Analytical data was gathered using a Tekmar LSC2 purge and trap concentrator with a 10-position autosampler. A gas chromatograph (Hewlett Packard 5890) with a flame ionization detector was used for identification and quantification. The column used was a DB624 (0.25 i.d. X 60 m) purchased from Supelco.

The field site used is located at Launch Complex 34 (LC34), Cape Canaveral Air Station, in east central Florida. Building 21900H, which is located on LC34, and is commonly referred to as the Engineering Support Building (ESB), was used to clean rocket parts (primarily with TCE) during the 1960s. Recent estimates indicate that over 20,000 kg of TCE remain beneath this building as DNAPL and continue to function as a contaminant source, fueling a greater than 1-km^2 plume. The PTW placed at this site was not intended to treat the entire plume but was instead intended for a research project testing the construction technique and the use of ultrasound.

Several observation and monitoring wells (10-cm diameter) were placed in front of, inside, and behind the PTW. Ultrasound is applied by use of a submergible 1,000-W resonator that could be lowered into the wells and held at various depths for the intended treatment period. Core samples were taken (within 30-cm of the well) before and after treatment and the iron was removed from these samples under a nitrogen atmosphere. The iron was then used in bag laboratory experiments to determine the 1st order rate constants for the disappearance of TCE.

RESULTS AND DISCUSSION

Acid washed vs unwashed iron: Experiments were conducted to determine the rate constants for the degradation of TCE with both types of iron (100-mesh and iron filings) in an unwashed (as received) condition, and after the iron was acid-washed and purposely oxidized. Table 1 shows the first order rate constants for the disappearance of TCE for both types of iron in acid-washed, unwashed and oxidized conditions. Acid-washed iron was more effective in the removal of TCE for both types of iron. This can be attributed to the increase in surface area of iron. Acid-washed 100-mesh was shown to be more reactive than acid-washed peerless iron. This is primarily due to increased surface area for 100-mesh iron and agrees with the findings of previous researchers that the reaction is surface area dependent (Johnson et al., 1996).

Table 1.   Comparison of first-order rate constants (hr^{-1}) for TCE disappearance before and after ultrasound treatment for iron under different conditions (in natural groundwater).

Iron	Condition	No US k_{obs}, 1st Order Rate hr^{-1}(x 10^{-3})	US k_{obs}, 1st Order Rate hr^{-1}(x 10^{-3})
100-mesh	Unwashed	3.75	5.00
	Acid-washed	6.42	6.54
	Oxidized	8.46	14.40
Peerless	Unwashed	2.70	3.36
	Acid-washed	4.74	4.68
	Oxidized	6.96	12.36
	AW-Air[a]	5.99	11.28

[a]AW-Air = acid-washed iron from previous experiment exposed to air then reused.

Effect of ultrasound on degradation of TCE: The experiments described in this paper used ultrasound prior to spiking with TCE so there was no impact to the TCE concentration or concentration of byproducts from ultrasound itself.

An important consideration was the time necessary to expose iron to ultrasound at a fixed energy level. Experiments were conducted using separate pairs of bags for each length of time examined and rate constants were determined for each set. The bags were exposed to ultrasound prior to the injection of TCE. The sets of bags were exposed to ultrasound according to the following times: zero minutes, 30 minutes, one hour, two hours and three hours. Figure 1 shows how the time exposure to ultrasound affects the rate of disappearance of TCE. The rate constant increases and then levels out after two hours. It can be assumed from this figure that there is an optimum time necessary for removing debris and corrosion products from the iron surface. For the types of iron and the conditions of this experiment, this is accomplished before two hours; therefore, two hours of ultrasound treatment was used for the remainder of the experiments described in this paper. Use of ultrasound beyond that point cannot be justified based on this work.

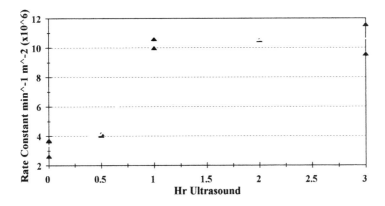

Figure 1. Effect of time of ultrasound treatment on reaction rates.

Effects of ultrasound on reaction rates: When acid-washed 100-mesh iron or iron filings was studied, it was found that the beneficial results of ultrasound were negligible (an improvement of only 2.0%). Table 1 lists the first order rate constants for the disappearance of TCE with time. Acid-washed, 100-mesh iron shows a very similar rate of disappearance whether it is treated with ultrasound prior to TCE injection or not treated with ultrasound at all. This is not surprising considering the acid washing process removes surface debris and oxidation products thus exposing fresh iron surface.

The influence of ultrasound on iron that was unwashed was more dramatic. As shown in Table 1, the rate constant for unwashed, 100-mesh iron increased from 3.750×10^{-3} hr^{-1} to 5.00×10^{-3} hr^{-1}, an improvement of 33.3%. Unwashed iron filings showed a similar change with an improvement of 24.1%. SEMs of unwashed, iron filings before and after sonication are shown in Figures 2(a) and 2(b). These figures show that the application of ultrasound removed corrosion products and surface debris revealing more iron surface area thus increasing reaction rates. The unwashed iron filings are somewhat uniformly coated with small debris and corrosion products before ultrasound. After ultrasound treatment, there are still some small areas of occluded surface but much more of the iron surface is free of debris.

Increased non-halogenated by-product formation: The rate of production of terminal by-products was greater in the bags exposed to ultrasound., for both the 100-mesh iron and iron filings. Figure 3 shows the rate of production of non-chlorinated, 2-carbon by-products (denoted as ethylene and expressed as equivalent TCE concentration) in bags that were not exposed to ultrasound compared to those that were, plotted against half-life reductions in TCE.

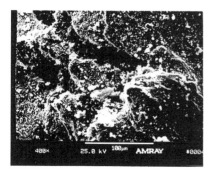 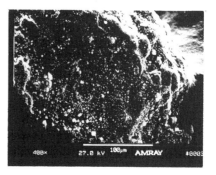

Figure 2. SEMs of (a) unwashed (as received) iron filings and (b) unwashed iron filings exposed to ultrasound.

The curves superimposed over the points are trend lines and are not intended to approximate best fits.

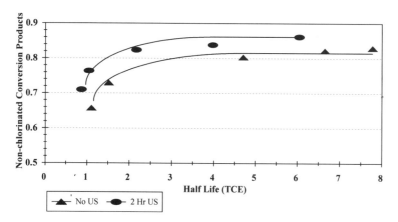

Figure 3. Non-chlorinated, two-carbon by-product formation as related to TCE half-life for iron filings with and without application of ultrasound.
It can be seen from Figure 3 that although TCE disappeared at approximately the same rates, the production of the non-chlorinated, 2-carbon by-products increased when the iron was exposed to two hours of ultrasound at the beginning of the experiments. This indicates that the fine etching and pitting that occurs on the iron surface may, in fact, be activating the iron surface in some way.

Surface Area: Because it was evident that the application of ultrasound altered the surface of the iron samples tested, surface area analysis was conducted to determine how the total surface area might have changed. Table 2 lists values obtained for 100-mesh iron and iron filings, in acid-washed, unwashed and oxidized conditions with no ultrasound and 2-hours of ultrasound exposure. For both the 100-mesh iron and the iron filings, treatment and sonication influenced specific surface area. Regardless of the initial specific surface area of the unwashed, unsonicated sample, washing in 5% sulfuric acid increased specific surface area for the two iron types an average of 1.17 m^2/g.

Table 2. Iron specific surface area resulting from pretreatment and sonication, m^2/g.

Iron Description	No Ultrasound	Ultrasound	Change, m^2/g	Percent Change
100-mesh unwashed	3.09	4.12	+1.03	+33.3
Acid-washed	4.28	5.40	+1.12	+26.2
Oxidized	49.16	22.33	-26.83	-54.6
Peerless - unwashed	0.77	2.07	+1.30	+1.69
Acid-washed	1.92	2.91	+0.99	+51.6
Oxidized	24.29	14.53	-9.76	-40.2
Several months in groundwater	11.31	7.49	-3.82	-33.8

Increases in surface area post-sonication varied somewhat by iron type and treatment, but averaged 1.11 m^2/g. Specifically, for unwashed samples, the specific surface areas for 100-mesh iron increased 1.03 m^2/g and the iron filings increased 1.30 m^2/g post-sonication. Considering acid-washed samples, sonication increased surface area 1.12 m^2/g and 0.99 m^2/g in the 100-mesh iron and Peerless iron, respectively. The changes in the surface areas reflect a cleaner, more open surface area but also could be related to, in some cases, the breakdown of the particle itself into smaller pieces with greater surface area to bulk ratio.

Field application of ultrasound: Iron taken from all depths showed an improvement after field application of ultrasound. Table 3 shows values for rate constants before and after ultrasound application at various depths of treatment. Rate constants increased over a range from 1.1 to 11.1 times the initial values. While these values are very encouraging, further study of the effects of ultrasound in field use is ongoing.

Table 3. Rate constant before and after ultrasound for field experiments.

Depth of Sample(meters)	Rate Constant X 10^4 (hr-g Fe)$^{-1}$
1.68	3.78
1.98	3.78.
1.98	2.21
2.13	4.36
2.29	25.5

CONCLUSIONS

The reaction of TCE with iron has been shown to be surface area dependent. As corrosion products and precipitates build up on the surface, the reaction with TCE is diminished because the unoccupied surface area is decreased. Ultrasound has been demonstrated to remove corrosion products, precipitates and other debris from the iron surface. This serves to increase the fresh iron surface, improve reaction rates and, in some cases, increase the concentration of terminal by-products indicating that the reaction is going to completion either at a faster rate or through a different mechanism.

REFERENCES

Agrawal, A., P. G. Tratnyek, P. Stoffyn-Egli, and L. Liang.1995. "Processes Affecting Nitro Reductions by Iron Metal: Mineralogical Consequences of Precipitation in Aqueous Carbonates Environments," *American Chemical Society Spring National Conference Extended Abstract*, Anaheim, CA, April 2-7.

Gillham, R. W. and S. F. O'Hannesin. 1994. "Enhanced Degradation of Halogenated Aliphatics by Zero-Valent Iron." *Ground Water*, 32, 958-967.

Johnson, T. L., M. M. Scherer, and P. G. Tratnyek. 1996. "Kinetics of Halogenated Organic Compound Degradation by Iron Metal." *Environmental Science and Technology*, 30, 2634-2640.

Johnson, T. L. and P. G. Tratnyek. 1995. "Dechlorination of Carbon Tetrachloride by Iron Metal: the Role of Competing Corrosion Reactions." *American Chemical Society Spring National Conference Extended Abstract*, Anaheim, CA, April 2-7.

Mackenzie, P. D., T. M. Sivavec, and D. P. Horney. 1997. "Extending Hydraulic Lifetime of Iron Walls." Presented at 1997 *International Containment Technology Conference and Exhibition Proceedings*, February 9-12, 1997. St. Petersburg, FL.

Matheson, L. J. and P. G. Tratnyek. 1994. "Reductive Dehalogenation of Chlorinated Methanes by Iron Metal." *Environmental Science and Technology*, 28, 2045-2053.

Orzechowska, G. E., E. J. Poziomek, V. F. Hodge, and W. E. Engelmann. 1995. "Use of Sonochemistry in Monitoring Chlorinated Hydrocarbons in Water." *Environmental Science and Technology*, 29, 1373-1379.

Vogan, J. L., J. K. Seaberg, B. Gnabasik, and S. O'Hannesin. 1994. "Evaluation of In situ Groundwater Remediation by Metal Enhanced Reductive - Dehalogenation." *87th Annual Meeting and Exhibition of the Air and Waste Management Association Proceedings*, Cincinnati OH, June 19-24.

INCORPORATING SURFACE SATURATION EFFECTS INTO
IRON WALL DESIGN CALCULATIONS

J.F. Devlin, University of Waterloo, Waterloo, Ontario, Canada N2L 3G1

M. Morkin, Geosyntec Consultants, Walnut Creek, California 94596

C. Repta, University of Waterloo, Waterloo, Ontario, Canada N2L 3G1

ABSTRACT
Previous research has shown that in the presence of high concentrations of reducible organics, reaction rates with granular iron tend to approach a maximum value. However, there is little field evidence of this reported in the literature. A field experiment in which saturation kinetics appeared to play a role was performed at the Alameda Naval Air Station located near San Francisco, between December, 1996, and January, 1998. The reactive barrier was constructed with granular iron from Peerless Metal Powders and Abrasive, Detroit, Michigan. Despite a conservative design, the barrier did not completely degrade cDCE or VC entering the system at peak concentrations in excess of 200 mg/L and 40 mg/L, respectively (removal was >95% but <100%). The barrier was initially designed on the basis of treatability studies with total concentrations (cDCE + VC) in the range of 30 to 50 mg/L, but was installed with a sufficient travel path length to transform the higher concentrations encountered, according to a pseudo-first-order model. It was hypothesized that the unexpected breakthrough was the result of slower reaction kinetics in the presence of high concentrations of contaminants, due to surface saturation effects.

Column experiments were performed to investigate the possibility of surface saturation and indicated that there was a notable effect of input concentration (20 mg/L to 300 mg/L) on transformation rates of cDCE in the Alameda water. Furthermore, a linear relationship was found to exist between half-lives (calculated from the pseudo-first-order rate constants) and the input concentrations, consistent with known saturation kinetic models. A new model, based on the assumption of first order reaction kinetics on the solid surface and a Langmuir isotherm (instantaneous, equilibrium sorption) described the data well and showed the importance of accounting for saturation kinetics in barrier design (wall thickness), rather than using the conventional pseudo-first-order model, when organic concentrations are high (>~20 mg/L in the case of cDCE).

INTRODUCTION
The granular iron barrier studied here was installed at the Alameda Naval Air Station, located on Alameda Island, near San Francisco. The barrier consisted of iron particles obtained from Peerless Metal Powders and Abrasive, Detroit, Michigan, and was found to have a surface area of 0.7 m^2/g, determined by the

BET method. The barrier measured about 4 m deep and 3 m wide and was instrumented with 3 transects of multilevel bundle piezometers. The first transect was located immediately upgradient of the barrier, the second within the barrier and the third immediately downgradient. Each bundle piezometer consisted of 3 stainless steel sampling tubes terminating in stainless steel mesh screens (0.3 mm ID). These were attached to PVC centre stocks with nylon straps. The barrier was part of a sequenced reduction-aerobic biodegradation treatment system for the removal   of   chlorinated   solvents,   including   trichloroethene   (TCE), *cis*-dichloroethene (cDCE), and vinyl chloride (VC), as well as petroleum hydrocarbons (benzene and toluene) from the groundwater. The contaminants were present in a shallow, sandy, unconfined aquifer about 6 metres deep and underlain by a Holocene mud that served as an aquitard. The organics in the plume ranged in their maximum concentrations (determined with multilevel sampling methods) from low mg/L (<10 mg/L hydrocarbons, TCE) to tens of mg/L (VC) and hundreds of mg/L (cDCE). The contaminant levels were found to vary seasonally with concentrations of individual compounds diminishing from the maximum by a factor of two or more. The groundwater was also characterised by elevated total dissolved solids (TDS = 800 to 8700 mg/L) dominated by calcium, magnesium, chloride and bicarbonate. Additional details concerning the Alameda site description and characterisation can be found in Barker et al. (2000).

Prior to installing the reactive barrier, groundwater was sampled from a fully screened well located just upgradient of the treatment gate and used in a laboratory column treatability study, to determine the residence times necessary for the dechlorination reactions. These studies indicated that a 37 cm thick barrier would be sufficient to accomplish the task. However, the concentration of cDCE in the test water was about 35 mg/L, significantly below the maximum concentration in the plume. A final design was adopted with a conservative barrier path length of 1.5 m to minimize the possibility of breakthrough in the gate. Despite this, cDCE and VC did breakthrough the iron barrier in low mg/L concentrations. These compounds were subsequently treated in the aerobic section of the sequence, but the causes of the breakthrough from the granular iron remained in question.

Two possible causes of the breakthrough were considered: channeled flow and surface saturation. Several attempts were made to characterize the groundwater velocities in the iron barrier, but these efforts were inconclusive. Therefore, the possibility that channeling played a role in the breakthroughs cannot be ruled out. However, additional laboratory work was undertaken to investigate the effect of cDCE concentrations on the reaction kinetics. These efforts are described below.

METHODS
Field Estimates of Pseudo-First-Order Rate Constants. As alluded to above, the plume was heterogeneous with respect to contaminant concentrations, both in

time and in space. This made the estimation of field rate constants rather difficult since, along some flow paths, in snapshot, concentrations actually rose through the barrier (Figure 1). To overcome this problem, it was decided to calculate bulk rate constants, reflecting overall barrier performance. This was accomplished by integrating the concentrations horizontally and vertically across Rows 1 and 3 (Figure 1) over a 4 month period to determine a 4 month mass entering and passing through the system

$$M = \int_{t_1}^{t_2} \int_0^z \int_0^x C\phi v \, dx \, dz \, dt \qquad (1)$$

where M= mass (M), C = solute concentration (M/L^3), ϕ = porosity, A = cross sectional area (L^2), v = average linear velocity (L/T), x, z = coordinates in the horizontal and vertical directions, respectively (L), t = time (T). The same procedure was applied to data from Row 4, at the downgradient end of the barrier, except that the time interval was offset by the residence time of water in the barrier, 1 month. The calculated masses were then fit with a first order kinetic equation to estimate the bulk pseudo-first-order rate constant, k_1. The rate constant was found to lie in the range 0.004 to 0.01 hr^{-1}.

Column Experiments. The first treatability column experiment was conducted in a fashion similar to those described in Gillham and O'Hannesin (1994). Briefly, the column was constructed of Plexiglas, 100 cm long, with a 3.8 cm internal diameter and 7 sampling ports positioned along its length. The column was packed with Peerless iron, as used in the field experiment. The grain size of the iron ranged from 0.25 to 2.0 mm diameter (-8 to +50 US Standard Sieve Mesh #). The BET surface area was 0.7 m^2/g and the particle density was 6.98 g/cm^3. While the original treatability study determined rate constants from steady state contaminant profiles in the column, we determined rate constants from breakthrough curves at the 16 cm sampling port, by fitting the data with the following equation from van Genuchten and Alves (1982),

$$C(x,t) = \frac{C_o}{2} \exp\left[\frac{vx}{2D}\left(1 - \sqrt{1 + \frac{4k_1 D}{v^2}}\right)\right] \mathrm{erfc}\left[\frac{Rx - v\sqrt{1 + \frac{4k_1 D}{v^2}}\, t}{2\sqrt{DRt}}\right] \qquad (2)$$

where, x = distance along the column (L), Co = concentration of solute entering the column (at x = 0), D = dispersion coefficient (L^2/T), R = retardation factor, k_1 = pseudo-first-order rate constant (T^{-1}). This approach has the advantage of providing an estimate of the retardation factor at the same time as the rate constant. In our experiments, cDCE was found to sorb minimally (R<2) to the granular iron, a result in agreement with the work of Allen-King et al. (1997).

Saturation Kinetics. A kinetic model, referred to as KIM, was derived assuming first order kinetics on the solid surface, a Langmuir isotherm (with instantaneous, equilibrium sorption). The model exhibits a functionality qualitatively similar to the hyperbolic expressions previously suggested for saturation kinetics (Scherer et al., 1999; Arnold and Roberts, 2000), but contains a fractional term in the denominator that accommodates a slightly extended range of concentration independence in the reaction rates of relatively dilute solutions. The significance of this feature of the model is currently under investigation.

$$\frac{dC}{dt} = \left[\frac{k(Fe/V_w)C_{max}}{\dfrac{1}{J} + \dfrac{C_{max}(Fe/V_w)}{1 + JC} + C} \right] C \tag{3}$$

where, k = the intrinsic first order rate constant for reaction on the surface of the granular iron (T^{-1}), Fe/Vw = the solid surface area to solution volume ratio (L^2/L^3), C_{max} = maximum concentration that can sorb to the solid (M/L^2), J = Langmuir isotherm constant, indicative of the affinity of the solute for the sorbing surface (M/L^3).

Results and Discussion. The column experiments provided a data set that confirmed the dependence of reaction rate on the concentration of cDCE in solution. Furthermore, there was reasonably good agreement between the estimated pseudo-first-order rate constants from the laboratory and those obtained from the field experiment. Converting rate constants to half lives ($t\frac{1}{2} = 0.693/k_1$) a linear dependence on concentration was observed (Figure 2). This is consistent with the kind of saturation kinetics previously described by Scherer et al. (1999) for carbon tetrachloride reacting on granular iron. In an effort to investigate the effects of the background TDS on reactivity, additional column experiments were performed with cDCE spiked deionized water as the influent. The same kind of relationship was found to exist between half life and concentration in the two solutions, but there was a noticeably lower slope to the line from the DI water column experiments (Figure 2). These data suggest that the inorganic constituents in the Alameda water may have been limiting access of cDCE to the reactive sites on the iron (note that the column experiments were performed in the absence of other competing organic solutes in the input water).

Experiments performed in our laboratory, using nitroaroamatic compounds as probes for iron reactivity, have indicated that carbonate, when present in sufficiently high concentration, can reduce iron reactivity (Allin and Devlin, 2000). This has also been noted by others (Agrawal and Tratnyek, 1996). Chloride caused much less noticeable changes to reactivity in our experiments, and has even been found to increase reactivity somewhat (with a borate background solution) by destabilizing the passivating oxide layer (Johnson et al., 1998).

The data from the column experiments were well described by the KIM equation; the fitted parameters, C_{max}, J and k, are summarized below in Table 1.

Table 1. Summary of KIM model fits, with 95% confidence intervals, and conditions for column experiments with deionized and Alameda Site waters.

Parameter	Deionized water	Alameda Site Water
C_{max} (mg/m^2)	31.4 ± 5.0	1.55 ± 0.44
J (mg/L)	0.0096 ± 0.004	0.011 ± 0.008
k (1/hr)	0.076 ± 0.01	0.85 ± 0.25
Fe/Vw (m^2/L) (range used)	0.379 - 2.5	1.36 - 1.52
Co (mg/L) (range used)	1.0 - 304	36 - 248

The decrease in C_{max} from deionized water to site water is consistent with expectations if sorption sites are lost to inorganic sorbates. The similarities in the J constants suggest that the affinity of the cDCE for the solid surface is virtually unaffected by the TDS, and the increase in k from low to high TDS waters may indicate that although some sorption sites are lost in the presence of the inorganic solutes, the remaining ones are, on average, more reactive. These interpretations are highly speculative given the limited data sets presented here. Further work is underway to elucidate these relationships in more detail.

IMPLICATIONS FOR BARRIER DESIGN

The effect of saturation kinetics on barrier design was considered by examining wall thickness as a function of plume concentration. Both pseudo-first-order and saturation models (KIM) were used to calculate reaction rates at various cDCE concentrations. The KIM parameters used were those from Table 1, Alameda site water case, while the k_1 value, representing the pseudo-first-order model, was calculated assuming equal rates for the two models at $C_o = 35$ mg/L. Bulk pseudo-first-order rate constants (k_{bulk}) were then calculated (k_{bulk} = rate/concentration) and used with equation 4 to estimate the wall thicknesses necessary for the required level of treatment (dispersion was assumed negligible in this analysis),

$$d = -\frac{v}{k_{bulk}} \ln\left(\frac{C_{lim}}{C_o}\right) \qquad (4)$$

where, d = wall thickness (L), C_{lim} = the concentration desired at the downgradient end of the reactive barrier (10 μg/L). The calculations show that inappropriate use of the pseudo-first-order model can result in significantly undersized barriers if plume concentrations ever exceed those used in the treatability studies (Figure 3). The implications for barrier design are that plumes should be characterized in considerable detail before finalizing plans, so that maximum concentrations are well known. To avoid breakthrough, plume cores should be identified and characterized at a relatively small scale, depending on the site. When dissolved chlorinated solvent concentrations rise into the tens of mg/L concentration levels, consideration should be given to the possibility that saturation kinetics apply. In some high TDS waters, this may be of concern for overall lower organic concentrations than in low TDS waters. In addition, inter-species competition, between organic compounds (such as VC and cDCE), would be expected to add to saturation related problems.

ACKNOWLEDGEMENTS
Funding for this work was provided by the NSERC/Motorola/ETI Industrial Research Chair in Groundwater Remediation held by Dr. R.W. Gillham and the AATDF Program administered by Rice University. Dr. J.F. Barker oversaw the installation of the Alameda iron barrier and assisted with the interpretation of the field data.

REFERENCES
Allen-King, R.M., R.M. Halket, and D.R. Burris. 1997. "Reductive transformation and sorption of cis- and trans-1,2- dichloroethene in a metallic iron-water system." *Environmental Toxicology and Chemistry.* 16(3): 424-429.

Allin, K., and J.F. Devlin. 2000. "Effect of water geochemistry on the reactivity of granular iron." Presented at the *Second International Conference on Remediation of Chlorinated and Recalcitrant Compounds*, May 22-25, Monterey, California.

Agrawal, A., and P.G. Tratnyek. 1996. "Reduction of nitroaromatic compounds by zero-valent iron metal." *Environmental Science and Technology.* 30(1): 153-160.

Arnold, W., and L. Roberts. 2000. "Pathways and kinetics of chlorinated ethylene and chlorinated acetylene reaction with Fe(0) particles." Submitted to *Environmental Science and Technology.*

Barker, J.F., B.J. Butler, E. Cox, J.F. Devlin, R. Focht, S.M. Froud, D.J. Katic, M. McMaster, M. Morkin, and J. Vogan. 2000. *Sequenced reactive barriers for groundwater remediation.* Stephanie Fiorenza, Carroll Oubre and Herb Ward (Eds.), Lewis Publishers, Boca Rotan, Florida.

Gillham, R.W., and S.F. O'Hannesin. 1994. "Enhanced degradation of halogenated aliphatics by zero-valent iron." *Ground Water.* 32(6): 958-967.

Johnson, T.L., W. Fish, Y.A. Gorby, and P.G. Tratnyek. 1998. "Degradation of carbon tetrachloride by iron metal: complexation effects on the oxide surface." *Journal of Contaminant Hydrology.* 29: 379-398.

Scherer, M.M., B. Balko, and P.G. Tratnyek. 1999. "The role of oxides in reduction reactions at the metal-water interface." Chapter 15 in *Kinetics and*

Mechanism of Reactions at the Mineral/Water Interface, T. Grundl and D. Sparks (Eds.), American Chemical Society, 301-322.

van Genuchten, M. Th., and W.J. Alves. 1982. *Analytical solutions of the one-dimensional convection-dispersion solute transport equation.* U.S. Department of Agriculture, Technical Bulletin no. 1661

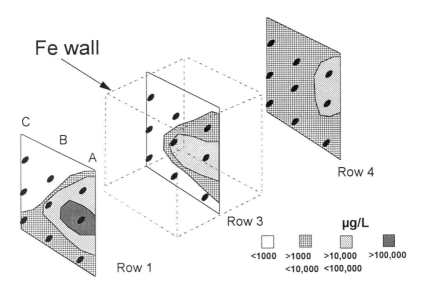

Figure 1: cDCE transects through the granular iron barrier. Data from May 22, 1997

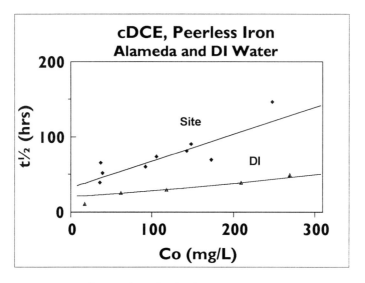

Figure 2: Comparison of saturation kinetics in site water and in deionized water.

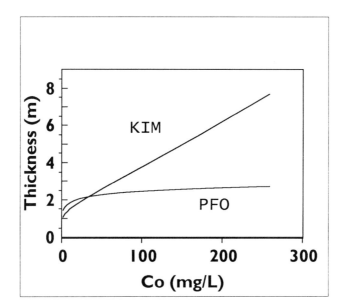

Figure 3: Wall thickness as a function of concentration for pseudo-first-order kinetics (PFO)and saturation kinetics as calculated with KIM (equation 3). Data used are for cDCE reaction with Peerless iron in Alameda site water.

REACTIVE BARRIER PERFORMANCE IN A COMPLEX CONTAMINANT AND GEOCHEMICAL ENVIRONMENT.

J. M. Duran (Woodward-Clyde Pty Limited, Sydney, NSW, Australia)
J. Vogan (EnviroMetal Technologies Inc, Waterloo, Ontario, Canada)
J. R. Stening (SHE Pacific Pty Ltd, Sydney, NSW, Australia)

ABSTRACT: A pilot scale zero valent iron reactive barrier test is currently being conducted downgradient of two former chlorinated solvent manufacturing plants which operated between the early 1950s and 1991. Site characterisation (Duran and Grounds, 1996) has demonstrated that extensive zones of DNAPL are present in the subsurface. The DNAPL includes the manufactured solvents trichloroethene (TCE), tetrachloroethene (PCE) and carbon tetrachloride (CTC) as well as a range of intermediary and byproducts of manufacturing including 1,1,2,2-tetrachloroethane (PCA), 1,1,1-trichloroethane, 1,2-dichloroethane, chloroform, chlorobenzenes, hexachlorobutadiene and hexachloroethane. Within the immediate pilot scale test zone the total dissolved phase concentration of volatile compounds ranges up to 220 mg/L.

The pilot scale system was installed in February 1999 using a sheet pile trench box method. Monitoring of the system after 30 days revealed that degradation of the volatile compounds ranged between 81% and 96%, with highest mass removal observed in zones where the DOC concentration was lowest. Complete mass removal was not expected, since compounds such as 1,2-dichloroethane and dichloromethane are known not to be significantly degraded by zero valent iron (Gillham and O'Hannesin, 1994). Monitoring after 9 months has revealed a decrease in the overall CHC mass reduction, primarily due to the decreased degradation rate of one of the principal components, PCE.

Monitoring of the system is expected to be undertaken for at least 12 months to allow sufficient groundwater pore volumes to pass through the barrier to assess the long term feasibility of a full scale system. However, present monitoring data has demonstrated that even within a complex contaminant and geochemical environment, zero valent iron provides a relatively robust passive system for the remediation of a wide range of chlorinated solvents.

INTRODUCTION

Site Description. The site is located in an area of coastal sand hills to the south of the City of Sydney, Australia. The site is underlain by unconsolidated sands, which are between 15 and 40 m thick and are underlain by sandstone basement rock. The unconsolidated sands are a high yielding unconfined and semi-confined aquifer system, which is used locally for industrial water, and turf irrigation supplies.

Site Characterisation. A hydraulic and geochemical study of the proposed test area was undertaken to design the pilot scale barrier and ensure that the barrier was installed perpendicular to the local groundwater flow direction. Geochemical characterisation of the test site has identified a complex geochemical environment, with reduced and highly acidic (pH<5) conditions. Significant concentrations of DOC (> 500 mg/L), COD (> 750 mg/L), BOD (> 600 mg/L) and sulphide (>30 mg/L) as well as the presence of volatile fatty acids. Dissolved volatile chlorinated hydrocarbons (CHCs) are found in concentrations of up to 200 mg/L, with predominant compounds being PCE (50 to 75 mg/L) and CTC (20 to 65 mg/L). The volatile chlorinated contaminants only partially contribute to the BOD loading in the system.

Column Studies. Prior to installing the pilot scale barrier, dynamic column tests were undertaken to establish the degradation rates of the volatile CHCs from the test site. The initial performance of the columns indicated that the rate of degradation was comparable to other studies. However, the performance of the system declined over time, in particular the degradation of PCE. Additional column studies concluded that elevated levels of organic carbon appeared to decrease the degradation rate of several CHCs including PCE. CTC did not appear to be affected. A half-life for PCE of 45hours was calculated from the column test, CTC half-lives were in the order of 1 to 2 hours or less.

METHODS
A zero valent iron barrier with dimension of 5 m in length, 3 m depth and 1.5–2.0 m wide was emplaced using a standard sheet pile trench box method. An extensive monitoring network consisting of 15 bundle piezometers was installed for the collection of groundwater samples to assess the performance of the barrier. Several 50 mm PVC monitoring wells were installed along with several multilevel 25 mm PVC piezometers for measurement of the horizontal and vertical gradients. A plan and cross-sectional view of the test area are shown on Figure 1.

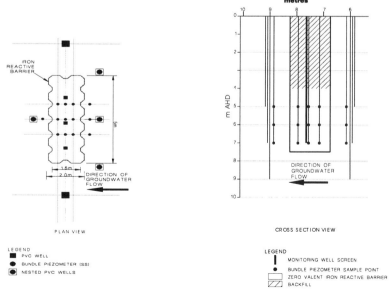

Figure 1 – Plan and Cross Section of Pilot Scale Barrier

Groundwater Sampling. Groundwater samples were collected from all the 15 bundle piezometers installed along the central alignment of the barrier (BP10, BP15, BP16, BP17 and BP08) in all monitoring events. Groundwater samples were collected after 1 month (March 1999), 3 months (June 1999) 6 months (September 1999) and 9 months (December 1999).

In situ groundwater flow velocity measurements were carried out using a KV-Associates Model 40 GeoFlo Groundwater Flowmeter. An average velocity of 0.6 m/day was measured within the barrier during the December 1999 monitoring event.

All groundwater samples were analysed for volatile CHCs, ethene, ethane, methane and a wide range of inorganic and aggregate organic parameters. Mercury and semivolatile CHCs were included in the first 6 months of monitoring and analyses for the presence of sulphate reducing bacteria (SRB) were included in the most recent monitoring event. All analyses were undertaken by a local commercial analytical laboratory, which is accredited by the National Accreditation and Testing Association (NATA) for all methods, which are based on SW-846 and APHA.

RESULTS AND DISCUSSION

Volatile Chlorinated Hydrocarbons (CHCs). Up to 17 individual volatile CHCs were detected during the monitoring of barrier performance. A summary of the principal components, degradation products and other significant analytes for Month 9 from the central transect through the test cell at an elevation of 5, 6 and 7 m is presented in Tables 1, 2 and 3 respectively.

TABLE 1 – Volatile CHCs (mg/L) – Central Transect December 1999 (Month 9) – Depth = 5 m. Groundwater flow direction from left to right.

Location	BP10	BP15	BP16	BP17	BP08	Decay	Half	Regression
Iron Thickness (m)	0	0.4	0.8	1.2	1.5	Constant	Life	Coefficient
Residence Time (days)	0	0.67	1.33	2.00	2.50	K	(hours)	
Tetrachloroethene	59	17.6	36.9	9.51	0.177	0.7	23.4	0.15
Trichloroethene	0.686	5.73	*3.32*	*0.909*	4.37	1.9*	8.6	1.0
cis-1,2-Dichloroethene	<0.2	*4.41*	*3.09*	3.52	8.55	0.5*	31.3	1.0
Vinyl chloride	3.09	6.75	*7.17*	*4.54*	6.73	0.7*	24.4	1.0
Carbon tetrachloride	*64.7*	*<0.2*	<0.2	<0.025	<0.025	>8.6*	<1.9	1.0
Chloroform	8.88	22.8	*22.8*	*10.8*	3.1	1.1*	14.8	1.0
Methylene chloride	<1	4.68	5.92	4.21	6.01	-	-	-
1,2-Dichloroethane	2.94	2.99	5.66	7.73	7.02	-	-	-
Other CHCs	0	0	1	1	1	-	-	-
Total CHCs	140	65	86	42	37	-	-	-
Mass reduction (%)	0	53	39	70	74	-	-	-

* Decay constant estimated from two point curve using maximum concentration difference over shortest travel distance. Data used shown in bold and italics in table.

The data indicate that the contaminant mass is comprised predominantly of PCE and CTC (approximately 88%). The monitoring data shows that although the principal components are almost completely degraded within the barrier, degradation products persist downgradient. In the case of the CTC, it was known that methylene chloride would be generated and would not be degraded by the zero valent iron (Gillham and O'Hannesin, 1994). However, degradation products of PCE such as TCE and vinyl chloride were expected to be readily degraded within the barrier. These compounds are emerging at the downgradient end of the barrier due principally to the slow degradation rate for PCE, and consequently, insufficient residence time for further degradation. The half-life for PCE was estimated using the first order kinetic model to be 23.4 hours. It is encouraging that PCE appears to degrade at a faster rate in the field than the laboratory column, although reasons for this are unclear. It is difficult to estimate the first order decay rate for compounds such as TCE, cis-1,2-DCE and vinyl chloride since these are present in the influent samples and are being formed and degraded concurrently within the barrier. A two-point curve was used to estimate the half-life for these compounds. All of the CTC was degraded at the first sample point (0.4 m) within the barrier, and the half-life estimate represents the maximum. This was expected based on column test results.

The estimates of the decay constant also indicate a significantly larger half-life for cis-1,2-DCE and VC, which tend to act as a rate-limiting steps for the complete degradation of PCE within the barrier.

TABLE 2 – Volatile CHCs (mg/L) – Central Transect December 1999 (Month 9) – Depth = 6 m. Groundwater flow direction from left to right.

Location	BP10	BP15	BP16	BP17	BP08	Decay	Half	Regression
Iron Thickness (m)]	0	0.4	0.8	1.2	1.5	Constant	Life	Coefficient
Residence Time (days)	0	0.67	1.33	2.00	2.50	K	(hours)	
Tetrachloroethene	79.3	4.27	2.88	4.01	0.301	2.5	6.7	0.84
Trichloroethene	1.99	4.22	*3.19*	*0.763*	1.3	2.1*	7.8	1.0
cis-1.2-Dichloroethene	0.578	10.8	*9.78*	*5.71*	4.85	0.8*	20.7	1.0
Vinyl chloride	5.63	7.49	*8.45*	*5.67*	3.59	0.6*	27.9	1.0
Carbon tetrachloride	*26.6*	*<0.1*	<0.1	<0.05	<0.05	>8.3*	<2	1.0
Chloroform	13.3	15.9	*14.8*	*7.34*	2.91	0.6*	28.7	1.0
Methylene chloride	1	3.43	4.95	3.85	3.62	-	-	-
1.2-Dichloroethane	9.98	8.68	7.06	7.44	6.42	-	-	-
Other CHCs	1	1	1	1	1	-	-	-
Total CHCs	140	56	52	36	24	-	-	-
Total Mass Reduction (%)	0	60	63	74	83	-	-	-

*Decay constant estimated from two-point curve using maximum concentration difference over shortest travel distance. Data used shown in bold and Italics in Table.

Data from 6 m (Table 2) shows a similar trend to 5 m albeit with a larger CHC mass reduction (83%) due principally to a faster degradation rate for PCE (half-life of 6.7 hours).

**TABLE 3 – Volatile CHCs (mg/L) – Central Transect December 1999
(Month 9) – Depth = 7 m. Groundwater flow direction left to right.**

Location	BP10	BP15	BP16	BP17	BP08	Decay	Half	Regression
Distance (m)	0	0.4	0.8	1.2	1.5	Constant	Life	Coefficient
Time	0.00	0.67	1.33	2.00	2.50	k	(hours)	
Tetrachloroethene	53.6	15.3	12.1	4.89	64	1.1	15.0	0.86
Trichloroethene	1.28	1.29	0.527	0.44	0.556	0.81	20.6	0.87
cis-1,2-Dichloroethene	0.204	1.73	2.57	3.9	<0.2	-	-	-
Vinyl chloride	2	1.12	<1	0.878	<2	-	-	-
Carbon tetrachloride	*15.8*	*<0.1*	<0.1	<0.05	53.5	>7.6*	<2.2	-
Chloroform	8.52	2.71	2.52	1.44	5.24	-	-	-
Methylene chloride	0.05	2.4	2.53	2.74	<0.2	-	-	-
1,2-Dichloroethane	5.77	3.44	3	3.79	1.43	-	-	-
Other CHCs	1	0	0	0	0	-	-	-
Total CHCs	89	28	23	18	125	-	-	-
Mass Reduction (%)	0	68	74	79	-	-	-	-

*Decay constant estimated from two-point curve using maximum concentration difference over shortest travel distance. Data used shown in bold and Italics in Table.

At a depth of 7 m, the concentrations of CHCs downgradient of the barrier (BP08) are similar to the influent concentration (BP10). Groundwater samples collected from this sampling port do not appear to flow through the barrier due to vertical hydraulic gradients. At this elevation the results from BP17 (1.2 m iron thickness) are used to assess the performance of the reactive barrier.

The data from a depth of 7 m (Table 3) shows that the concentration of total volatile CHCs at a distance of 1.2 m is lower than the corresponding distance at an elevation of 5 and 6 m. This is principally due to the lower influent concentration of CHCs at this elevation. The estimated half life for PCE (15.0 hours) at this elevation is larger than at 6 m (6.7 hours). This suggests a slower reaction rate in a zone with the lowest DOC concentration (see Table 6), which is contrary to the findings of the column studies (which indicated an inhibition in degradation rates when DOC was introduced into the columns). The data from earlier monitoring events (Table 4) shows that the PCE half life at a depth of 7 m has decreased significantly over the 9 months, whereas rates at 5 m have remained relatively constant. Table 4 indicates an increasing trend at a depth of 6 m.

TABLE 4. Estimated Half Life (Hours) for PCE

	5 m	6 m	7 m
Month 3	23.3	27.1	3.0
Month 6	19.7	20.8	6.7
Month 9	23.4	6.7	15.0

The performance of the wall as measured by the mass reduction of the total volatile CHCs over the monitoring period are presented in Table 5. The data suggests a decreasing performance at a depth of 5m even though the PCE and CTC degradation rates have not shown a decreasing trend. The data shows that the influent concentration has varied significantly during the monitoring events and indicates that some uncertainty exists in the estimate of the total volatile CHC mass reduction based solely on the influent and effluent concentrations.

TABLE 5 – Mass Reduction Performance - Total Volatile CHCs.

	5 m			6 m			7 m		
Month	BP10 (mg/L)	BP17 (mg/L)	Mass Reduction (%)	BP10 (mg/L)	BP17 (mg/L)	Mass Reduction (%)	BP10 (mg/L)	BP17 (mg/L)	Mass Reduction (%)
1	150	24	85	81	25	69	69	10	86
3	193	46	76	111	38	66	87	11	88
6	202	35	83	99	29	72	72	14	81
9	140	42	70	140	36	75	89	18	79

Note BP10 = influent samples and BP17 = iron thickness of 1.2 m.

Semivolatile Chlorinated Hydrocarbons. Although the data has not been presented herein, a range of semivolatile CHCs were detected in all influent samples and included hexachloroethane (HCE), hexachlorobutadiene (HCBD) and a range of di- and trichlorobenzene isomers. The analytical results from the effluent samples collected downgradient of the barrier at BP08 indicate that all of these compounds are being completely degraded within the barrier.

Aggregate Organic Compounds. The concentration of the DOC, TOC, BOD and COD remains relatively constant along the flow path. The average concentrations for these aggregate organic parameters for the December 1999 (month 9) sampling event are shown on Table 6. As indicated earlier, the concentrations of these parameter decrease with depth.

Table 6 – Average Concentration of Aggregate Organic Parameters – December 1999

	5 m	6 m	7 m
BOD (mg/L)	652	582	134
COD (mg/L)	773	665	174
TOC (mg/L)	511	484	102
DOC (mg/L)	501	469	97

The column studies, which were performed prior to the installation of the pilot scale barrier, indicated that the high concentration of DOC may inhibit the degradation rate of PCE within the barrier. The early monitoring data (Month 3) tended to confirm that reaction rates were faster at a depth of 7 m where the DOC concentrations are significantly lower. However as noted earlier, the PCE degradation rates after 9 months indicate that the degradation rates at 7 m have decreased below those estimated at 6 m where the DOC concentration is higher.

Inorganic Parameters. The field parameters (pH and Eh) measured at the time of sampling are presented in Table 7. The data shows that the influent groundwater (BP10) is reduced and has a very low pH (<5) at all sample depths. As the groundwater enters the barrier, the pH increases by approximately 2 pH units and the redox decreases further. In all monitoring events it was noted that the Eh at a depth of 7 m was lower than at 5 and 6 m. The trend is also evident in the TDS and ferrous iron within the barrier, which reveals a significantly lower concentration of these parameters at a depth of 7 m. Although the magnitude of the pH increase (2 to 3 pH units) is similar to that observed at other sites, the influent pH at this site is significantly lower. Therefore the pH does not increase to the levels which trigger significant iron hydroxide precipitation, leading to an increase in dissolved ferrous iron concentration in solution relative to most sites.

TABLE 7.   Inorganic analyses (mg/L) in influent samples (BP10) and at an iron thickness of 1.2 m (BP17).

Depth	5 m		6 m		7 m	
Location	BP10	BP17	BP10	BP17	BP10	BP17
Iron Thickness (m)	0	1.2	0	1.2	0	1.2
pH	4.82	6.82	4.6	6.95	4.71	6.74
Eh (mV)	-136	-214	-115	-217	-105	-162
TDS	1610	2550	3410	2790	702	712
Bicarbonate as CaCO3	46	173	<1	226	<1	38
Carbonate as CaCO3	<1	<1	<1	<1	4	<1
Sulphate	53	21	159	11	140	9
Sulphide	34.2	<0.1	31.9	0.2	18.8	0.4
Iron(II)	7.5	267	41.4	265	5.2	71.9

The influent waters do not contain any carbonate or bicarbonate because of the very low pH of the groundwater. The carbonate in the groundwater upgradient of the barrier is present as dissolved carbon dioxide (CO_2) and is normally referred to as carbonic acid (H_2CO_3).

Sulphate and sulphide decrease significantly in the reactive barrier. Sulphide in particular is completely removed in a short distance within the iron barrier and is most likely complexing with iron to form iron sulphide (FeS). The sulphate (SO_4^{2+}) is being reduced within the wall where it may possibly be forming either hydrogen sulphide (H_2S) or iron sulphide. Sulphate may also be incorporated into the structure of certain iron hydroxide precipitates.

The reduction of sulphate within the barrier may be biologically mediated (Benner, Blowes and Ptacke,1997) and samples of groundwater were analysed for the presence of Sulphate Reducing Bacteria (SRBs) . Only two samples detected the presence of SRBs (BP17 – 6 m and BP16 – 7 m), however the numbers present are indicative of very minor populations of SRBs which suggest that most of the sulphate reduction is not being biologically mediated.

The loss of porosity from precipitation of FeS, based on the maximum measured difference of the sulphate in the influent and effluent concentration, is estimated to be approximately 1.3%/annum.

CONCLUSIONS

The installation of a pilot scale zero valent iron barrier in an environment containing high concentrations of a mixture of chlorinated solvents has achieved a mass reduction of approximately 70 to 80% of the total volatile CHCs in the influent groundwater within the first 1.2 m of the barrier. The original influent and subsequent effluent includes approximately 10% of compounds that are known not to be degraded by zero valent iron. Some reduction in the degradation rate of PCE was observed in the zone where the DOC concentration is lowest. This is contrary to the findings of the column studies and will be investigated further.

Sulphide and sulphate reduction within the barrier is likely to be precipitated as iron sulphide and does not appear to be biologically mediated at this site. It is estimated that this will have only a minor impact on the performance of the wall with respect to hydraulic performance rather than mass reduction of CHCs.

Further monitoring of adjacent sample ports in the barrier will be undertaken at the 12 month sampling period to assess the lateral extent of the vertical variations which have been observed in the central alignment of the barrier. In particular, the observed decline in the degradation rate of PCE at 7 m to rates observed in zones of higher DOC concentration at shallower depths will be investigated. However, the monitoring data collected to date has demonstrated that even within a complex contaminant and geochemical environment, zero valent iron provides a relatively robust passive system for the remediation of a wide range of chlorinated solvents.

REFERENCES

Benner, S. G., D. W. Blowes and C. J. Ptacke. "A Full–Scale Porous Reactive Wall for the Prevention of Acid Mine Drainage." *Ground Water Monitoring Review*. Fall 1997, pp 99-107.

Duran, J. M., and J. A. Grounds (1996). "Site Characterisation of a Complex DNAPL Site – An Australian Experience." In L. N. Reddi (Ed.), *Non-Aqueous Phase Liquids (NAPLs) in Subsurface Environment: Assessment and Remediation.* ASCE National Convention, Washington D.C., November 12-14, 1996

Gillham, R. W., and S. F. O'Hannesin (1994). "Enhanced Degradation of Halogenated Aliphatics by Zero Valent Iron." *Ground Water 32*(6): 958-967.

INNOVATIVE CONSTRUCTION AND PERFORMANCE MONITORING OF A PERMEABLE REACTIVE BARRIER AT DOVER AIR FORCE BASE

Sam Woong-Sang Yoon, Arun Gavaskar, Bruce Sass, Neeraj Gupta, Robert Janosy, Eric Drescher, Lydia Cumming, and James Hicks (Battelle Memorial Institute, Columbus, Ohio, USA) and Alison Lightner (Air Force Research Laboratory, Tyndall AFB, Florida, USA)

ABSTRACT: A pilot-scale funnel-and-gate permeable reactive barrier (PRB) with two gates was installed in December 1997 to remediate dissolved chlorinated solvents (CVOCs) at Area 5, Dover Air Force Base (AFB), Delaware. A caisson was used to install the two gates, which were keyed into an aquitard located approximately 40 feet below ground surface (bgs). A pre-treatment zone (PTZ) was placed in each gate to remove dissolved oxygen (DO) in the influent water in order to promote degradation rates and extend the life of the reactive cell material. Two types of pretreatment media (one composed of pyrite and sand and the other composed of iron and sand) were used in the demonstration.

The field assessment of the hydrogeology and the geochemistry was performed from January 1998 to December 1999. The hydrogeologic evaluation, including continuous and periodic groundwater level surveys, slug tests, in-situ groundwater velocity probe data analyses and colloidal borescope tests, showed that groundwater was captured as designed and sufficient residence time was available in the reactive cell to degrade the influent CVOCs. The geochemical evaluations included long-term monitoring of CVOC, field parameters, inorganic groundwater constituents at or near the PRB, and analysis of iron cores from the barrier. All dissolved CVOCs in the influent groundwater appeared to be degraded to below the target levels (U.S. EPA maximum contaminant levels [MCLs] for trichloroethene [TCE] = 5 μg/L, tetrachloroethene [PCE] = 5μg/L, *cis*-1,2-dichloroethene [*cis*-1,2-DCE] = 70 μg/L, and vinyl chloride = 2 μg/L). Both PTZs succeeded in removing DO from the groundwater prior to the reactive cell, but the expected pH reduction in the pyrite could not be sustained as the groundwater flowed into the 100% iron reactive cell. The results of the field assessment demonstrate that the PRB is capturing the plume and is treating the contaminated groundwater at Dover AFB to target levels.

INTRODUCTION

The Area 5 aquifer at Dover Air Force Base (AFB), DE is contaminated with dissolved chlorinated volatile compounds (CVOCs): primarily perchloroethene (PCE), trichloroethene (TCE) and *cis*-1,2-dichlroethene (DCE). The exact source of the plume is not known but CVOC solvents have been used for aircraft maintenance at the base. A pilot-scale funnel-and-gate permeable reactive barrier (PRB) was installed to capture and treat the dissolved CVOC plume. A two-year field assessment of the PRB performance was conducted to demonstrate the effectiveness of the technology to treat the contaminated groundwater to target levels.

Objective. The objectives of this study are to design, install, and monitor the performance of a funnel-and-gate PRB system with two gates containing different pre-treatment zone (PTZ) reactive media. The performance evaluation consisted of the hydrogeology and geochemistry aspects of the application.

Site Description. The site lies in the Atlantic Coastal Plain Physiographic Province consisting of Cretaceous to Recent sedimentary deposits of gravel, sand, silt, clay, and limestone. The two uppermost geologic units consist of the Columbia Formation, which overlies the Calvert Formation of the Chesapeake Group. The Columbia Formation consists of sand with gravel 35 to 40 feet below ground surface (bgs) and the Calvert Formation consists of gray to blue gray silt with little fine to medium grained gray sand interbeded (Battelle, 1997). Groundwater elevation is seasonally fluctuating from 17 feet bgs during winter to as much as 5 to 7 feet higher during spring to summer.

Detailed site characterization, using cone penetrometer test (CPT) and groundwater level measurements from 30 CPT-installed temporary points at discrete levels, indicates the presence of clayey lenses in the Columbia Formation at the depth of 17 to 18 feet bgs ranging from 2 to 8 feet thick across the study area. Groundwater samples collected during the CPT investigation was used to delineate the plume, which was estimated to be about 50 feet wide at right angle to the flow path, with two areas of high concentrations of PCE.

DESIGN AND CONSTRUCTION
Design. Two gates, connected by impermeable sheet pile, were placed at the leading edges of the plume (Figure 1). Both gates consist of a pre-treatment zone (PTZ), an iron granular reactive cell, and an exit zone. For Gate 1 in the east of the PRB system, a mixture of 10 % of iron and 90% sand was used in the pre-treatment zone; for Gate 2 in the west, the mixture of 10 % of pyrite and 90% sand was used. Both mixtures were designed to remove dissolved oxygen in the influent groundwater prior to the iron cell to reduce rust and precipitate that may form in the presence of high dissolved oxygen (DO) and high pH.

The media used in the pre-treatment zones were selected based on long-term field column tests conducted by the U.S. EPA-NRMRL (U.S. EPA,1997). Connelly™ -8+50 mesh-size granular iron was used in the reactive iron cell in both gates, which extended from 10-ft bgs to 40-ft bgs (Battelle, 1997).

Gate Construction. A 8-ft diameter caisson in each gate location was used to drive down to 40 ft to the Calvert Formation. A vibratory hammer was used to push the bottom of the caisson 2 ft into the clay aquitard. A 5-ft diameter auger was used to remove the soil inside the caisson. During the soil removal from caisson, water was filled at the bottom of the caisson to prevent the inward pressure from outside the caisson.

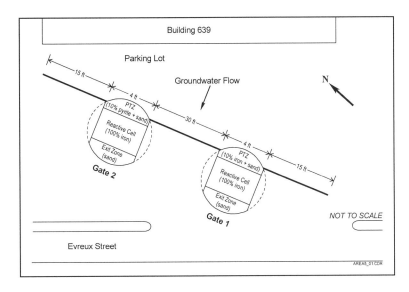

Figure 1. Schematic View of the PRB at Dover AFB

A frame of two parallel steel plates with 14 monitoring well clusters was inserted into the caisson in each gate. The steel plates are permanent dividers between the PRB area and the surrounding formation.

The granular iron was poured down in the iron reactive cell compartment in the frame by 2-ton-size of super sack and was filled up to 10-ft bgs to accommodate variable groundwater elevations. About 54,000 lbs of granular iron was required to pack each reactive cell. At the top of the iron fillings, high-density polyethylene layer was placed to provide a cap. In the PTZ, a mixture of 10 % of iron or pyrite (Gates 1 or 2, respectively) and 90% sand was placed. The same particle sized sand was also tremied in the exit zone. Once the emplacement of the gate media was completed, the caisson was pulled out of the ground.

Funnel Construction. Sealable-joint steel sheet piles (Waterloo Barrier™) patented by University of Waterloo were used as the funnel. A vibratory hammer was used to insert sheet piles into the ground. After all the sheet piles were connected and pushed into the ground, sealant was grouted in the joint to create a continuous impermeable barrier. The methods and materials used allowed for rapid installation and minimal disturbance at the installation site (Gavaskar *et al.*, 1998).

MONITORING
Well design. 2 in- and 1 in-ID schedule-40 PVC multi-level well clusters were installed to create a strategic hydrogeologic and geochemical monitoring network in the aquifer and PRB. The screen intervals are generally spaced at 15 to 20 ft, 20 to 25 ft, or 31 to 36 ft bgs for the shallow, middle, or deep depth level, respectively.

Hydrogeologic Monitoring. Groundwater levels were surveyed periodically to investigate the hydraulic gradients in the upgradient aquifer, the Gates 1 and 2, the downgradient aquifer, and background wells. Over 20 water level surveys have been conducted since the completion of the PRB. Figure 2 shows a contour map based on one of the groundwater surveys of the shallow wells. The regional hydraulic gradient is 0.002 and the general groundwater flow direction is to the southwest; the extremely low gradient and velocity of the groundwater made the determination of groundwater flow direction in the immediate vicinity of the PRB difficult. The water level maps do indicate hydraulic capture along expected portions of the aquifer, although seasonal variability makes the capture zone asymmetric, with more capture probably occurring on the north end of the PRB.

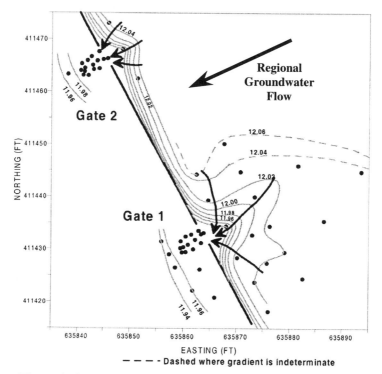

Figure 2. Map of Water Levels in Shallow Wells near the PRB

To measure the hydraulic conductivity of the materials in and around the barrier, slug tests were conducted in surrounding wells during the performance monitoring in the first year. The conductivity from all aquifer slug tests averaged 7.4 ft/day with the range between 1.8 and 101 ft/day. The geometric mean slug test results of Gate 1 vicinity upgradient aquifer wells and Gate 1 wells were 5.7 ft/day and 812 ft/day, respectively. The geometric mean of slug test results of Gate 2 vicinity upgradient wells and Gate 2 wells were 21.9 ft/day and 234 ft/day, respectively. The aquifer conductivity at the PRB site is in the range of 10 to 50 ft/day. Based on the slug test results, the flow velocity in the aquifer was estimated in a range from 0.026 to 0.16 ft/day.

Two in-situ HydroTechnics™ sensors were installed in the reactive cells and two sensors were installed in the Gate 1 upgradient aquifer to measure the flow direction and the velocity in and around the PRB. Overall, the probes indicated that flow conditions varied with climate and precipitation conditions. The readings of the sensors in June 1999 are indicated in Table 1. The general direction is to the southwest. The results of Probes 1 and 2 show velocity ranges that are very similar to the regional water flow velocity estimated from the modeling.

Table 1. HydroTechnics™ Sensor Measurements in June 1999

Probe	Location	Groundwater Velocity (ft/day)			Direction
		Min	Max	Mean	
1	Upgradient Aquifer of Gate 1	0.010	0.045	0.04	SSW
2	Upgradient Aquifer of Gate 1	0.050	0.095	0.09	SW
3	In Gate 1	0.005	0.075	0.03	S to SW
4	In Gate 2	0.010	0.600	0.03	SSW

Colloidal borescope tests were conducted at selected 2 in.-ID wells in the aquifer and in the reactive cells in collaboration with Oak Ridge National Laboratory (ORNL) (Korte, 2000). The borescope sensor searches colloidal particles and records the particle movements within a measurable period. While recording the particle movements, the flow direction and orientation are being calculated. The flow velocity from the borescope test is generally higher than that of HydroTechnics™ sensors or estimated from the slug tests because the borescope measures very localized preferential flow in the water column. The velocities from the colloidal borescope ranged from 1 to 15.7 ft/day in the Gate 1 upgradient aquifer and 3.6 to 15.5 ft/day in the Gate 2 upgradient aquifer. The velocity results from the borescope test were much higher compared to a regional flow velocity estimated by the other conventional methods. The overlay of the water level survey and the colloidal borescope and HydroTechnics™ results is shown in the Figure 3. The flow directions indicated by the borescope in wells F1S (in the Gate 1 PTZ) and U9D (in upgradient aquifer near the Gate 2 wing wall) appear to be the only significantly anomalous readings that did not match the directions indicated by water level maps. It is unclear whether the preferential flow patterns at these locations are complex or there are transient periods of back-flow at these locations.

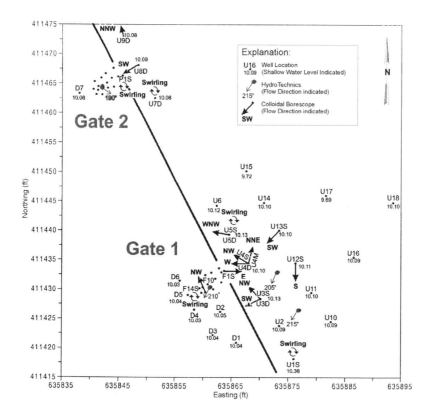

Figure 3. Colloidal Borecope and HydroTechnics™ Sensor Readings and Water Levels from Shallow Wells in December 1999

Geochemical Monitoring. A low-flow pump (peristaltic pump) was used to collect groundwater samples. Two comprehensive sampling events and several smaller scale sampling events at selected wells were conducted to delineate the dissolved CVOC plume in conjunction with field parameter measurements in the upgradient aquifer, two PTZs, Gates 1 and 2, and the downgradient aquifer.

PCE, TCE, and *cis*-1,2-DCE along with other VOC compounds were analyzed. The summary of the analysis results for Gates 1 and 2 is listed in Table 2. CVOCs were not detected in downgradient wells to the reactive cells or in the Gates 1 and 2 exit zone wells. The results indicate that the CVOCs degraded to the target levels as the plume passes through the reactive cell. Vinyl chloride was negligible at Dover site.

The performance in two different media in the PTZs was evaluated to the degree of oxygen scrubbing and the reactivity with the groundwater prior to flow into the reactive cells. The field parameters such as ORP, pH, and DO and the levels of inorganic constituents, sulfate and TDS, were measured. Although there was some difference between the pyrite mixture (Gate 2) and the iron mixture (Gate 1), the result shows that the dissolved oxygen in both PTZs is removed before entering into the reactive cells.

Table 2. CVOC Degradation Along Flow Path

Target Level Zone	PCE (μg/L) MCL = 5 μg/L		TCE (μg/L) MCL = 5 μg/L		*cis*-1,2-DCE (μg/L) MCL = 70 μg/L	
	July 98	June 99	July 98	June 99	July 98	June 99
Gate 1 and Vicinity Aquifer						
Immediate Upgradient Aquifer	ND to 155-334	ND to 210-520	ND to 11-22	ND to 34	ND to 20-69	ND to 44-130
PTZ	ND	8.2	ND	ND	ND	ND
Reactive Cell	ND to 7.0	ND	ND to 6.0	ND	ND	ND
Exit Zone	ND	ND	ND	ND	ND	ND
Downgradient	ND to 110	ND to 73	ND	ND to 19	ND	ND to 16
Gate 2 and Vicinity Aquifer						
Immediate Upgadient Aquifer	ND to 47-275	ND to 5.6-480	ND to 9-24	ND to 5.9-65	ND to 6-52	ND to 17-140
PTZ	ND	ND	ND	ND	ND	ND
Reactive Cell	ND	ND	ND	ND - 7.6	ND	ND
Exit Zone	ND	ND	ND	ND	ND	ND
Downgradient	ND	ND	ND to 10	ND	ND to 9	ND

ND: Non-detect and below a reporting detection limit.

To observe precipitation and corrosion of the iron media in the reactive cell, several iron core samples were collected and tested. The analyses employed for these analyses were X-Ray Diffraction (XRD), scanning electron microscopy (SEM), Raman spectroscopy, infrared spectroscopy, carbon analysis and microbiological evaluations (Battelle, 1999). The results of the analyses and evaluations showed that there was little precipitation and corrosion built up on the iron surface in both reactive cells in the 18 months since installation in the low-alkalinity aquifer and with the PTZ to remove DO from the groundwater.

CONCLUSIONS

The contaminant levels of the plume, field parameters, and hydraulic behavior in the barrier were monitoring during a two-year period. Concentrations of PCE, TCE, and *cis*-1,2-DCE in the groundwater at Dover AFB Area 5 were degraded to below the target levels. TCE and *cis*-1,2-DCE levels are comparatively lower than the levels of PCE. The results of hydrogeologic modeling based on the hydraulic tests show that the dissolved CVOC plume is captured and enters Gates 1 and 2 to react in the reactive barrier. No significant difference was noted between the two different media used in the PTZs with respect to the removal of dissolved oxygen in the groundwater. Both PTZs succeeded in removing DO from the groundwater prior to the reactive cell, but the expected pH reduction in the pyrite could not be sustained as the groundwater flowed into the 100% iron reactive cell. The use of caissons is a viable option for installing reactive media in relatively deep aquifer or at sites with a lot of underground utility lines.

ACKNOWLEDGEMENTS

This project was funded by DoD's Strategic Environmental Research and Development Program (SERDP) through the Air Force Research Laboratory (AFRL). Authors would like to acknowledge the support of the base Civil Engineering department at Dover AFB and the staff at the Dover National Test Site (DNTS) AFB for their on-site support during this project. DNTS is one of four national environmental technology test site locations established and managed by U.S. DoD SERDP. We would also like to thank the staff from ORNL for their cooperation during the colloidal borescope measurements.

REFERENCES

Battelle. 1999. *Design, Construction, and Monitoring of the Permeable Reactive Barrier in Area 5 at Dover Air Force Base.* Prepared for Air Force Research Laboratory, Florida. November.

Battelle. 1997. *Design and Test Plan: Permeable Barrier Demonstration at Area5, Dover AFB* prepared for AFRL, Tyndall Air Force Base, Floria. November.

Gavaskar, A.R., N. Gupta, B.M. Sass, R.J. Janosy, and D. O'Sullivan. 1998. *Permeable Barriers for Groundwater Remediation: Design, Construction and Monitoring.* Battelle Press, Columbus, Ohio.

Korte, N. and R. Schlosser. 2000. *Application of the Colloidal Borescope at the Dover AFB.* Oak Ridge National Laboratory, Grand Junction, Colorado.

United States Environmental Protection Agency. 1997. *Selection of Media for the Dover AFB Field Demonstration of Permeable Barriers to Treat Groundwater Contaminated with Chlorinated Solvents.* Preliminary report to U.S. Air Force for SERDP Project 107. August 4.

EVALUATION OF THE KANSAS CITY PLANT IRON WALL

Alan D. Laase and Nic E. Korte
(Oak Ridge National Laboratory, Grand Junction, Colorado, USA)
Joseph L. Baker and Paul D. Dieckmann
(Honeywell Federal Manufacturing & Technology, Kansas City, Missouri, USA)
John L. Vogan and Robert L. Focht
(Envirometal Technologies, Inc., Waterloo, Ontario, Canada)

ABSTRACT: A 130-ft long by 6-ft wide passive iron treatment wall was installed at the U. S. Department of Energy Kansas City Plant to treat groundwater contaminated with trichloroethylene and associated degradation products. Water quality data collected from performance monitoring wells installed in and around the iron wall showed contaminated groundwater flowing around the southern end of the wall. The bypass is caused by inadequate iron wall length. The iron wall was designed to treat a narrower plume representative of pumping conditions rather than the actual, wider ambient plume. Recent characterization identified a previously unknown more permeable sandy gravel zone south of the wall that may also be a cause of contaminant bypass. Incomplete contaminant degradation was observed in the southern end of the wall. Increased groundwater velocities associated with the sandy gravel zone and increasing basal gravel thickness may be the cause of the incomplete contaminant degradation. A previously unknown groundwater mound located immediately downgradient of the iron wall was also identified as potentially hindering the performance. In portions of the iron wall where there is sufficient residence time, groundwater contamination is effectively degraded and all applicable groundwater standards are achieved.

INTRODUCTION

As a result of manufacturing activities at the US Department of Energy (DOE) Kansas City Plant (KCP) shallow groundwater is contaminated with trichloroethylene (TCE) and associated degradation products 1,2-dichloroethene (DCE) and chloroethene (vinyl chloride [VC]). In 1990 an interceptor trench was installed to prevent one of KCP two plumes from reaching a nearby river (Figure 1). At the time of installation it was recognized that because of the presence of dense non-aqueous phase liquids the interceptor trench would have to operate in perpetuity and its operation was considered temporary until different remedial strategies could be further evaluated and more efficient replacement technologies became available (Korte and Kearl, 1990). In 1996 KCP began exploring use of an iron wall as a replacement remedial technology for the interceptor trench.

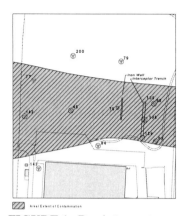

FIGURE 1. Pre-interceptor system plume configuration.

A number of participants have been involved directly or indirectly with the KCP iron wall. URS Greiner Woodward Clyde Federal Services (URSGWWCFS) performed initial characterization activities associated with the iron wall, prepared construction documents, and completed some testing of the hydraulic and remedial effectiveness of the iron wall. Envirometal Technologies Inc. (ETI), as a subcontractor to URSGWWCFS, prepared iron wall design specifications.

Two KCP entities were involved, Facility Engineering tasked with overseeing preparation of the design specifications and installation of the iron wall, and the Environmental Compliance group tasked with long-term performance. Oak Ridge National Laboratory (ORNL) is currently performing research regarding appropriate methods for evaluating the performance of in situ permeable reactive barriers and in that capacity has assisted in the evaluation of the KCP iron wall (Korte et al.1999).

HYDROGEOLOGY

At the KCP the 25 to 45-foot thick alluvial aquifer consists of two distinct hydrologic zones, an upper clayey-silt zone overlying a semi-confined basal gravel zone. The upper clayey-silt consist of thin-bedded clayey-silt with minor amounts of sand. The basal gravel ranges in thickness from a few inches to 8 feet and consists of fragments of eroded bedrock in a sand-silt-clay matrix. Basal gravel hydraulic conductivities are typically an order of magnitude or more greater than upper clayey-silt hydraulic conductivities. Underlying the alluvial aquifer are relatively impermeable shale and sandstone.

PRE-DESIGN CHARACTERIZATION, INITIAL DESIGN AND CONSTRUCTION

A location 175 ft west of the interceptor trench was selected as the preferred site for installation of the reactive permeable barrier (Figure 1). In the fall of 1996, to obtain site- specific design data, URSGWWCFS conducted a field investigation consisting of soil borings, groundwater sampling and aquifer testing. To facilitate the investigation, the interceptor trench was turned off 16 days prior to the start of the field program to allow re-establishment of the natural flow pattern (WCFS 1997). Based on groundwater sampling, the study concluded that at the proposed wall location the plume was 120 ft wide (WCFS 1997). However, since the interceptor trench had only been off a short time prior to sample collection after having been in continuous operation for seven years, the characterized contaminant distribution reflected a plume that had narrowed in response to pumping and was not representative of ambient (pre-interceptor trench) conditions. Prior to operation of the interceptor trench, the plume at the same location was depicted as being in excess of 200 feet wide (Figure 1) (DOE 1990). However, the plume width was extrapolated and not measured due to a limited number of sampling points. In the absence of exact pre-interceptor trench historical plume measurements, the 120 ft width was assumed more representative than the estimated 200 ft width.

Hydrologic and chemical data collected during characterization and used in the permeable barrier design are presented in Table 1. Based on these data and bench-scale testing, a 130 ft long iron wall with widths of two feet and six feet for the upper clayey-silt and basal gravel zones, respectively, was specified (WCFS 1997).

TABLE 1. Hydrologic and chemical data used in the iron wall design.

Saturated Thickness	21 ft
Hydraulic Gradient	0.01 ft/ft
Thickness of the Basal Gravel	3 ft
Hydraulic Conductivity of the Basal Gravel	34 ft/d
Hydraulic Conductivity of the Upper Silty-Clay	0.75 ft/d
Porosity	0.3
Groundwater Velocity in Basal Gravel	1.13 ft/d
Groundwater Velocity in the Upper Silty-Clay	0.025 ft/d
Maximum 1,2 DCE Concentration	1,337 µg/L
Maximum Vinyl Chloride Concentration	291 µg/L
Plume Width	120 ft

Source: URSGWWCFS 1999

Construction of the KCP iron wall was completed in May 1998 at a cost of $1.2 million. Originally the wall was to have been installed using a single pass trencher (URSGWWCFS 1999). However, the trenching machine proved incapable of excavating through the clayey-silt. Next trench box construction techniques were attempted but that too failed. Finally, the iron wall was constructed in 26-ft segments using braced sheet piling, which was driven to depth using a vibratory hammer. The construction cells were dewatered prior to placement of the iron.

PERFORMANCE MONITORING

The current performance evaluation monitoring well network consists of 39 wells, all located in the immediate vicinity of the iron wall (Figure 2). The majority of the wells are located in three clusters (north, central and south) that transect the width of the wall. In a typical cluster one well is located within three feet of the upgradient side of the iron wall, two more are placed at two foot intervals within the wall, and a fourth well is located within three feet of the down gradient side of the wall. Some of the clusters have additional up and downgradient-monitoring wells located within 25 feet of either side of the iron wall. Finally, additional monitoring wells are located at the north and south ends of the wall to evaluate the potential for contamination bypass around the ends of the wall. Many of the wells located outside the wall are dual completion wells having upper and lower completions screened in the clayey-silt and basal gravel, respectively, which are separated by a bentonite seal.

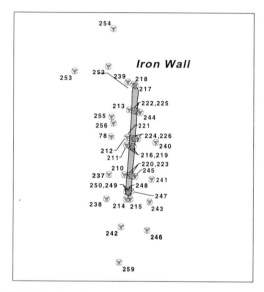

FIGURE 2. Performance monitoring well network.

Slug tests were performed on a number of the performance monitoring wells located in and outside of the iron wall. Tests performed by ORNL and URSGWWCFS within the wall yielded hydraulic conductivities between 300 ft/d and 500 ft/d (DOE 1999a and URSGWWCFS 1999). Tests performed by both ORNL and URSGWWCFS in native material at the southern end of the iron wall yielded hydraulic conductivities between three and 12 times greater than the hydraulic conductivity value of 34 ft/d used in the design (DOE 1999a and URSGWWCFS 1999). The 34 ft/d hydraulic conductivity value was obtained from a pumping test conducted near the center of the iron wall (WCFS 1997). Lithologic logs for wells at the southern end of the iron wall indicate the presence of a sandy gravel zone that is absent in the central and northern portions of the iron wall (URSGWWCFS 1999). The supervising engineer also noted the presence of the same sandy gravel zone during construction of the iron wall. Also, the basal gravel is thicker in this area allowing for the transmission of greater volumes of contaminated groundwater (URSGWWCFS 1999). At this time it is not known whether the sandy gravel is a local feature or is extensive, and whether the sandy gravel and the basal gravel are hydrologically connected. In addition to the slug tests, a pumping test performed 50 ft north of the north end of the iron wall produced an average hydraulic conductivity value of 21 ft/d, similar to the hydraulic to the 34 ft/d obtained from the pumping test conducted near the center of the iron wall (DOE 1999b).

The iron wall design was partially based on a silty-clay and basal gravel zone potentiometric surface constructed from water-level elevations from 19 monitoring well and 11 temporary piezometers. The upper clayey-silt and basal gravel zone potentiometric surface showed primarily easterly groundwater flow in the in the vicinity of the proposed iron wall (Figure 3) (WCFS 1997). The upper silty-clay and basal gravel zone potentiometric surfaces (Figure 4) constructed using the current performance monitoring well network have the same general geometry as the design potentiometric surfaces (Figure 3) with some subtle differences. A groundwater mound, thought to be associated with infiltration from a nearby surface water drainage ditch, is present in the basal gravel zone immediately downgradient of the iron wall. The density of the original monitoring well and piezometer network was not sufficient to identify the mound during the design phase. For the performance evaluation, the mound was excluded from the potentiometric surface evaluation (URSGWWCFS 1999) (Figure 4). Rationale for excluding the mound from the evaluation is that the well defining the mound is completed in a lower permeable zone relative to the surrounding wells. As a result of the conductivity contrast, the water level in the well is considerably higher than in the surrounding wells. Therefore, the water level from the well was not considered representative and was not used in constructing the basal gravel zone potentiometric surface. When the groundwater mound is excluded, the basal gravel potentiometric surface shows groundwater contamination flowing through the center portion of the iron wall (Figure 4). When the groundwater mound is included, the basal gravel potentiometric surfaces sometimes show groundwater being diverted away from the central portion of the iron wall and flowing to the northeast and southeast (as shown in

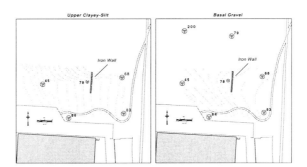

FIGURE 3. Potentiometric surfaces used to design the iron wall.

Figure 4). Other times, because of reduction in size of the groundwater mound, the potentiometric surface shows that the diversion occurs after groundwater flows through the central portion of the iron.

Water levels in the confined basal gravel zone increase rapidly in response to precipitation events. Because of the much greater storage capacity associated with unconfined relative to confined conditions, water levels in the fully

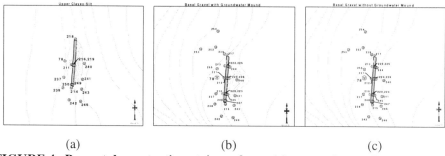

(a)   (b)   (c)

FIGURE 4. Present day potentiometric surfaces, (a) upper clayey-silt, (b) basal gravel with mound, (c) basal gravel without mound.

penetrating unconfined iron wall, respond much more slowly to precipitation events. Consequently, immediately following precipitation events, groundwater in the basal gravel zone enters the trench from all sides (URSGWWCFS 2000). This condition is relatively short-lived and persists until hydrologic equilibrium is achieved.

There is a significant downward vertical flow in the vicinity of the iron wall as evidenced by the two-foot difference in water-level elevations between the upper clayey-silt and the basal gravel zones. Because there is no resistance to vertical flow in the iron wall, the water-level elevation in the iron wall is equal to the water-level elevation in the basal gravel. Consequently, with respect to the upper clayey-silt zone, the iron wall is a sink and contaminated groundwater flows into the wall from all sides (Figure 4). Thus, contaminated groundwater does not flow horizontally through the wall in the upper clayey-silt zone. Rather, treatment is affected as contaminated groundwater originating in the clayey-silt zone migrates vertically through the iron to the basal gravel zone.

Chemical sampling of the performance monitoring well network shows contaminated groundwater extending 20 ft north and 40 ft south of the iron wall (Figure 5). Groundwater contamination north of the iron wall is reported captured by the iron wall (URSGWWCFS 2000). However, the analysis was based on a basal gravel zone potentiometric surface that did not include the groundwater mound. When the groundwater mound is included in the basal gravel zone potentiometric surface, some contamination north of the wall appears to bypasses the wall. Groundwater contamination concentrations in wells located south of the wall, including those completed in the sandy-gravel zone, are increasing with time suggesting a widening of the plume in that direction. Sampling results suggest that the iron wall is effectively reducing contaminant levels except at the southern end (wells 247 and 248) where DCE and VC have repeatedly been detected at approximately 40 µg/L and 15 µg/L, respectively (URSGWWCFS 2000). Based on upgradient concentrations, contamination is being degraded at this location, just not below the maximum concentration limit (MCL) for VC. The presence of contamination within the south end of the iron wall is likely a result of groundwater velocities within the sandy gravel zone that exceed design specification (Table 1). Low levels of VC (<5 µg/L) have been sporadically detected in the upper reaches of well 223, located in the south-central portion of the iron wall along the downgradient edge of the iron wall (Figure 5) (URSGWWCFS 2000). Contamination has never been detected in companion well 220 located two feet further upgradient within the iron wall. The presence of VC is reportedly the result of groundwater flowing into the iron wall from the downgradient side during periods of rapid rise in the potentiometric surface (URSGWWCFS 2000). Contamination is present downgradient of the southern half of the iron wall (Figure 5). Given that groundwater contamination is absent within the iron wall, the contamination is surmised to be residual contamination that has not been flushed from the aquifer (URSGWWCFS 2000). Potentiometric surface evaluation that includes the groundwater mound suggests that the some of this contamination enters the trench and is treated.

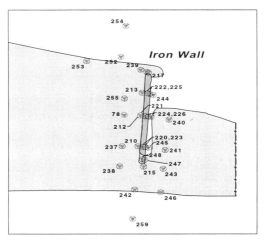

FIGURE 5. Groundwater contamination distribution in the vicinity of the iron wall.

PROPOSED MODIFICATIONS TO THE IRON WALL

To treat groundwater contamination that is flowing around and through the southern end of the iron wall a 10 ft wide and 48 ft long extension has been proposed as a worst-case design (URSGWWCFS 2000). Additionally, the same worst-case design suggests widening the southern most 30 ft of the existing iron wall from 6 ft to 10 ft so that complete contaminant degradation can be achieved in the vicinity of wells 247 and 248. Modifying the iron wall is estimated to cost between $350,000 and $475,000.

CONCLUSIONS

Better recognition of the variability in hydraulic conductivity at the iron wall location and the temporal changes that occurred to the plume geometry as a result of interceptor trench operation would have led to a more robust design. Where groundwater flow velocities are below the design velocities, the iron wall effectively degrades groundwater contamination below MCLs. In retrospect, the pre-design characterization effort should have been more extensive. Rather than a single pumping test, multiple pumping tests, or other hydraulic conductivity measurements (e.g. slug tests), should have been performed along the length of the proposed iron wall to better characterize lateral variations in hydraulic conductivity. Additionally, more soil borings were needed to characterize horizontal and vertical variations in lithology. Finally, because the groundwater plume narrowed with time as it converged on the interceptor trench, consideration should have been given to using historic information on plume width rather than the width determined during the pre-design characterization effort.

REFERENCE

DOE. 1999a. *Preliminary report-comparison of hydraulic assessment methods for the permeable barrier at the Kansas City Plant.* U. S. Department of Energy, EM-50 Subfocus Contaminant Focus Area, Washington, DC.

DOE. 1999b. *Abandoned Blue River Channel Pumping Test.* U. S. Department of Energy, Kansas City Plant, Kansas City, Missouri.

DOE. 1990. *Groundwater quality assessment plan for the northeast area.* U. S. Department of Energy, Kansas City Plant, Kansas City, Missouri.

Korte N., L. Liang, R. Puls, and C. Reeter. 1999. *Interagency Program for Long-Term Performance Monitoring of Permeable Reactive Barriers. Abiotic In Situ Technologies for Groundwater Remediation.* U. S. Environmental Protection Agency. Dallas, Texas, September 1999.

Korte, N. E. and P. M. Kearl. 1990. "Plume management. An inexpensive long-term strategy for groundwater contamination at low-risk sites." *Waste Management & Research* , 8, 246-248.

URSGWWCFS. 2000. *Iron wall demonstration montoring report- addendum no. 1.* URS Greiner Woodward Clyde Federal Services, Overland Park, Kansas.

URSGWWCFS. 1999. *Iron wall demonstration montoring report.* URS Greiner Woodward Clyde Federal Services, Overland Park, Kansas.

WCFS. 1997. *Pre-design field investigation report-funnel and gate passive groundwater treatment systems.* Woodward Clyde Federal Services, Overland Park, Kansas.

CHARACTERIZING A CULTURE THAT DECHLORINATES TCE WITH FE(0)

Pei C. Chiu, Minho Lee, Daniel K. Cha, Mark Radosevich, and E. Danielle Rhine
University of Delaware, Newark, Delaware

ABSTRACT: A mixed culture, which was previously shown to dechlorinate TCE to ethene in the presence of Fe(0), was studied. TCE was also dechlorinated to ethene when H_2 replaced Fe(0) as the sole reducing agent, suggesting H_2 produced from anaerobic Fe(0) corrosion was the actual electron donor. When receiving lactate, the culture reduced TCE to ethane via DCEs, VC and ethene and formed high quantities of methane. In contrast, when 2-bromoethane sulfonate was added or when H_2 or Fe(0) was used as electron source, ethene was the end product. Reduction of ethene to ethane appeared to depend on methanogens, although they were probably not involved in the dechlorination reaction. Digital confocal images of Fe(0) surface taken after 10-day incubation with the culture showed sporadic but significant colonization. DGGE analysis of the PCR products from the culture gave several distinct bands using universal primers P63f and P518r.

INTRODUCTION

Trichloroethene (TCE) is a prevalent groundwater contaminant that often exists in the subsurface as dense non-aqueous phase liquid (DNAPL). Due to the low efficiency and high cost associated with conventional pump-and-treat method to remediate DNAPLs, permeable reactive barrier (PRB), a passive technology using zero-valent iron, was developed (Tratnyek, 1996; Vidic and Pohland, 1996). Dozens of full- and pilot-scale iron PRBs have been installed to date (RTDF, www.rtdf.org). However, the long-term performance of these barriers remains uncertain and has been an area of active research (Gu et al., 1998; O'Hannesin and Gillham, 1998). The complex biogeochemical processes involved in iron PRBs are not well understood, particularly with respect to the role(s) of microorganisms in groundwater-iron systems containing dissolved chlorinated ethenes.

Zero-valent iron in PRBs creates an anaerobic environment by depleting dissolved O_2 and nitrate. In addition, anaerobic iron corrosion (i.e., with water being the electron acceptor) produces H_2, which is an excellent electron donor and energy source for certain autotrophic anaerobes including methanogens, acetogens, sulfate reducers, and TCE dehalogenators (Holliger et al., 1998; Löffler et al. 1999). Weathers et al. (1997) and Novak et al. (1998) showed that methanogens can couple the oxidation of H_2 from iron corrosion to the reductive dechlorination of $CHCl_3$ and CCl_4. Lampron et al. (1998) reported that a mixed culture mediated TCE dechlorination in a batch aqueous system containing Fe(0) as the sole electron source. Gu et al. (1999) demonstrated that a sulfate reducer population developed in iron columns receiving synthetic groundwater, but the organisms involved were not identified.

In addition to other potential effects such as changing PRB permeability, microbial activities, especially dechlorination activities, may be detrimental to the effectiveness and longevity of iron PRBs. Unlike abiotic TCE reduction by Fe(0), which produces ethene and ethane as end products with insignificant formation of chlorinated intermediates, microbial TCE reduction often results in accumulation of vinyl chloride (VC, a carcinogen), which is mobile and unreactive with iron.

To better assess the long-term performance of iron PRBs, it is necessary to obtain information about the microorganisms that can derive reducing equivalents from iron for energy and for the dechlorination of chlorinated ethenes. The objective of this study is to characterize a microbial culture which has been shown to dechlorinate TCE in the presence of Fe(0). We will identify the dominant members in the consortium and elucidate their roles in TCE transformation. The results presented here are preliminary data from the first phase of the study.

MATERIALS AND METHODS

Chemicals. All chemicals used for preparing the culture medium were obtained from Sigma (St. Louis, MO) or Aldrich (Milwaukee, WI) and used as received. TCE (>99.5%), 1,1-DCE (99%), *cis*-1,2-DCE (97%), and *trans*-1,2-DCE (98%) were purchased from Aldrich. VC (0.1% in N_2), ethene (99.5%), ethane (99%), methane (99.99%), and H_2 (99.99%) were purchased from Scott Specialty Gases (San Bernardino, CA). Iron chips were obtained from either Alfa Aesar (99.97%, Ward Hill, MA) or Master Builders, Inc. (approximately 95%, Cleveland, OH).

Culture Maintenance. An anaerobic enrichment culture used in this study was originally isolated from a landfill site at Dover Air Force Base (Dover, DE). The culture was maintained in 120-mL serum bottles (Supelco, Bellefonte, PA) containing ammonium acetate (0.3 g/L), monopotassium phosphate (0.5 g/L), magnesium acetate (0.21 g/L), calcium acetate (0.12 g/L), 4-(2-hydroxyethyl)piperazine-1-ethanesulfonic acid buffer (HEPES, 0.02 M), sodium bicarbonate (1.2 g/L), 10% yeast extract (1 mL/L, 0.2 μm filter-sterilized, Millipore, Bedford, MA), 60% sodium lactate (0.72 mL/L), mineral salt solution (6.7 mL/L), filter-sterilized vitamin solution (0.1 mL/L), and TCE-saturated deionized water (4.5 mL/L). The compositions of the mineral salt solution and vitamin solution have been described elsewhere (Lampron et al., 1998). Five mL HEPES buffer (1.0 M) was added to assist in dissolving the chemicals. The final pH of the growth medium was 7.0 ± 0.1. The stock culture bottles were sealed with Teflon-lined rubber septa and aluminum crimp caps, foil-wrapped, and stored at 22 ± 2°C in an anaerobic glove bag (I^2R, Cheltenham, PA) filled with N_2 (Keen, Wilmington, DE). Once a month, 50 mL of liquid from each stock culture bottle was replaced by 50 mL of freshly prepared culture medium to replenish the nutrients and TCE.

Two sub-cultures derived from the stock culture were also used for the study. One sub-culture has been maintained using the same medium without TCE for the past 6 months. This is to eliminate obligate TCE dehalogenators in the culture, if any. The other sub-culture has been receiving 3 mM 2-bromoethane sulfonate (BES, a methanogenesis inhibitor) for the past 18 months.

Batch Experiment. Dechlorination experiments were conducted in 63-mL amber bottles (Alltech, Deerfield, IL) at room temperature under anoxic and light-excluded conditions. The growth medium used in all experiments was identical to the stock culture medium with modifications to eliminate any potential electron sources: lactate was omitted, HEPES was replaced with phosphate, acetate was replaced with chloride, and reactors were prepared under H_2-free (N_2) atmosphere. Each reactor contained 16.5 mL of electron donor-free medium, 15 mL of 30-day old culture, and 31.5 mL of N_2 headspace. All bottles were sealed with Mininert™ valves (Precision, Baton Rouge, LA) and low-permeability vinyl tape (3M, St. Paul, MN). Two mL of H_2 gas was added to each reactor but not to the controls. All bottles were then spiked with 1.0 µL neat TCE (~11 µmol) and incubated statically at room temperature in an inverted position. At predetermined times, headspace samples (100µL) were withdrawn with a gas-tight syringe and analyzed using a Hewlett Packard (Wilmington, DE) 6890 gas chromatograph with a flame-ionization detector and a 30-m GS-GasPro capillary column (J&W, Folson, CA).

Confocal Images. Digital confocal images of microbial colonies on iron surface were acquired with a laser scanning confocal microscopy (Zeiss LSM 510). Iron chips from incubation bottles, with and without culture, were collected at 0 and 10 days. The microbial colonies on the surfaces were labeled with a live-cell nucleic acid stain SYTO 13 (Molecular Probes, Eugene, OR), which has an excitation and emission spectra similar to fluorescent isothiocyanates (FITC) dye. The stained iron chips were observed under 63x C-Apochromat water immersion objective

(NA 1.2) on a Zeiss LSM 510 laser scanning confocal microscope using the 488 nm excitation line of an Omnichrome Krypton-Argon laser. The reflection (iron) and fluorescence (cells) signals from the surface were simultaneously collected by separating the signals with a 570 nm dichroic and a 505 nm longpass filter for the fluorescence channel and no barrier filter for the reflection channel.

DNA Extraction. Fifty mL stock culture was centrifuged (x12,000 g) for 10 min to collect cells. Most of the supernatant was removed and the cells were vortexed with the remaining liquid (approximately 3 mL). The cell suspension was then transferred to a 1.5 mL microtube and centrifuged for 2 min (x10,000 g) and the supernatant was discarded. This step was repeated until the resuspended liquid was depleted. One hundred μL of Tween 20 (0.1 %, Bio-Rad, Hercules, CA) was added to the microtube and vortexed to promote cell lysis and the mixture was heated to 100°C for 10 min in a thermocycler (MJ Research, Inc). The cells were resuspended by vortexing for 10 s and centrifuged for 1 min. Eighty μL of InstaGene™ Matrix (Bio-Rad), while being mixed on the stirrer, was added to remove the cell lysis products in the tube and the mixture was then vortexed. The sample was incubated at 56°C for 20 min, vortexed, and incubated again at 100°C for 8 min. After incubation, the sample was vortexed (10 s) again and centrifuged for 2 min and the supernatant containing DNA was kept at -20°C before use for the PCR. In addition, 5 μL of the supernatant was stained with ethidium bromide (Sigma) and run in 1% agarose gel at 120 V for 30 min.

PCR. A modified PCR from previously reported procedure (EL Fantroussi et al., 1999) was used to amplify the DNA. The forward and reverse primers used were P63f (5'CAG GCC TAA CAC ATG CAA GTC 3') and P518r (5'ATT ACC GCG GCT GCT GG 3'), respectively. P63f had been shown to be applicable to a wide range of bacteria (Marchesi et al., 1998). P518r was based on a universally conserved region of bacteria (Øvreås et al., 1997). These primers were expected to give PCR products of approximately 495 base pair (EL Fantroussi et al., 1999), a size suitable for denaturing gradient gel electrophoresis (DGGE) analysis. A GC clamp of 40 bases was added to the forward primer.

The PCR mixture consisted of 3.5 μL 10x Taq buffer, 0.28 μL 100 mM dNTP mix (0.25 mM/base), 0.18 μL Taq polymerase (5U/μL), and 7 μL Q solution (Qiagen PCR Core Kit). Primers P63f and P518r (Operon) were added in the amounts equivalent to 50 pmol per 100 μL reaction. One μL extracted DNA was used as template for the PCR. Sterile deionized water was added to make up the mixture volume to 35 μL. PCR was performed using an iCycer (Bio-Rad) and the temperature program: 95°C for 5 min with hot start (initial denaturation step); 30 cycles of 92°C for 1 min, 55°C for 1 min, and 72°C for 1 min; and 72°C for 10 min (final extension step). Where the amounts of PCR products are low, the PCR cycle was repeated with addition of 0.13 µL Taq polymerase in 1x buffer. Five µL of the PCR products was stained with ethidium bromide and run in 1% agarose gel at 120 V for 30 min to ensure the DNA fragments had the same length of approximately 495 base pairs.

DGGE. DGGE was performed using a Bio-Rad DCode System. Twenty-five µL PCR products was loaded into an 8% (wt/vol) polyacrylamide gel in 1x TAE running buffer (40 mM Tris base, 20 mM acetic acid, and 1 mM EDTA; pH 8.0). The linear 25-75% and 50-85% denaturing gradients were created by mixing 7M urea (Fisher) and 40% deionized formamide (EM Science). The electrophoresis was performed at 60°C at 75 V for 12 h. After electrophoresis, the gels were soaked for 1 h in a vessel containing ethidium bromide solution (0.5 g/L). The stained gel was viewed on a UV transilluminator (Fisher) and photographed with a polaroid camera (Fotodyne).

RESULTS AND DISCUSSION

This culture has been shown to dechlorinate TCE to ethene in the presence of Fe(0) as the sole electron source (Lampron et al., 1998). Result from the batch experiment shows that H_2 alone can support complete TCE dechlorination to ethene (Figure 1). This suggests that cathodic H_2 produced during anoxic Fe(0) corrosion was probably the actual electron donor for TCE reduction. After a short lag, TCE was dechlorinated to DCEs and VC until the initial H_2 was depleted. Significant CH_4 production indicates the presence of methanogens in the culture. Re-spiking of H_2 stimulated DCE and VC dechlorination to ethene without immediate CH_4 formation, which suggests that the methanogens were probably not responsible for the observed dechlorination.

The original culture and its BES sub-culture both dechlorinated TCE completely but showed different product distribution. With lactate as the substrate, the original culture reduced TCE to ethane via *cis*-DCE, VC, and ethene as intermediates. Much methane was formed and no 1,1-DCE was observed. In contrast, the BES sub-culture reduced TCE to ethene (but not ethane) and both 1,1-DCE and *cis*-DCE were formed in significant amounts (data not shown). The result suggests that either the methanogens play a role in 1,1-DCE and ethene reduction or BES has an inhibitory effect on non-methanogenic organisms (e.g., dehalogenators) that are capable of reducing 1,1-DCE and ethene. BES, which

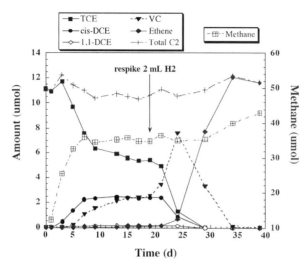

Figure 1. TCE reduction by the mixed culture with 2 mL H₂ as sole electron donor (another 2 mL H₂ added at 19d).

was thought to be a methanogen-specific inhibitor, has been shown to inhibit TCE dehalogenation (Löffler et al., 1997).   Interestingly, black precipitates (presumably metal sulfides) were formed and a strong sulfide smell was detected in the BES sub-culture. Since the medium does not contain sulfate, BES was probably metabolized to produce sulfides. The effect of BES on microbial populations is being investigated in our laboratory using PCR-DGGE.

Digital confocal images of Fe(0) surface taken before and after 10-day incubation with the culture showed significant difference in surface coverage by microorganisms (only the 10-d image is shown in Figure 2). The fluorescence signals (nucleic acid stain SYTO 13) on the surface show that microbes were able to colonize the iron surface. The confocal images of z-axis profiles also indicate that some colonies were formed within the crevices. The 0-day iron sample showed no fluorescence signals of colonies (only few individual cells).

The original culture and the sub-culture receiving no TCE gave essentially identical DGGE profiles (Figure 3). Using a denaturant gradient range of 25% to 75%, a strong (at approximately 55% denaturant) and a minor (at approximately 40%) band were obtained (Figure 3a). Attempt was made to further resolve the strong band using a narrower denaturant range. Several distinct bands were resolved in the denaturant range of 50% to 55%, which were followed by a series of smeared bands (Figure 3b). We are in the process of optimizing the denaturing conditions to enhance resolution for sequencing of the DNA fragments and for identifying the dominant organisms. No PCR products were obtained from the BES sub-culture using the PCR protocol and primers P63f and P518r.

Figure 2. Digital confocal image of Fe(0) surface after 10-day incubation with the culture. Fluorescence signals (white regions) represents bacteria (1 mm = 2.5 μm). Two z-axis profiles are shown at top and right edges.

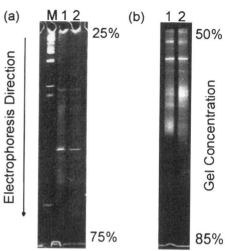

Figure 3. DGGE profiles of the original culture (2) and the sub-culture which did not receive TCE (1). M = DNA marker.

ACKNOWLEDGMENTS

The authors thank Michael D. Lee for sharing the mixed culture and Kirk Czymmek for assistance in acquiring the digital confocal image of iron surface.

REFERENCE

EL Fantroussi, S., L. Verschuere, W. Verstraete and E. A. Top. 1999. "Effect of Phenylurea Herbicides on Soil Microbial Communities Estimated by Analysis of 16S rRNA Gene Fingerprints and Community-Level Physiological Profiles." *Appl. Environ. Microbiol.* 65: 982-988.

Gu, B., T. J. Phelps, L. Liang, M. J. Dickey, Y. Roh, B. L. Kinsall, A. V. Palumbo and G. K. Jacobs. 1999. "Biogeochemical Dynamics in Zero-Valent Iron Columns: Implications for Permeable Reactive Barriers." *Environ. Sci. Technol.* 33: 2170-2177.

Holliger, C., D. Hahn, H. Harmsen, W. Ludwig, W. Schumacher, B. Tindall, F. Vazquez, N. Weiss and A. J. B. Zehnder. 1998. "*Dehalobacter restrictus gen. nov.* and *sp. nov.*, a Strictly Anaerobic Bacterium that Reductively Dechlorinates Tetra- and Trichloroethene in an Anaerobic Respiration." *Arch. Microbiol.* 169: 313-321.

Lampron, K. J., P. C. Chiu and D. K. Cha. 1998. "Biological Reduction of Trichloroethene Supported by Fe(0)." *Bioremediation.* 2: 175-181.

Löffler, F. E., K. M. Ritalahti and J. M. Tiedje. 1997. "Dechlorination of Chloroethenes Is Inhibited by 2-Bromoethanesulfonate in the Absence of Methanogens." *Appl. Environ. Microbiol.* 63: 4982-4985.

Löffler, F. E., J. M. Tiedje and R. A. Sanford. 1999. "Fraction of Electrons Consumed in Electron Acceptor Reduction and Hydrogen Thresholds as Indicators of Halorespiratory Physiology." *Appl. Environ. Microbiol.* 65: 4049.

Marchesi, J. R., T. Sato, A. J. Weightman, T. A. Martin, J. C. Fry, S. J. Hiom and W. G. Wade. 1998. "Design and Evaluation of Useful Bacterium-Specific PCR Primers that Amplify Genes Coding for Bacterial 16S rRNA." *Appl. Environ. Microbiol.* 64: 795-799.

Novak, P. J., L. Daniels and G. F. Parkin. 1998. "Enhanced Dechlorination of Carbon Tetrachloride and Chloroform in the Presence of Elemental Iron and *Methanosarcina barkeri, Methanosarcina thermophila,* or *Methanosaeta concillii*." *Environ. Sci. Technol.* 32: 1438-1443.

O'Hannesin, S. F. and R. W. Gillham. 1998. "Long-Term Performance of an In Situ "Iron Wall" for Remediation of VOCs."*Ground Water.* 36: 164-170.

Øvreås, L., L. Forney, F. L. Daae and V. Torsvik. 1997. "Distribution of Bacterioplankton in Meromictic Lake Sælenvannet, as Determined by Denaturing Gradient Gel Electrophoresis of PCR-Amplified Gene Fragments Coding for 16S rRNA." *Appl. Environ. Microbiol.* 63: 3367-3373.

Tratnyek, P. G. 1996. "Putting Corrosion to Use: Remediationg Contaminated Groundwater with Zero-Valent Metals." *Chem. Ind.* 499-503.

Vidic, R. D. and F. G. Pohland. 1996. *Treatment Walls.* GWRTAC Technology Evaluation Report.

Weathers, L. J., G. F. Parkin and P. J. Alvarez. 1997. "Utilization of Cathodic Hydrogen as Electron Donor for Chloroform Cometabolism by a Mixed, Methanogenic Culture." *Environ. Sci. Technol.* 31: 880-885.

HYDROGEN EVOLUTION FROM ZERO-VALENT IRON IN BATCH SYSTEMS

Christian Bokermann[1], Andreas Dahmke[2] & Martin Steiof[1]
([1] Technical University of Berlin, Germany
[2] Christian-Albrechts-University of Kiel, Germany)

ABSTRACT: While pilot and field–scale applications of zero-valent iron (Fe^0) permeable reactive barriers are performing successfully, there is still not much known about their long-term performance. We conducted experiments with two types of iron in water-vapor batch systems. The liquid-vapor ratio was kept constant (50/50 v/v-%), the mass of iron, the bicarbonate concentration (HCO_3^-) and the gas phase composition (100% nitrogen [N_2]; 80/20 v/v-% nitrogen/carbon dioxide [CO_2]) were varied.

Depending on the gas and liquid phase composition, we found hydrogen (H_2) evolution rates in the range from 0.2 µmol H_2 * g^{-1} Fe * d^{-1} to 50 µmol H_2 * g^{-1} Fe * d^{-1}. The presence of CO_2 and HCO_3^- clearly increased H_2 evolution rates. Since the bicarbonate system is the prevailing buffer system in groundwater we also expect the presence of significant amounts of H_2 under field conditions. Consequently the barrier performance, especially the hydraulic permeability, might be influenced by physical effects (e.g. clogging with gas bubbles) or growth of microorganisms supported by H_2.

INTRODUCTION

The application of permeable reactive barriers containing zero-valent iron for the reductive dechlorination of chlorinated compounds such as tetrachloroethylene (PCE) has gained increasing importance (e.g. LfU 1996). However, data about the long-term performance are still rare. Especially the influence of H_2 evolution and the resulting possibility of microbiological effects on the performance of reactive systems under field conditions have to be examined further. Several corrosion reactions occur when iron is in contact with a water phase containing chlorinated compounds under anaerobic conditions:

$$Fe^0 + XCl + H_2O \Leftrightarrow Fe^{2+} + OH^- + XH + Cl^- \qquad (1)$$

$$Fe^0 + 2 H_2O \Leftrightarrow Fe^{2+} + H_2 + 2 OH^- \qquad (2)$$

Equation (1) describes the dechlorination of a chlorinated compound (e.g. Matheson & Tratnyek 1994), which results in the corrosion of Fe^0. H_2 is produced by anaerobic iron corrosion according to equation (2) (Andrzejaczek 1984). The process is pH dependent, slowing down with increasing pH. It has been reported that carbon dioxide (CO_2), carbonic acid (H_2CO_3) and bicarbonate (HCO_3^-) enhance anaerobic corrosion (De Waard & Milliams 1975). The enhancement of the corrosion rate can not be attributed entirely to the buffer mechanism. It appears that CO_2 and HCO_3^- are electroactive species that promote corrosion through a different mechanism. The following equations describe the cathodic steps of carbonic acid corrosion (Gray et al. 1990):

$$H_2CO_3 + e^- \Leftrightarrow H_{(ads)} + HCO_3^- \tag{3}$$

$$HCO_3^- + e^- \Leftrightarrow H_{(ads)} + CO_3^{2-} \tag{4}$$

After CO_2 has been hydrated to H_2CO_3, reduction to HCO_3^- and finally to CO_3^{2-} takes place. Adsorbed hydrogen ($H_{(ads)}$) combines to H_2 and is released from the surface. The CO_2/HCO_3^- buffer system is encountered in the majority of groundwaters and its effect on the H_2 evolution of zero-valent iron might be substantial. Here we report results of H_2 evolution by two different types of iron under different environmental conditions, focusing on the CO_2/HCO_3^- system.

MATERIAL AND METHODS

Nitrogen gas (purity 5.0), carbon dioxide (purity 4.5) and a nitrogen/carbon dioxide mixture (80/20 v/v-%) were purchased from Messer-Griesheim (Berlin, Germany), $NaHCO_3$ (p.A.) from Merck (Darmstadt, Germany).

Two different types of iron were examined. The first iron was delivered in pellets with diameters ranging from 1 mm to 1.6 mm. Specific surface area determined by BET was 0.043 m^2/g, the iron mass fraction was determined to be 95%. This iron is denoted „pellet-iron" further on. The second type of iron had the form of splinters with 0.3 mm to 3 mm grain size and a BET specific surface area of 0.5 m^2/g. The iron mass fraction was given with 92%, this iron is referred to as „splinter-iron". Iron was purchased from the companies Würth (pellet form, Germany) and Maier Metallpulver GmbH (splinter form, Rheinfelden, Germany).

Batch experiments were prepared in 118 ml brown glas bottles. Millipore™ (deionized water additionally purified through an active carbon filter and ion-exchange cartridges) was filled into the bottles and purged with N_2 (100%) for 30 minutes before use. Zero-valent iron was transferred into the bottles, and the systems were purged with N_2 for another 15 minutes. The bottles were sealed with two septa (butyl rubber and rubber with PTFE coating) and screw tops. This procedure guaranteed sufficiently low oxygen levels ($< 0.5\%$) at the beginning of the experiments. Different amounts of HCO_3^- solution and/or CO_2 gas were injected with a gas-tight syringe. The aqueous to gas phase ratio was 50/50 v/v-%. The specifications of the conditions for different batch series are given in the following table:

TABLE 1: Composition of batch experiments

Series	Gas phase (v/v-%)	Liquid phase
Series A	N_2 (100)	Millipore
Series B	N_2/CO_2 (80/20)	Millipore
Series C	N_2/CO_2 (80/20)	Millipore, 250 mg/L HCO_3^-
Series D	N_2 (100)	Millipore, 1 g/L HCO_3^-
Series E	N_2 (100)	Millipore, 7.5 g/L HCO_3^-

Each series consisted of two identical batches with 0.5; 1 and 2.5 g splinter-iron and 1; 5 and 10 g pellet-iron

Batches were incubated at 25°C in the dark on a horizontal shaker at 120 rpm. Samples (100µL, duplicate determinations) for H_2 and $N_2/O_2/CO_2$ respectively were withdrawn with a gas-tight syringe.

Analytical Methods. Gas samples were analyzed on a Shimadzu GC14A gas chromatograph equipped with a thermal conductivity detector and a combined porapak/molecularsieve 13X column (CTR I, Alltech, Unterhaching, Germany). For H_2 detection nitrogen was used as the carrier gas, while helium was the carrier gas for determination of the other compounds. N_2 and O_2 were determined for control purposes (data not shown).

RESULTS AND DISCUSSION

In all batch systems, H_2 evolution was found to depend strongly on the iron mass present, as well as on the composition of the liquid and gas phases. Figure 1 summarizes the H_2 evolution in the batch systems containing 0.5 g splinter-iron. Displayed values are the means of duplicates and reflect the total H_2 amount in the gas phase. Liquid phase H_2 concentrations were negligible due to the low Henry-constant.

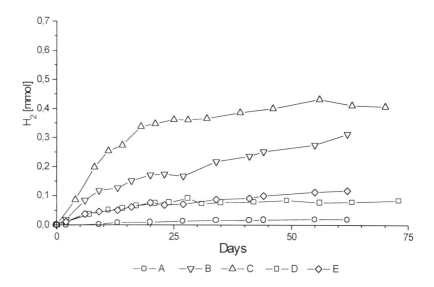

FIGURE 1: Hydrogen evolution in batch systems containing 0.5 g splinter-iron and different gas and liquid phases (see Table 1)

The lowest H_2 evolution was found in batch systems containing only Millipore and N_2 (100%). It is assumed that H_2 production in these batches occured only due to anaerobic corrosion according to equation (2). H_2 evolution ceased completely after 75 days (data not shown). The presence of HCO_3^- (series D and E) resulted in increased H_2 production, apparently due to the buffer effect and the mechanism of carbonic acid corrosion (equation (4)). Batch systems with CO_2 (series B and C) showed the highest H_2 evolution, which can be explained with the mechanisms according to equations (3) and (4). The additional presence of HCO_3^- increased H_2 production further (series C).

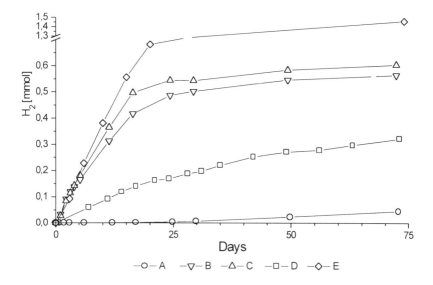

FIGURE 2: Hydrogen evolution in batch systems containing 5 g pellet-iron and different gas and liquid phases (see Table 1)

Experiments with 5g pellet-iron overall showed a similar pattern regarding the influence of different gas and liquid phases. Again the lowest H_2 evolution was observed with a N_2-atmosphere and Millipore (series A). A concentration of 1 g/L HCO_3^- resulted in moderate H_2 production (series D), while the presence of CO_2 caused a significantly higher H_2 evolution (series B and C). Batches with 7.5 g/L HCO_3^- (series E) showed a different behavior. While the H_2 evolution rate was in the same range as for batches from series C and D, the absolute amount of H_2 produced was nearly 3 times higher.

All batches, except those from series A, showed a decrease of H_2 evolution rates with time, where the decrease occured sooner and more clearly in batches with high inital rates. We attribute this effect to the experimental setup, as conditions like concentration of bicarbonate species, pressure and pH change with time.

Our further analysis of experimental data will therefore focus on initial H_2 evolution rates, since relatively stable rates were determined through the first phase of the experiments. The following table summarizes the absolute and relative (normalized to 1 g Fe) H_2 evolution rates (denoted as rX_{Batch}, S: splinter-iron, P: pellet-iron), determined by linear regression of at least 4 consecutive data points.

TABLE 2: Initial absolute and relative (normalized to 1 g Fe^0) hydrogen evolution rates (rX_{Batch}, S: splinter-iron, P: pellet-iron),

Series A: N_2-atmosphere (100%); Millipore						
	$rA_{S;0.5 g}$	$rA_{S;2.5 g}$	$RA_{S;5 g}$	$rA_{P;1 g}$	$rA_{P;5 g}$	$rA_{P;10 g}$
$\mu mol * d^{-1}$	0,55	1,39	3,17	0,96	1,02	1,94
$\mu mol * d^{-1} * g^{-1}$ Fe	1,09	0,56	0,63	0,96	0,20	0,19
Series B: N_2/CO_2-atmosphere (80/20 v/v-%); Millipore						
	$rB_{S;0.5 g}$	$rB_{S;2.5 g}$	$RB_{S;5 g}$	$rB_{P;1 g}$	$rB_{P;5 g}$	$rB_{P;10 g}$
$\mu mol * d^{-1}$	13,54	39,97	56,03	22,32	32,39	47,28
$\mu mol * d^{-1} * g^{-1}$ Fe	27,09	15,99	11,21	22,32	6,48	4,73
Series C: N_2/CO_2-atmosphere (80/20 v/v-%), 250 mg/L HCO_3^-						
	$rC_{S;0.5 g}$	$rC_{S;2.5 g}$	$RC_{S;5 g}$	$rC_{P;1 g}$	$rC_{P;5 g}$	
$\mu mol * d^{-1}$	24,71	47,77	54,48	21,57	35,62	
$\mu mol * d^{-1} * g^{-1}$ Fe	49,42	19,11	10,90	21,57	7,12	
Series D: N_2-atmosphere (100%), 1 g/L HCO_3^-						
	$rD_{S;0.5 g}$	$rD_{S;2.5 g}$	$RD_{S;5 g}$	$rD_{P;1 g}$	$rD_{P;5 g}$	$rD_{P;10 g}$
$\mu mol * d^{-1}$	4,16	12,62	21,17	5,98	7,90	10,34
$\mu mol * d^{-1} * g^{-1}$ Fe	8,31	5,05	4,23	5,98	1,58	1,03
Series E: N_2-atmosphere (100%); 7.5 g/L HCO_3^-						
	$rE_{S;05 g}$	$rE_{S;2.5 g}$	$RE_{S;5 g}$	$rE_{P;1 g}$	$rE_{P;5 g}$	$rE_{P;10 g}$
$\mu mol * d^{-1}$	5,27	23,59	46,67	12,42	38,76	50,19
$\mu mol * d^{-1} * g^{-1}$ Fe	10,54	9,43	9,33	12,42	7,75	5,02

Absolute and relative H_2 evolution rates were lowest in batches from series A, where only anaerobic corrosion according to equation (2) occured. Higher amounts of Fe^0 resulted in higher rates, but a strict linear relationship was not observed, as can be seen in the decrease of relative rates.

The presence of a bicarbonate species (series B to E) drastically enhanced H_2 evolution rates, supporting the assumption that carbonic acid corrosion became the dominant mechanism for H_2 production.

The pattern that increased Fe^0 amounts and therefore increased available surface area resulted in higher H_2 evolution rates was observed throughout all experiments. It was also evident, that for most series the relative H_2 evolution rates decreased with increasing Fe^0 surface area, indicating that factors other than the available surface area limited the overall H_2 production. Only for splinter-iron in series E relative rates for batches with different amounts of iron were nearly identical. While $rD_{S;0.5\ g}$ and $rE_{S;0.5\ g}$ lie in the same range, $rD_{S;2.5\ g}$, $rE_{S;2.5\ g}$ and also $rD_{S;5\ g}$ and $rE_{S;5\ g}$ are significantly lower. The decrease in relative rates can be attributed to the lower buffer capacity of the liquid phase in series D compared to series E. The same conclusion can be made for batches with pellet-iron in series E and D, although even 7.5 g/L HCO_3^- in series E was not sufficient to exclude other limiting factors than available surface area.

For both types of iron the presence of CO_2 (series B and C) resulted in higher H_2 evolution rates compared to the rates that were determined in batches with HCO_3^- (series D and E). This observation is more distinct for batches with low iron content, because in batches with higher iron mass present other factors than the available surface area restricted the H_2 evolution. The lower pH in batches with CO_2 might have caused an increased hydrogen production through anaerobic corrosion (equation (2)). With the current experimental setup it is not possible to estimate to which degree the higher H_2 evolution rate was caused by increased anaerobic corrosion.

The pellet-iron supported a more drastic increase of H_2 evolution rates in the presence of CO_2 than the splinter-iron did, which is shown by a comparison of $rD_{P;x}$ with $rB_{P;x}$ and $rD_{S;x}$ with $rB_{S;x}$. When additional HCO_3^- was present (series C) no substantial change of $rC_{P;x}$ was observed, while $rC_{S;x}$ increased significantly.

The two types of iron showed a different potential to produce H_2. Based on the mass of iron that was used, splinter-iron developed more H_2. Based on specific surface area the pellet-iron was substantially more reactive with respect to H_2 evolution. This corresponds with results from Alter (1998), who showed that PCE degradation rates, normalized to surface area, were one order of magnitude higher for the pellet-iron compared to the splinter-iron.

Batches from series B and C were treated further (day 75) after H_2 evolution had leveled off to a low rate (1-2 µmol H_2^* g^{-1} Fe $*$ d^{-1}). In some batches the gas phase was exchanged with N_2 (100%) in order to examine whether the elevated H_2 partial pressure in the gas phase had an effect on the H_2 evolution. Equally low H_2 evolution rates were determined after the exchange, indicating that the H_2 partial pressure had no significant influence on H_2 evolution at a late stage of the initial experiments. Furthermore we investigated whether the H_2 evolution capacity of the iron changed during the course of the experiments. In order to achieve conditions comparable to those at the beginning of the experiments, the gas phase was exchanged with N_2/CO_2 (80/20 v/v-%). The exchange restored H_2 evolution rates to a level comparably high to the initial ones.

These results support the assumption that the decrease of H_2 evolution rates in initial experiments was caused by the depletion of CO_2 and HCO_3^-, and that e.g. surface coatings did not significantly alter the reactivity of the iron during the course of the experiments.

CONCLUSION

Our results indicate that H_2 evolution can play an important role in zerovalent iron permeable reactive barriers. Especially when CO_2 and/or HCO_3^- are present in significant concentrations, elevated H_2 levels can be expected. The H_2 evolution might influence the long-term performance of permeable reactive barriers, either directly due to clogging with gas bubbles or indirectly via the support of growth of microbiological communities.

Since H_2 production was limited in different batch systems to varying extents by factors other than available surface area, a generalization of results from our batch experiments is not advisable. Current experiments involve anaerobic column systems to achieve data more closely related to conditions in a reactive barrier. Furthermore the effect of different mircoorganism groups on H_2 evolution and dechlorination of PCE will be studied.

ACKNOWLEDGEMENTS

Part of this experimental work was funded by a research grant from the Deutsche Forschungsgemeinschaft (DFG) (STE 765/1-1).

REFERENCES

Alter, M. 1998. "Reduktive Dechlorierung von PCE in Systemen mit Fe^0 und einer methanogenen Mischkultur." M.S. Thesis, Institute for Environmental Engineering, Technical University of Berlin, Germany.

Andrzejaczek, B. J. 1984. "Der Einfluß des Sauerstoffs auf die Kinetik der Korrosionsprozesse am Eisen in Wasser." *Korrosion.* 15(4): 239-256.

de Waard, C. & D. E. Milliams. 1975. "Carbonic Acid Corrosion of Steel." *Corrosion*, 31(5): 177-181.

Gray, L. G. S., B. G. Anderson, M. J. Danysh & P. R. Tremaine. 1990. "Effect of pH and Temperature on the Mechanism of Carbon Steel Corrosion by Aqueous Carbon Dioxide". *Corrosion/90*, NACE, Houston, paper no. 40.

LFU. 1996. *Literaturstudie Reaktive Wände - ph-Redox-reaktive Wände.* Texte und Berichte zur Altlastenbearbeitung, Nr. 24/96, Landesanstalt für Umweltschutz Baden-Württemberg, Germany.

Matheson, L. J. & P. G. Tratnyek. 1994. "Reductive Dehalogenation of Chlorinated Methanes by Iron Metal." *Environ. Sci. Technol.*, 28(12): 2045-2053.

2000 AUTHOR INDEX

This index contains names, affiliations, and book/page citations for all authors who contributed to the seven books published in connection with the Second International Conference on Remediation of Chlorinated and Recalcitrant Compounds, held in Monterey, California, in May 2000. Ordering information is provided on the back cover of this book.

The citations reference the seven books as follows:

2(1):   Wickramanayake, G.B., A.R. Gavaskar, M.E. Kelley, and K.W. Nehring (Eds.), *Risk, Regulatory, and Monitoring Considerations: Remediation of Chlorinated and Recalcitrant Compounds.* Battelle Press, Columbus, OH, 2000. 438 pp.

2(2):   Wickramanayake, G.B., A.R. Gavaskar, and N. Gupta (Eds.), *Treating Dense Nonaqueous-Phase Liquids (DNAPLs): Remediation of Chlorinated and Recalcitrant Compounds.* Battelle Press, Columbus, OH, 2000. 256 pp.

2(3):   Wickramanayake, G.B., A.R. Gavaskar, and M.E. Kelley (Eds.), *Natural Attenuation Considerations and Case Studies: Remediation of Chlorinated and Recalcitrant Compounds.* Battelle Press, Columbus, OH, 2000. 254 pp.

2(4):   Wickramanayake, G.B., A.R. Gavaskar, B.C.Alleman, and V.S. Magar (Eds.) *Bioremediation and Phytoremediation of Chlorinated and Recalcitrant Compounds.* Battelle Press, Columbus, OH, 2000. 538 pp.

2(5):   Wickramanayake, G.B. and A.R. Gavaskar (Eds.), *Physical and Thermal Technologies: Remediation of Chlorinated and Recalcitrant Compounds.* Battelle Press, Columbus, OH, 2000. 344 pp.

2(6):   Wickramanayake, G.B., A.R. Gavaskar, and A.S.C. Chen (Eds.), *Chemical Oxidation and Reactive Barriers: Remediation of Chlorinated and Recalcitrant Compounds.* Battelle Press, Columbus, OH, 2000. 470 pp.

2(7):   Wickramanayake, G.B., A.R. Gavaskar, J.T. Gibbs, and J.L. Means (Eds.), *Case Studies in the Remediation of Chlorinated and Recalcitrant Compounds.* Battelle Press, Columbus, OH, 2000. 430 pp.

Bergersen, Ove (SINTEF Oslo/NOR-
WAY) *2(7):*385
Berini, Christopher M. (U.S. Army
Corps of Engineers/USA) *2(6):*109
Beyke, Gregory (Current Environmental
Solutions, LLC/USA) *2(5):*183, 191
Bienkowski, Lisa A. (IT Corporation/
USA) *2(4):*229
Binard, Kinsley (Geomatrix Consultants,
Inc./USA) *2(4):*485
Bjerg, Poul L (Technical University of
Denmark/DENMARK) *2(3):*9
Blanchet, Denis (Institut Francais Du
Petrole/FRANCE) *2(7):*205
Blickle, Frederick W. (Conestoga
Rovers & Associates/USA)
*2(1):*133, 295; *2(2):*133
Blowes, David (University of
Waterloo/CANADA) *2(6):*361
Blum, Brian A. (McLaren/Hart, Inc./
USA) *2(2):*25
Boettcher, Gary (ARCADIS Geraghty &
Miller, Inc./USA) *2(1):*311
Boggs, Kevin G. (Wright State
University/USA) *2(5):*253
Bokermann, Christian (Technical
University of Berlin/GERMANY)
*2(6):*433
Bollmann, Dennis D. (City and County
of Denver/USA) *2(5):*113
Booth, Robert (XCG Consultants Ltd./
CANADA) *2(5):*135
Borch, Robert S. (USA) *2(7):*93
Borchert, Susanne (CH2M Hill/USA)
*2(5):*19
Borden, Robert C. (North Carolina State
University/USA) *2(4):*47, 421
Bosma, Tom N.P. (TNO Institute of
Environmental Sciences/THE
NETHERLANDS) *2(4):*63
Boulicault, Kent J. (Parsons Engineering
Science/USA) *2(4):*1
Bow, William (CADDIS Inc./USA)
*2(4):*15
Bowen, William B. (Advanced
GeoServices Corp/USA) *2(1):*231
Boyd, Thomas J. (U.S. Navy/USA)
*2(7):*189

Boyle, Susan L. (Haley & Aldrich,
Inc./USA) *2(4):*255
Bradley, Paul M. (U.S. Geological
Survey/USA) *2(3):*169; *2(7):*17
Brady, Warren D. (IT Corporation/
USA) *2(3):*201, 209
Brauning, Susan (Battelle/USA)
*2(1):*245
Brenner, Richard C. (U.S. EPA/USA)
*2(7):*393
Bricelj, Mihael (National Institute of
Biology/SLOVENIA) *2(4):*123
Bridge, Jonathan R. (HSI GeoTrans,
Inc./USA) *2(2):*149
Briseid, Tormod (SINTEF Oslo/NOR-
WAY) *2(7):*385
Brooker, Daniel (Applied Power
Concepts, Inc./USA) *2(4):*101
Brourman, Mitchell D. (Beazer East,
Inc./USA) *2(2):*1
Brown, Richard A. (ERM/USA)
*2(6):*125
Brown, Susan (Environment Canada/
CANADA) *2(5):*261
Bryant, J. Daniel (Geo-Cleanse Interna-
tional, Inc./USA) *2(5):*307
Buchanan, Ronald J. (DuPont Co./USA)
*2(4):*77
Buggey, Thomas R. (Dames & Moore/
USA) *2(2):*141
Burdick, Jeffrey S. (ARCADIS Ger-
aghty & Miller, Inc./USA) *2(4):*263
Burken, Joel G. (University of Missouri-
Rolla/USA) *2(7):*25
Burnett, R. Donald (Morrow Environ-
mental Consultants Inc./CANADA)
*2(5):*35
Burwinkel, Stephen (University of
Central Florida/USA) *2(6):*385
Butler, David (Applied Engineering &
Sciences, Inc./USA) *2(5):*127
Caffoe, Todd M. (New York State-
DEC/USA) *2(4):*255

Campbell, Ted R. (U.S. Geological
Survey/USA) *2(1):*349
Cannata, Marc A. (Parsons Engineering
Science/USA) *2(5):*9; *2(6):*385

Urynowicz, Michael A. (Envirox
LLC/USA) *2(6):*75, 117
Utgikar, Vivek P. (U.S.EPA/USA)
*2(7):*307

Vail, Christopher H. (Focus Environ-
mental, Inc./USA) *2(1):*207
Vainberg, Simon (Envirogen, Inc./USA)
*2(4):*165
VanBriesen, Jeanne Marie (Carnegie
Mellon University/USA) *2(3):*25
Vancheeswaran, Sanjay (CH2M
Hill/USA) *2(4):*303
Vandecasteele, Jean-Paul (Institut
Francais du Petrole/FRANCE)
*2(7):*205
Vartiainen, Terttu (National Public
Health Institute & University of
Kuopio/FINLAND) *2(7):*343
Venosa, Albert (U.S. EPA/USA)
*2(4):*191
Vermeul, Vince R. (Pacific Northwest
National Laboratories/USA)
*2(6):*369
Vierkant, Gregory P. (Lucent
Technologies, Inc./USA) *2(6):*217
Vilardi, Christine L. (STV
Incorporated/USA) *2(1):*199
Vincent, Jennifer C. (Earth Tech/USA)
*2(4):*437
Vinegar, Harold J. (Shell E&P
Technology/USA) *2(5):*197
Vogan, John L. (EnviroMetal
Technologies Inc/CANADA)
*2(6):*401, 417
Vogt, Carsten (UFZ-Centre for
Environmental Research/
GERMANY) *2(4):*133
Vroblesky, Don A. (U.S. Geological
Survey/USA) *2(1):*349; *2(7):*17

Waisner, Scott A. (TA Environmental,
Inc./USA) *2(7):*213
Wallace, Mark N. (U.S. Army Corps of
Engineers/USA) *2(7):*269
Wallis, B. Renee Pahl (U.S. Navy/USA)
*2(4):*467
Wallis, F.M. (University of Natal/REP
OF SOUTH AFRICA) *2(7):*131

Walsh, Matthew (Envirogen, Inc./USA)
*2(4):*157
Walti, Caryl (Northgate Environmental
Mgt, Inc./USA) *2(6):*49, 57
Wanty, Duane (The Gillette Company/
USA) *2(4):*405
Warburton, Joseph M. (Metcalf &
Eddy, Inc./USA) *2(4):*221
Ware, Leslie (Anniston Army
Depot/USA) *2(6):*153
Warith, Mustafa A. (University of
Ottawa/CANADA) *2(4):*381
Warner, Scott D. (Geomatrix Consul-
tants, Inc./USA) *2(4):*361
Warren, Randall J. (Shell Canada
Products, Ltd./USA) *2(5):*207
Watkinson, Robert J. (Sheffield Univer-
sity/UNITED KINGDOM) *2(4):*183
Weeber, Phil (HSI Geotrans/USA)
*2(4):*429
Weiss, Holger (Centre for Environmental
Research/GERMANY) *2(6):*331
Wellendorf, William G. (Southwest
Ground-water Consultants, Inc./
USA) *2(7):*161
Wells, Samuel L. (Golder Sierra
LLC/USA) *2(6):*307
Werner, Peter (Technische Universitat
Dresden/GERMANY) *2(7):*205
West, Brian (U.S. Army Corps of
Engineers/USA) *2(7):*269
Westerheim, Michael (Unisys Corpor-
ation/USA) *2(1):*387
Weston, Alan (Conestoga-Rovers &
Associates/USA) *2(6):*161; *2(7):*301
Wharry, Stan (U.S. Army/USA) *2(7):*81
Whiter, Terri M. (Focus Environmental,
Inc./USA) *2(1):*207
Wickramanayake, Godage B.
(Battelle/USA) *2(1):*339
Widdowson, Mark A. (Virginia
Polytechnic Inst & State Univ/USA)
*2(4):*493
Wiedemeier, Todd H. (Parsons Engin-
eering Science, Inc./USA) *2(1):*357;
*2(3):*81; *2(4):*1
Wildenschild, Dorthe (Lawerence
Livermore National Laboratory/
USA) *2(5):*277

2000 KEYWORD INDEX

This index contains keyword terms assigned to the articles in the seven books published in connection with the Second International Conference on Remediation of Chlorinated and Recalcitrant Compounds, held in Monterey, California, in May 2000. Ordering information is provided on the back cover of this book.

In assigning the terms that appear in this index, no attempt was made to reference all subjects addressed. Instead, terms were assigned to each article to reflect the primary topics covered by that article. Authors' suggestions were taken into consideration and expanded or revised as necessary to produce a cohesive topic listing. The citations reference the seven books as follows:

2(1): Wickramanayake, G.B., A.R. Gavaskar, M.E. Kelley, and K.W. Nehring (Eds.), *Risk, Regulatory, and Monitoring Considerations: Remediation of Chlorinated and Recalcitrant Compounds.* Battelle Press, Columbus, OH, 2000. 438 pp.

2(2): Wickramanayake, G.B., A.R. Gavaskar, and N. Gupta (Eds.), *Treating Dense Nonaqueous-Phase Liquids (DNAPLs): Remediation of Chlorinated and Recalcitrant Compounds.* Battelle Press, Columbus, OH, 2000. 256 pp.

2(3): Wickramanayake, G.B., A.R. Gavaskar, and M.E. Kelley (Eds.), *Natural Attenuation Considerations and Case Studies: Remediation of Chlorinated and Recalcitrant Compounds.* Battelle Press, Columbus, OH, 2000. 254 pp.

2(4): Wickramanayake, G.B., A.R. Gavaskar, B.C.Alleman, and V.S. Magar (Eds.) *Bioremediation and Phytoremediation of Chlorinated and Recalcitrant Compounds.* Battelle Press, Columbus, OH, 2000. 538 pp.

2(5): Wickramanayake, G.B. and A.R. Gavaskar (Eds.), *Physical and Thermal Technologies: Remediation of Chlorinated and Recalcitrant Compounds.* Battelle Press, Columbus, OH, 2000. 344 pp.

2(6): Wickramanayake, G.B., A.R. Gavaskar, and A.S.C. Chen (Eds.), *Chemical Oxidation and Reactive Barriers: Remediation of Chlorinated and Recalcitrant Compounds.* Battelle Press, Columbus, OH, 2000. 470 pp.

2(7): Wickramanayake, G.B., A.R. Gavaskar, J.T. Gibbs, and J.L. Means (Eds.), *Case Studies in the Remediation of Chlorinated and Recalcitrant Compounds.* Battelle Press, Columbus, OH, 2000. 430 pp.

A

α-ketoglutarate-dependent cleavage **2(7):**229
abiotic release date **2(7):**181
acetate **2(4):**23, 107, 389, 437
acid mine drainage, *see* mine waste
acid-enhanced degradation **2(7):**33
acridine orange (AO) **2(6):**233
actinomycetes **2(4):**455

activated carbon **2(6):**257, 315
advanced oxidation technology (AOT) **2(6):**201, 209, 217, 225, 233, 241, 249; **2(7):**25
aeration **2(5):**237
air monitoring **2(1):**207
air sparging, *see* sparging
air stripping **2(5):**293